The Laser Cutting Process

The Laser Cutting Process

Analysis and Applications

Bekir Sami Yilbas

ELSEVIER

Elsevier
Radarweg 29, PO Box 211, 1000 AE Amsterdam, Netherlands
The Boulevard, Langford Lane, Kidlington, Oxford OX5 1GB, United Kingdom
50 Hampshire Street, 5th Floor, Cambridge, MA 02139, United States

Library of Congress Cataloging-in-Publication Data
A catalog record for this book is available from the Library of Congress

British Library Cataloguing-in-Publication Data
A catalogue record for this book is available from the British Library

ISBN: 978-0-12-812982-1

Publisher: Matthew Deans
Acquisition Editor: Christina Gifford
Editorial Project Manager: Lindsay Lawrence
Production Project Manager: Sruthi Satheesh
Cover Designer: Miles Hitchen

Typeset by SPi Global, India

Dedication

To my mother Ayse, my wife Zahide, and
my daughters Ayse and Merve

Contents

Preface

Laser cutting is one of the important processes of laser machining, which has several advantages over the conventional machining methods. Some of these advantages include precision of operation, short processing time, local treatment, and low cost. Although the laser- machining technologies are have been known for decades, improvements in these technologies and development of new materials such as metallic foams, super alloys, ceramics, and composite materials, which are difficult to cut, keep the laser machining research current. The laser cutting process is involved with high-temperature phenomena, which are associated with multiphysics processes. These processes may be categorized into two main parts: key-hole and kerf formations. A key-hole formation is the initial step of the cutting process and is related to solid heating, melting, evaporation, cavity formation, and hole forming. Although key-hole formations are completed within a short time period, they have a significant effect on kerf formation with the progressing cutting process. In the early periods of laser research, the key-hole formation and its modeling was the primary interest. As the computational power growth increased significantly over the years, the research focus has been tilted toward the comprehensive modeling of the process, including multiphysics incorporating phase change, thermal stresses, and quality assessments. However, capturing the complete picture of the physical situation still needs tremendous computational efforts as it forms the bases for the current research directions.

The efforts for the analytical formulation of the laser cutting process falls into two main categories: the prediction of thermal and stress fields, and kerf width size formulation in terms of laser cutting parameters. Because the laser cutting process involves multiphysics, the analytical solutions to the problems are difficult to achieve mainly because of the involvement of a high degree of nonlinearity, and a simultaneous coupling of multi processes. The simplification of the physical problem after introducing useful assumptions enables the formulaic reduction of the process, which can then be tackled analytically. Some of these assumptions are necessary to reduce the degree of nonlinearity to achieve possible closed-form solutions to the heating and stress problems. Although incorporating many useful assumptions simplifies these problems, the actual physical process may be overlooked, and physical details may not be captured with accuracy in the formulations. Nevertheless, the analytical solutions to the heating problem provide a functional relationship between the independent variables and the controlling parameters while allowing an improved physical understanding of the laser cutting process. On the other hand, numerical treatment of the process can require high computational efforts and becomes expensive to capture the physical insight. In the physical frame of the laser cutting process, the accuracy of analytical and numerical solutions remains in question because of the limitations in model studies that feature the complete physical aspects of the real phenomena. This gives rise to the importance of experimental studies toward understanding of the laser cutting process and the capturing of the actual physical insight of the process. Therefore, the experimental efforts to identify the affecting parameters of laser cutting and the assessment of end-product quality are still irreplaceable.

In this book, the laser cutting process is presented and model studies associated with the analytical and numerical treatment of the thermal and stress fields are incorporated. The closed-form solutions prior to key-hole formation are demonstrated in terms of temperature and thermal stress fields using the convective and insulated boundary conditions. The lump-parameter analysis predicting the kerf width size is also presented, and various examples are provided with the experimental data to demonstrate the accuracy of the formulation. An analytical approach for a two-dimensional heating situation is also provided. Numerical simulations predicting the influence of an assisting gas jet on the heat transfer and skin friction at the kerf surfaces are presented for various geometric configurations of the kerf, which are possibly formed during the laser cutting of solid substrates. The model and experimental studies for laser cutting of high thermal conductive materials are included for various cutting configurations, including straight, circular, and triangular/rectangular geometries. In addition, experimental and model studies pertinent to laser cutting of difficult-to-cut materials, including ceramics, Kevlar, and carbon nano-composites, are also included. Working examples are provided and the findings are discussed in detail to gain physical insight into the laser cutting process. However, some of the other issues related to laser cutting processes are not presented in this book due to space limitations. Future treatments will review these issues and related processes.

Acknowledgments

I would like to acknowledge the role of King Fahd University of Petroleum & Minerals (KFUPM) in extending strong support from beginning to end facilitating every means during the preparation of this book. The author also thanks Deanship of Scientific Research at KFUPM for supporting the writing of this book through the project #IN141006. The author wishes to thank the colleagues who contributed to the work presented in the book through previous cooperation of the author. In particular, thanks to Drs. Haider Ali, Shahzada Zaman Shuja, Sohail Akhtar, Faleh Al-Sulaiman, Nasser Al-Aqeeli, and all my graduate students.

Introduction

1

Lasers are widely used in industry as machine tools due to their precision of operation, low cost, localized processing, and high speed of operation. In laser machining applications, a laser beam is used as a heat source, rapidly increasing the temperature to the melting and evaporation temperature of the substrate material. Because the optical settings for the laser beam are very precise, the localized heating can be controlled easily. Laser machining can be categorized into several groups based on the type of processing being involved during the machining: drilling, cutting, welding, alloying, and others. Moreover, the laser process can involve pre- or posttreatment operations, such as duplex treatment for coatings and scribing after coatings. In order to optimize the laser-cutting process and reduce the experimental time and cost, the model studies receive considerable attention. Moreover, the model studies give insight into the physical processes that take place during the heating process making them easier to accomplish compared to experimental studies. The measurement of physical properties during laser workpiece interaction is difficult and costly because the process involves high temperatures, short durations, and localized heating.

When a high-powered laser beam is focused onto a substrate surface, the beam of energy is partially absorbed by the substrate material. Depending on the focused beam diameter at the surface, laser power intensity (combining the laser output energy and pulse length), and reflectivity of the surface, the substrate material undergoes solid heating, melting, and evaporation. In the case of evaporation process, the evaporating front detaches from the liquid surface generating a recoil pressure across the vapor-liquid interface. As the evaporation of the surface progresses, the recoil pressure increases considerably influencing the evaporation rate. As the heating progresses further, the liquid surface recesses toward the solid bulk forming a cavity in the substrate material. Depending on the pulse length and power intensities, liquid ejection from the cavity occurs, which is particularly true for the long pulses ($\sim$ ms pulse lengths); however, the surface ablation without liquid ejection takes place for short pulses ($\sim$ ns pulse lengths). Moreover, the liquid ejection improves the material removal rates from the cavity. In the case of laser shortpulse processing, the recoil pressure increases substantially due to high rates of momentum exchange during the evaporation process. In this case, high pressure at the vapor/liquid interface acts as a pressure force, generating surface stresses at the liquid/solid interface. This, in turn, results in a pressure wave propagating into the substrate material. Depending on the magnitude of the pressure wave, plastic deformation through dislocations in the surface region of the substrate material takes place. The depth of the deformed region is limited with the interaction of loading (plastic wave) and unloading (elastic wave) waves, that is, as the loading phase is completed (when the evaporation is completed, the recoil pressure diminishes), the unloading wave (elastic wave) from the liquid/solid interface initiates. Because the unloading wave travels faster than the loading wave, both waves meet

The Laser Cutting Process. https://doi.org/10.1016/B978-0-12-812982-1.00001-1

at some depth below the surface. It should be noted that the wave motion in the substrate material is complicated and requires comprehensive investigation.

From the modeling point of view, the laser machining can be classified into two categories: (i) laser conduction limited heating, and (ii) laser nonconduction limited heating. In the laser conduction limited heating situation, the substrate surface is heated up to the melting temperature of the substrate material; in this case, substrate material remains in the solid state during the process. One of the laser conduction limited applications is the laser quenching of the surfaces. In the laser nonconduction limited heating situation, the substrate surface undergoes a phase change during the processing, resulting in melting and subsequent evaporation. The laser drilling, cutting, and welding are typical examples of laser nonconduction limited heating situations.

In the laser cutting process, the end-product quality is very important from the manufacturing point of view. Consequently, the optimization studies improve the laser cutting process considerably. The quality assessment of the end product, such as drilled hole, cut surface, etc., can be possible through examining the geometric features of the laser-machined section. One of the methods associated with the quality assessment is factorial analysis. In this case, the affecting factors are varied randomly or with increments to generate random blocks. Through the statistical testing of the measurable responses, the significant levels of the affecting parameters can be identified.

The rapid growth of the use of high-powered lasers for profiling sheet metals has generated a great deal of scientific and commercial interest in the quality of the cut edges produced [1–10]. In recent years, some problems relevant to laser cutting quality have been encountered and the process parameters are adjusted and fine-tuned to provide the quality of cut desired [11–15]. However, this consumes a considerable amount of time and effort, and still the optimal cutting parameters may not be found. If a different substance is used for cutting, all these parameters may require readjustment. This has been recognized as a major shortcoming, and this problem of optimized process needs to be addressed accordingly. Deficiencies in the cutting quality may be related to the slow process drifts and disturbances. The main causes for this drift may include velocity fluctuations, variation in power and spatial intensity distribution, and optical integrity perturbations.

A jet of gas, capable of taking part in exothermic reaction, is directed onto the moving region of the workpiece at which a laser beam is concentrated, to provoke an exothermic reaction and thereby increase the heating of the workpiece. The momentum of the jet sweeps away the products of combustion. This oxidation front comes to rest just outside the laser material interaction area and the subsequent encroachment of the laser beam on the new cutting front repeats the oxidation initiation process. This initiation/propagation event is repeated as the laser beam moves across the material's surface resulting in a characteristic cut edge consisting of regularly spaced striations. The formation of striations has received considerable attention, because it is closely related to the laser-cutting quality. Explanations given on striation are, in general, quantitative and some little work has been carried out on classification of striation patterns. In addition, an attempt was made to calculate the striation frequency, but the predictions are limited in its use for characterizing and predicting surface roughness or a striation depth, which is considered as an important parameter in judging the laser cut quality [15]. However, the results predicted for striation frequency agreed well with the experimental data.

Some problems may be encountered during laser cutting with difficult-to-cut materials such as ceramics (due to low thermal conductivity) and copper alloys (due to high thermal conductivity). These problems are mainly associated with: (1) the formations of cracks in the cutting sections due to high thermal strains for low thermal-conductivity materials and (2) low-intensity cutting that can cause sideways burning around the cut edges or incomplete cutting in high-thermal conductivity materials. Although altering the laser power intensity during the cutting process is one of the alternative solutions to avoid cutting defects and deficiencies, it is hard to achieve because of the difficulties faced with online control of the laser output power during the cutting process. However, repetitive pulsing of the laser output power can improve the cutting defects and may reduce the thermal strain and stress levels formed in the cut section vicinity [16–18]. Consequently, various laser-cutting parameters have individual and combined effects on the resulting cut geometry and its quality.

References

[1] B.S. Yilbas, A.Z. Sahin, Oxygen assisted laser cutting mechanism – a laminar boundary layer approach including the combustion process, Opt. Laser Technol. 27 (1995) 175–184.

[2] B.S. Yilbas, A.Z. Sahin, Turbulent boundary layer approach allowing chemical reactions for CO_2 laser oxygen-assisted cutting process, Proc. Inst. Mech. Eng. C J. Mech. Eng. Sci. 208 (1994) 275–284.

[3] B.S. Yilbas, Effect of process parameters on the kerf width during the laser cutting process, Proc. Inst. Mech. Eng. C J. Mech. Eng. Sci. 215 (2001) 1357–1365.

[4] B.S. Yilbas, M. Rashid, CO_2 laser cutting of Incoloy 800 HT alloy and quality assessment, Opt. Lasers Eng. 12 (2002) 135–145.

[5] B.S. Yilbas, The analysis of CO_2 laser cutting, Proc. Inst. Mech. Eng. B 211 (1997) 223–232.

[6] J.N. Gonsalves, W.W. Duley, Cutting thin metals with the cw CO_2 laser, J. Appl. Phys. 43 (1972) 4684–4687.

[7] A. Kar, J.A. Rothenflue, W.P. Latham, Scaling laws for thick-section cutting with a chemical oxygen-iodine laser, J. Laser Appl. 9 (1997) 279–286.

[8] B.S. Yilbas, R. Davies, A. Gorur, Z. Yilbas, F. Begh, M. Kalkat, N. Akcakoyun, Study into the measurement and prediction of penetration time during CO_2 laser cutting process, Proc. Inst. Mech. Eng. B 204 (1990) 105–113.

[9] B.S. Yilbas, B.J. Abdul Aleem, Dross formation during laser cutting process, J. Phys. D. Appl. Phys. 39 (2006) 1451–1461.

[10] B.S. Yilbas, Laser cutting of thick sheet metals: effects of cutting parameters on kerf size variations, J. Mater. Process. Technol. 201 (1–3) (2008) 285–290.

[11] B.S. Yilbas, Z. Yilbas, Effects of plasma on CO_2 laser cutting quality, Opt. Lasers Eng. 9 (1988) 1–12.

[12] B.S. Yilbas, Z. Yilbas, Effects of plasma in CO_2 laser cutting quality, Opt. Lasers Eng. 9 (1988) 1–12.
[13] B.S. Yilbas, Laser cutting quality assessment and thermal efficiency analysis, Int. J. Mater. Prod. Technol. 155–156 (2004) 2106–2115.
[14] B.S. Yilbas, C. Karatas, O. Keles, I. Uslan, Y. Usta, Z. Yilbas, M. Ahsan, Laser cutting of corners: assessment of cutting quality, Opt. Lasers Eng. 16 (1–2) (2006) 157–171.
[15] F. Al-Sulaiman, B.S. Yilbas, C. Karatas, O. Keles, I. Uslan, Y. Usta, M. Ahsan, A. Bazoune, Laser cutting of Kevlar and mild steel composite structure: end product quality assessment, J. Mater. Eng. Perform. 161 (2007) 22–29.
[16] B.S. Yilbas, A.F.M. Arif, Modelling of residual stress during laser cutting of small diameter holes, Proc. IMechE B J. Eng. Manuf. 222 (12) (2008) 1577–1587.
[17] B.S. Yilbas, A.F.M. Arif, B.J. Abdul Aleem, Laser cutting of sharp edge: thermal stress analysis, Opt. Lasers Eng. 48 (1) (2010) 10–19.
[18] B.S. Yilbas, S.S. Akhtar, Laser cutting of alloy steel: 3-dimensional modeling of temperature and stress fields, Mater. Manuf. Process. 26 (1) (2011) 104–112.

Thermal analysis of the laser cutting process

2

2.1 Introduction

The laser cutting process is initiated with the absorption of the laser beam by the substrate material through which heat is generated within the absorption depth of the irradiated surface. Although surface reflection plays an important role on the peak intensity absorbed by the irradiated surface, the influence of surface reflection reduces as the surface temperature increases [1]. Because the absorption depth is shallow in the surface vicinity, heat generation becomes significantly high during high-intensity laser irradiation. In this case, a melting temperature is reached rapidly and surface evaporation takes place almost immediately after the melting process. The evaporated front detaches from the surface while the melted zone recesses towards the substrate material, forming a cavity during the heating process. As the heating progresses, the cavity formed becomes deeper, causing the formation of a key-hole prior to cutting along the workpiece movement. The vapor pressure in the cutting section diminishes once the key-hole is formed. Hence, the vapor and liquid phases of the substrate material remain in the cutting section causing burrs and dross attachments around the cut edges. In addition, the presence of oxygen causes the initiation of high-temperature oxidation reactions at the melt/vapor surface. This results in additional heat generation in the cutting region. However, the irregular behavior of the oxidation reactions causes oscillatory behavior of heat input in the cutting section. Consequently, irregular thermal erosion causes large variation in the kerf width size while reducing the cut section quality. In order to avoid irregular behavior of the high temperature exothermic oxidation reactions, an assisting gas is used in the cutting section. The assisting gas is usually inert, such as argon, helium, nitrogen, etc., and prevents or minimizes the oxidation reactions taking place in the cutting section. On the other hand, the assisting gas causes convection cooling in the cutting section while enhancing the temperature gradients in the cutting section vicinity. This, in turn, enhances the thermal strain and stress levels causing an adverse effect on the cutting section quality. Therefore, the velocity of the assisting gas and its pressure needs to be regulated to minimize the adverse effect on the cutting quality. The mathematical analysis pertinent to absorption, temperature, and stress fields during the key-hole formation, and the assisting gas effect, is presented.

2.2 Absorption of laser beam intensity

The absorption of laser intensity at any point in a substrate material can be written in terms of the Beer-Lambert's law [2], which is:

$$\frac{dI}{dx} = -\alpha I \tag{2.1}$$

The Laser Cutting Process. https://doi.org/10.1016/B978-0-12-812982-1.00002-3

where x is the depth along the laser beam direction, in units of m, $I = I(x)$ is the laser intensity at a depth x in units of W/m^2, and α is the absorptivity or the absorption coefficient of the material, in units of m^{-1}. The surface of the substrate material lies at $x = 0$. The absorption coefficient can be variable within the substrate material; in which case, the laser intensity can be a function of x. It can be obtained after integrating Eq. (2.1), which is:

$$\frac{dI}{I} = -\alpha(x)dx$$

or integrating both sides yields

$$\Rightarrow \int_{I_o}^{I} \frac{dI}{I} = -\int_0^L \alpha(x)dx$$

Due to the integration purpose, x variable may be changed to another variable u. Also, to find the laser intensity at any point x, we may replace L by x in the above integration. Therefore,

$$\Rightarrow \int_{I_o}^{I} \frac{dI}{I} = -\int_0^x \alpha(u)du$$

or

$$\Rightarrow I = I_o e^{-\int_0^x \alpha(u)du} \tag{2.2}$$

Hence, the laser intensity at any point x in the substrate material is given by

$$I = I_o e^{-\int_0^x \alpha(u)du} \tag{2.3}$$

However, when the thin film absorbs the laser beam energy, it generates heat. The heat generation rate is defined as

$$S = -\frac{dI}{dx} \tag{2.4}$$

Substituting Eq. (2.3) in Eq. (2.4) and using Leibniz rule yields,

$$S = -\frac{d}{dx}\left\{ I_o e^{-\int_0^x \alpha(u)du} \right\}$$

or

$$= -I_o e^{-\int_0^x \alpha(u)du} \frac{d}{dx}\left\{ -\int_0^x \alpha(u)du \right\}$$

or

$$= I_o e^{-\int_0^x \alpha(u)du} \left\{ \int_0^x 0.du + \alpha(x).1 - \alpha(0).0 \right\}$$

$$S = I_o \alpha(x) e^{-\int_0^x \alpha(u) du}$$

Hence, the volumetric heat generation becomes:

$$S = I_o \alpha e^{-\int_0^x \alpha(u) du} \tag{2.5}$$

The variable absorption coefficient can be determined after considering the absorption coefficient in the mathematical form of:

$$\alpha_1 = \alpha_{\max} \exp\left[-\left(\frac{x-b}{d}\right)^2\right] \tag{2.6}$$

where b and d are the constants. Eq. (2.3) yields

$$I = I_o e^{-\int_0^x \alpha_{\max} \exp\left[-\left(\frac{u-b}{d}\right)^2\right] du} \tag{2.7}$$

Let $g(x) = \int_0^x \alpha_{\max} \exp\left[-\left(\frac{u-b}{d}\right)^2\right] du$

then,

$$g(x) = \frac{\sqrt{\pi}}{2} \alpha_{\max} d \left[\operatorname{erf}\left(\frac{x-b}{d}\right) - \operatorname{erf}\left(\frac{b}{d}\right)\right] \tag{2.8}$$

The laser intensity and the volumetric heat generation are then given by

$$I = I_o e^{-g(x)} \tag{2.9}$$

and

$$S = I_o \alpha e^{-g(x)} \tag{2.10}$$

Eq. (2.10) can be used in the analysis to incorporate the heat source generation due to the absorption of the laser beam by the substrate material.

In the laser heating process, irradiated energy is absorbed in the surface skin of the substrate material. This results in excess computational efforts due to grid arrangements in the irradiated region, and the remaining region in the solution domain are due to the fine grid spacing in the irradiated region. However, consideration of surface heat source minimizes this problem, because it does not require fine-grid spacing in the surface skin.The comparison of the surface temperature difference due to the volumetric heat source and the surface heat source is shown in Fig. 2.1 [3]. It is evident that the surface heat source model predicts higher values of temperature than those corresponding to the volumetric heat source at the surface vicinity. However, as the depth increases, temperature distributions predicted from both models become almost identical.

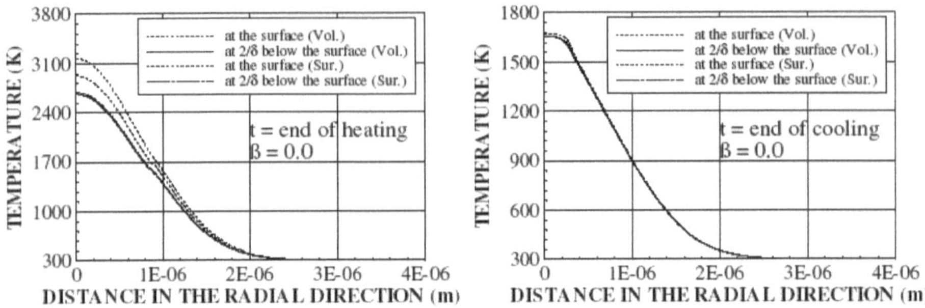

Fig. 2.1 Temperature variation in the radial direction for surface and volumetric heat source models [3].

2.3 Temperature and stress fields in relation to key-hole formation

Surface evaporation is the dominant mechanism in terms of the thermal analysis during the key-hole formation. This is because of the fact that the latent of evaporation is much higher than the latent heat of melting most of the substrate materials, particularly engineering materials. Moreover, the key-hole formation is involved with a nonconduction limited heating situation, due to the fact that during the key-hole formation, phase change takes place. Although key-hole formation is transient in nature, the surface recession velocity can be considered to attain a steady value during the laser nonconduction heating process. This assumption simplifies the problem and provides the formulation of the recession velocity at the evaporating surface. After considering a recessing front, which is traveling steadily into the material with an instantaneous velocity V, fixing a coordinate reference frame on this front gives the steady-state equation for the laser evaporation process:

$$k\frac{\partial^2 T}{\partial x^2} + \rho C_p V\frac{\partial T}{\partial x} + I_0\delta\exp(-\delta x) = 0 \tag{2.11}$$

with boundary conditions:

$$\left.\frac{\partial T}{\partial x}\right|_{x=0} = \rho\frac{VL}{k} \tag{2.12}$$

and

$$T(\infty) = 0.$$

Because T is a function of x only and V is a constant, Eq. (2.11) may be rewritten:

$$\frac{d^2T}{dx^2} + \frac{V}{\alpha}\frac{dT}{dx} + \frac{I_0\delta}{k}\exp(-\delta x) = 0 \tag{2.13}$$

Eq. (2.13) yields a solution:

$$T = A\exp\left(-\frac{Vx}{\alpha}\right) + B - \frac{I_0}{k}\frac{\exp(-\delta x)}{\delta - \frac{V}{\alpha}} \tag{2.14}$$

where A and B are constants.

Introducing the boundary conditions in Eq. (2.14) gives:

$$B = 0$$

and

$$A = \frac{\alpha}{V}\left[\frac{I_0\delta}{k}\frac{1}{\delta - \frac{V}{\alpha}} - \rho\frac{VL}{k}\right] \tag{2.15}$$

The complete solution for the temperature is therefore:

$$T(x) = \frac{\alpha}{V}\left[\frac{I_0\delta}{k}\frac{1}{\delta - \frac{V}{\alpha}} - \rho\frac{VL}{k}\right]\exp\left(-\frac{Vx}{\alpha}\right) - \frac{I_0}{k}\frac{\exp(-\delta x)}{\delta - \frac{V}{\alpha}} \tag{2.16}$$

Although Eq. (2.15) is the solution based on the steady evaporation, it is worthwhile to consider, in more detail, its implications. The surface temperature T_s can be obtained after setting $x = 0$ in Eq. (2.16), i.e.,

$$T(0) = T_s = \frac{\alpha}{V}\left[\frac{I_0\delta}{k}\frac{1}{\delta - \frac{V}{\alpha}} - \rho\frac{VL}{k}\right] - \frac{I_0}{k}\frac{1}{\delta - \frac{V}{\alpha}} \tag{2.17}$$

or

$$T_s = \frac{I_0}{k}\frac{\alpha}{V} - \frac{\alpha\rho L}{k} \tag{2.18}$$

After rearrangement Eq. (2.18) gives:

$$I_0 = \rho V\left(C_p T_s + L\right) \tag{2.19}$$

Eq. (2.19) represents the energy balance at the surface.

However, another approach is possible to formulate the recession velocity at the surface after making the assumption that:

$$V \gg \alpha\delta$$

This assumption is valid for most applications such as for nickel with a surface temperature of 5×10^3 K, $\alpha = 10^{-4}$ m^2/s, and $\delta = 10^7$ 1/m, the velocity of the evaporating front is about 60 m/s. This corresponds to a laser beam with incident intensity of approximately 10^{12} W/m^2. This assumption then leads from Eq. (2.16) to the following form of the temperature distribution:

$$T(x)=\frac{\alpha}{V}\left[\frac{I}{k}-\rho\frac{VL}{k}\right]\exp\left(-\frac{Vx}{\alpha}\right)-\frac{I}{k_o}\exp(-\delta x) \tag{2.20}$$

It is obvious that this solution is still valid, because $x=0$ the surface temperature is given as:

$$T(0)=T_s=\frac{I}{k}\left[\frac{\alpha}{V}-\frac{1}{\delta}\right]-\frac{L}{C_p} \tag{2.21}$$

or

$$T(0)=\frac{I\alpha}{kV}--\frac{L}{C_p}\quad \text{since } \frac{\alpha}{V}\gg\frac{1}{\delta} \tag{2.22}$$

or

$$I_0=\rho V\left(C_pT_s+L\right) \tag{2.23}$$

which is the same as that presented in Eq. (2.19).

Thermal analysis pertinent to the key-hole formation can be further simplified after considering the formation of a key hole at a constant rate. In this case, temperature distribution at the recessed surface can be readily obtained, that is, the expression for the surface temperature can be obtained after setting $x=0$ in Eq. (2.16). This results:

$$T=T_s\exp\left(-\frac{Vx}{\alpha}\right)+\frac{I_0}{k\delta}\left[\exp\left(-\frac{Vx}{\delta}\right)-\exp(-\delta x)\right] \tag{2.24}$$

The solution in this form clearly shows that the temperature inside the material will rise from the surface to a maximum at some point below the surface before decaying to zero at infinity. This is because the last two exponential terms of the solution are of the form $\exp(-\beta x)-\exp(-\gamma x)$ where $\beta<\gamma$. It can be shown that this function initially rises to a maximum before decaying away towards zero.

The position at which the maximum temperature can be obtained through differentiating Eq. (2.24) with respect to x, gives:

$$\frac{\partial T}{\partial x}=-\frac{VT_s}{\alpha}\exp\left(-\frac{Vx}{\alpha}\right)-\frac{VI_0}{\alpha k\delta}\exp\left(-\frac{Vx}{\alpha}\right)+\frac{I_0}{k}\exp(-\delta x) \tag{2.25}$$

The condition for a maximum temperature is:

$$\frac{\partial T}{\partial x}=0.$$

Hence:

$$\frac{V}{\alpha}\left[T_s \exp\left(-\frac{Vx}{\alpha}\right)+\frac{I_0}{k\delta}\exp\left(-\frac{Vx}{\alpha}\right)\right]=\frac{I_0}{k}\exp(-\delta x) \quad (2.26)$$

or

$$x=-\frac{1}{\delta-\frac{V}{\alpha}}\ln\left[\frac{kV}{I_0\alpha}\left(T_s+\frac{I_0}{k\delta}\right)\right] \quad (2.27)$$

However,

$$\frac{V}{\alpha}\ll\delta,$$

Therefore, the location of the maximum temperature is:

$$x_{\max}\simeq-\frac{1}{\delta}\ln\left[\frac{\rho C_p V T_s}{I_0}\right] \quad (2.28)$$

or, substituting for I_0 from Eq. (2.23) gives:

$$\delta x_{\max}=-\ln\left[\frac{\rho C_p V T_s}{\rho V(C_p T_s+L)}\right] \quad (2.29)$$

Finally,

$$\delta x_{\max}=\ln\left[1+\frac{L}{C_p T_s}\right] \quad (2.30)$$

The steady state analysis provides useful information on the recession velocity and the location of maximum temperature in the key hole; however, the phase change process is transient and temperature should be obtained from the solution of the transient heat equation.

The liquid surface layer formed during a laser pulse moves into the metal at a rate determined by the quantity of vapor expelled. As the temperature of the liquid molecules is increased the additional energy needed to free them from the binding forces decreases. The latent heat of vaporization therefore decreases with temperature until, at the critical temperature and above, it remains zero. It is an elliptical function of temperature, which appears adequate, for the accuracy sought [4].

$$L(T)=L_o\left[1-\left(\frac{T_c}{T_s}\right)^2\right]^{1/2} \quad (2.31)$$

where L_o is latent heat of vaporization at absolute zero.

The rate of change of latent heat with temperature can be expressed as:

$$\frac{dL}{dt}=\frac{L}{T}+\left(C_{p_2}-C_{p_1}\right)-\frac{L}{v_2-v_1}\left[\left(\frac{\partial v_2}{\partial T}\right)_p-\left(\frac{\partial v_1}{\partial T}\right)_p\right] \tag{2.32}$$

where C_{p_1} and C_{p_2} are specific heats at constant pressure, v_1 and v_2 are specific volumes, and 1,2 subscripts present for liquid and vapor states, respectively.

Although integration of the latent heat over the temperature range 0 degree to T_c is impossible, it can be used to show that little inaccuracy is involved in taking the room temperature latent heat as the latent heat at absolute zero. Because the specific volume of gas is much greater than the condensed liquid and its rate of change with temperature, at constant pressure, is corresponding greater, it results $v_2 \gg v_1$ and $\left(\frac{\partial v_2}{\partial T}\right)_p \gg \left(\frac{\partial v_1}{\partial T}\right)_p$.

Therefore, Eq. (2.32) yields:

$$\frac{dL}{dt}=\frac{L}{T}+\left(C_{p_2}-C_{p_1}\right)-\frac{L}{v_2}\left(\frac{\partial v_2}{\partial T}\right)_p \tag{2.33}$$

Applying the equation of state yields $v_2=\frac{RT_2}{P}$ and $\frac{\partial v_2}{\partial T}=\frac{P}{R}$

Thus, Eq. (4.35) becomes:

$$\frac{dL}{dt}=\frac{L}{T}+\left(C_{p_2}-C_{p_1}\right)-\frac{L}{T}=\Delta C_p \tag{2.34}$$

C_p is extremely small for temperatures up to the room temperatures [5] and so little error will result in taking L_o as latent heat at normal temperatures.

According to Maxwell's law the velocity distribution of molecules is [6],

$$f(V_z)dV_z=\left(\frac{m}{2\pi kT}\right)^{1/2}\exp\left(-\frac{mV_z^2}{2\pi kT}\right)dV_z \tag{2.35}$$

where V_z is the velocity in the direction normal to the surface, T is the temperature of solid, liquid, or gas, k is the Boltzmann's constant, and m is the mass of the atom.

Only those molecules whose velocity is greater than that given by:

$$\frac{1}{2}mV_{\min}^2=L(T) \tag{2.36}$$

where $V_{\min}$ lies in the direction normal to the surface and molecules with $V_{\min}$ will escape from the retaining potential. If n is the number of atoms per unit volume, then the number of atoms with these velocities passing a unit area per unit time is $nf(V_z)dV_z$. All these atoms satisfying the condition $V_z > V_{\min}$ do not return to their equilibrium position and are evaporated. If G is the number of atoms evaporated per unit time per unit area, then,

$$G = \int_{V_{\min}}^{\infty} nf(V_z)V_z dV_z \tag{2.37}$$

or

$$G = n\left(\frac{m}{2\pi k_B T}\right)^{1/2} \int_{V_{\min}}^{\infty} \exp\left(-\frac{mV_z^2}{2\pi k_B T}\right) V_z dV_z \tag{2.38}$$

After integration yields:

$$G = n\left(\frac{m}{2\pi k_B T}\right)^{1/2} \left\{ \frac{\exp\left(-\frac{mV_z^2}{2\pi k_B T}\right)}{mkT} \Bigg|_{V_{\min}}^{\infty} \right\} \tag{2.39}$$

or

$$G = n\left(\frac{m}{2\pi k_B T}\right)^{1/2} \exp\left[-\frac{L(T)}{kT}\right] \tag{2.40}$$

If the atoms are equally spaced within the lattice, a surface layer would consist of $n^{2/3}$ with an evaporation time $n^{2/3}/G$. The average velocity of the surface V_s is, therefore,

$$V_s = \frac{1}{n^{1/3}} \frac{G}{n^{2/3}} = \left(\frac{k_B T}{2\pi m}\right)^{1/2} \exp\left[-\frac{L(T)}{k_B T}\right] \tag{2.41}$$

Consider the laser irradiation of the solid substrate, the transient Fourier heat transfer equation governing the evaporation at the surface, including the absorption term, can be written as:

$$k\frac{\partial^2 T}{\partial x^2} + \rho C_p V \frac{\partial T}{\partial x} + I_0 \delta \exp(-\delta x) = \rho C_p \frac{\partial T}{\partial t} \tag{2.42}$$

where V is the same as V_s in Eq. (2.35), in which is $V = \left(\frac{k_B T_s}{2\pi m}\right)^{1/2} \exp\left(-\frac{L}{k_B T_s}\right)$

It can be observed that the transient heating problem is nonlinear, because the velocity is surface temperature (T_s) dependent. Consequently, complete solution to transient heat transfer equation (Eq. 2.42) is extremely difficult, but quasi-steady solution is feasible. The set of boundary conditions relevant to Eq. (2.42) are:

$$k\frac{\partial T}{\partial x}\bigg|_{x=0} = \rho VL; \ \ T(\infty, t) = 0 \ \text{and} \ T(x, 0) = 0. \tag{2.43}$$

The solution of Eq. (2.42) with the appropriate boundary conditions can be obtained using a Laplace transformation with respect to time (t), that is:

$$\frac{d^2\overline{T}(x,p)}{dx^2}+\frac{V}{\alpha}\frac{d\overline{T}(x,p)}{dx}-\frac{p}{\alpha}\overline{T}(x,p)=-\frac{I_0\delta}{kp}\exp(-\delta x) \tag{2.44}$$

The inversion of boundary conditions gives:

$$T(x,0)=0;\ \left.\frac{dT(x,p)}{dx}\right|_{x=0}=\frac{\rho VL}{kp};\ T(\infty,p)=0 \tag{2.45}$$

where p is the Laplace transform variable.

The solution to Eq. (2.42) gives the result in the Laplace domain:

$$\begin{aligned}\overline{T}(x,p) &= A\exp\left[-\frac{x}{\sqrt{\alpha}}\left(b-\sqrt{b^2+p}\right)\right]\\ &\quad +B\exp\left[-\frac{x}{\sqrt{\alpha}}\left(b+\sqrt{b^2+p}\right)\right]\\ &\quad -\frac{I_0\delta\alpha}{kp}\left[\frac{\exp(-\delta x)}{c^2-(b^2+p)}\right]\end{aligned} \tag{2.46}$$

where $b=\frac{V}{2\sqrt{\alpha}}$, $c=b-\delta\sqrt{\alpha}$ and A and B are constants of integration. Using the boundary condition (Eq. 2.45), it yields:

$$A=0 \tag{2.47}$$

Furthermore, substitution of boundary condition (Eq. 2.45) gives:

$$B=\frac{\sqrt{\alpha}}{b+\sqrt{b^2+p}}\left[\frac{I_0\alpha\delta^2}{kp(c^2-b^2-p)}-\frac{\rho VL}{kp}\right] \tag{2.48}$$

Therefore, the complete solution to the transformed equation (Eq. 2.46) is:

$$\begin{aligned}\overline{T}(x,p)&=\frac{\sqrt{\alpha}}{b+\sqrt{b^2+p}}\left[\frac{I_0\alpha\delta^2}{kp(c^2-b^2-p)}-\frac{\rho VL}{kp}\right]\exp\left[-\frac{x}{\sqrt{\alpha}}\left(b+\sqrt{b^2+p}\right)\right]\\ &\quad -\frac{I_0\delta\alpha}{kp}\left[\frac{\exp(-\delta x)}{c^2-(b^2+p)}\right]\end{aligned} \tag{2.49}$$

Inversion of the transformed solution, the difficulty arises due to the first term, which is a rather complicated function of the subsidiary variable p. However, use of the observation that the first term may be written as the indefinite integral, that is:

$$f(x,p)=-\int_0^x\left[\frac{I_0\alpha\delta^2}{kp(c^2-b^2-p)}-\frac{\rho VL}{kp}\right]\exp\left[-\frac{x}{\sqrt{\alpha}}\left(b+\sqrt{b^2+p}\right)\right]dx \tag{2.50}$$

or

$$f(x,p) = -\int_0^x g(x,p)dx \tag{2.51}$$

The inverse transformation of $f(x,p)$ function may be carried out in the following manner:

$$L^{-1}[f(x,p)] = -L^{-1}\left[\int_0^x g(x,p)dx\right] = -\int_0^x L^{-1}[g(x,p)]dx \tag{2.52}$$

where L^{-1} is the inverse Laplace transformation. The function $g(x,p)$ is easier to invert than the function $f(x,p)$, but involves indefinite integration after the inversion process. The result of this procedure for inverting the solution is the same as that obtained in the following method of expansion into partial fractions. Using the relationship:

$$L^{-1}[\phi(p+a)] = e^{-at}L^{-1}[\phi(p)] \tag{2.53}$$

it yields:

$$\begin{aligned} &L^{-1}\left[\frac{\sqrt{\alpha}}{b+\sqrt{b^2+p}}\frac{I_0\alpha\delta^2}{kp(c^2-b^2-p)}\exp\left[-\frac{x}{\sqrt{\alpha}}\left(b+\sqrt{b^2+p}\right)\right]\right] \\ &= -\exp\left[-\left(\frac{bx}{\sqrt{\alpha}}+b^2t\right)\right]L^{-1}\left[\frac{I_0\alpha\delta^2\exp(-qx)}{k(p-b^2)(\rho-c^2)(q+b/\sqrt{\alpha})}\right] \end{aligned} \tag{2.54}$$

where $q^2 = p/\alpha$, $b = V/(2\sqrt{\alpha})$, and $c = b - \delta\sqrt{\alpha}$.

This expression may be expanded into partial fractions using the residual theorem (Eq. 2.7):

$$\begin{aligned} &-\frac{I_0\delta^2}{k\alpha}\exp\left[-\left(\frac{bx}{\sqrt{\alpha}}+b^2t\right)\right]L^{-1}\left\{-\frac{\alpha\sqrt{\alpha}}{2b(b^2-c^2)}\frac{\exp(-qx)}{(q+b/\sqrt{\alpha})^2}\right. \\ &\left.-\frac{\alpha^2(5b^2-c^2)}{4b^2(b^2-c^2)}\frac{\exp(-qx)}{(q+b/\sqrt{\alpha})}+\frac{\alpha^2\exp(-qz)}{4b^2(b^2-c^2)(q+b/\sqrt{\alpha})}\right) \\ &+\frac{\alpha^2\exp(-qz)}{4b^2(b^2-c^2)(q-b/\sqrt{\alpha})}+\frac{\alpha^2\exp(-qx)}{2c(b+c)(b^2-c^2)(q-c/\sqrt{\alpha})}\Bigg) \\ &\left.+\frac{\alpha^2\exp(-qx)}{2c(b-c)(b^2-c^2)(q+c/\sqrt{\alpha})}\right\} \end{aligned} \tag{2.55}$$

which gives on inversion and after many algebraic manipulations:

$$\begin{aligned}
&\frac{I_0\delta\sqrt{\alpha}}{2\rho C_p(\alpha\delta-V)}\left\{4t\ \mathrm{ierfc}\left(\frac{x}{2\sqrt{\alpha t}}+b\sqrt{t}\right)\right.\\
&+\frac{3b^2+c^2}{2b(b^2-c^2)}\mathrm{erfc}\left(\frac{x}{2\sqrt{\alpha t}}+b\sqrt{t}\right)\Bigg)\\
&+\frac{1}{2b}\exp\left(-\frac{2bx}{\sqrt{\alpha}}\right)\mathrm{erfc}\left(\frac{x}{2\sqrt{\alpha t}}-b\sqrt{t}\right)\Bigg)\\
&-\frac{1}{b-c}\exp\left[-(\delta x+b^2-c^2)t\right]\mathrm{erfc}\left(\frac{x}{2\sqrt{\alpha t}}+c\sqrt{t}\right)\Bigg)\\
&\left.-\frac{1}{b+c}\exp\left[\frac{x}{\sqrt{\alpha}}(b+c)+(b^2-c^2)t\right]\mathrm{erfc}\left(\frac{x}{2\sqrt{\alpha t}}-c\sqrt{t}\right)\right\}
\end{aligned} \tag{2.56}$$

The second part of the term in the transformed solution may be inverted in a similar manner:

$$\begin{aligned}
&L^{-1}\left\{\frac{\rho VL\sqrt{\alpha}}{kp}\frac{\exp\left[-(x/\sqrt{\alpha})\left(b+\sqrt{p+b^2}\right)\right]}{b+\sqrt{p+b^2}}\right\}\\
&=\frac{\rho VL}{4kb^2}\exp\left[-\left(b^2t+\frac{bx}{\sqrt{\alpha}}\right)\right]\\
&\times L^{-1}\left[\frac{\exp(-qx)}{q-b/\sqrt{\alpha}}-\frac{\exp(-qx)}{q+b/\sqrt{\alpha}}-\frac{2b}{\sqrt{\alpha}}\frac{\exp(-qx)}{(q+b/\sqrt{\alpha})^2}\right]
\end{aligned} \tag{2.57}$$

which, after transformation, gives:

$$\begin{aligned}
&\frac{\rho VL}{4bk}\left[4b\sqrt{\alpha t}\ \mathrm{ierfc}\left(\frac{x}{2\sqrt{\alpha t}}+b\sqrt{t}\right)-\sqrt{\alpha}\left(\frac{x}{2\sqrt{\alpha t}}+b\sqrt{t}\right)\right.\\
&\left.+\sqrt{\alpha},\exp\left(-\frac{2bx}{\sqrt{\alpha}}\right)\mathrm{erfc}\left(\frac{x}{2\sqrt{\alpha t}}-b\sqrt{t}\right)\right]
\end{aligned} \tag{2.58}$$

Finally, the term:

$$L^{-1}\left[\frac{I_0\alpha\delta}{k}\frac{\exp(-\delta x)}{p(p+b^2-c^2)}\right]=\frac{I_0\delta\sqrt{\alpha}}{\rho C_p(\alpha\delta-V)}\times\left\{\frac{\exp[-[\delta x+(b^2-c^2)t]]}{b-c}-\frac{\exp(-\delta x)}{b-c}\right\} \tag{2.59}$$

where $b-c=\delta\sqrt{\alpha}$.

Substitution of all these terms (Eqs. 2.56, 2.58, 2.59) into Eq. (2.49) gives the complete solution, that is:

$$
\begin{aligned}
T(x,t) = {} & \frac{I_0\delta\sqrt{\alpha}}{2\rho C_p(\alpha\delta - V)}\left\{4\sqrt{t}\,\text{ierfc}\left(\frac{x}{2\sqrt{\alpha t}} + b\sqrt{t}\right)\right. \\
& + \frac{3b^2+c^2}{2b(b^2-c^2)}\,\text{erfc}\left(\frac{x}{2\sqrt{\alpha t}} + b\sqrt{t}\right)\Bigg) \\
& + \frac{1}{2b}\exp\left(-\frac{2bx}{\alpha}\right)\text{erfc}\left(\frac{x}{2\sqrt{\alpha t}} - b\sqrt{t}\right)\Bigg) \\
& + \frac{1}{(b-c)}\exp\left[-\left[\delta x + (b^2-c^2)\right]t\right]\text{erfc}\left[-\left(\frac{x}{2\sqrt{\alpha t}} + c\sqrt{t}\right)\right]\Bigg) \\
& - \frac{1}{(b+c)}\exp\left[\frac{x}{\sqrt{\alpha}}(b+c) + (b^2-c^2)t\right]\text{erfc}\left(\frac{x}{2\sqrt{\alpha t}} - c\sqrt{t}\right)\Bigg) \\
& \left. - \frac{2}{(b-c)}\exp(-\delta x)\right\} - \frac{\rho VL}{4bk}\left\{4b\sqrt{\alpha t}\,\text{ierfc}\left(\frac{x}{2\sqrt{\alpha t}} + b\sqrt{t}\right)\right. \\
& - \sqrt{\alpha}\,\text{erfc}\left(\frac{x}{2\sqrt{\alpha t}} + b\sqrt{t}\right)\Bigg) \\
& \left. + \sqrt{\alpha}\exp\left(-\frac{-2bx}{\sqrt{\alpha}}\right)\text{erfc}\left(\frac{x}{2\sqrt{\alpha t}} - b\sqrt{t}\right)\right\}
\end{aligned}
\tag{2.60}
$$

Setting $x = 0$ in Eq. (2.60) results in the surface temperature, that is:

$$
\begin{aligned}
T(0,t) = {} & \frac{I_0\delta\sqrt{\alpha}}{2\rho C_p(\alpha\delta - V)}\left[4\sqrt{t}\,\text{ierfc}(b\sqrt{t}) + \frac{(b^2+c^2)}{b(b^2-c^2)}\,\text{erfc}(b\sqrt{t}) + \frac{1}{b}\right. \\
& \left. + \frac{\exp\left[-(b^2-c^2)t\right]}{(b-c)}\,\text{erfc}(-c\sqrt{t}) - \frac{\exp\left[-(b^2-c^2)t\right]}{(b+c)}\,\text{erfc}(-c\sqrt{t}) - \frac{2}{b-c}\right] \\
& - \frac{\rho VL}{4bk}\left[4b\sqrt{\alpha t}\,\text{ierfc}(b\sqrt{t}) - \sqrt{\alpha}\,\text{erfc}(b\sqrt{t}) + \sqrt{\alpha}\,\left(2 - \text{erfc}(b\sqrt{t})\right)\right]
\end{aligned}
\tag{2.61}
$$

rearrangement gives:

$$
\begin{aligned}
T(0,t) = {} & \frac{I_0\delta\sqrt{\alpha}}{2\rho C_p(\alpha\delta - V)}\left[4\sqrt{t}\,\text{ierfc}(b\sqrt{t}) + \frac{(b^2+c^2)}{b(b^2-c^2)}\,\text{erfc}(b\sqrt{t})\right. \\
& \left. + \frac{2c}{b^2-c^2}\exp\left[-(b^2-c^2)t\right]\,\text{erfc}(-c\sqrt{t}) - \frac{(b+c)}{b(b-c)}\right] \\
& - \frac{\rho VL}{2bk}\left[2b\sqrt{\alpha t}\,\text{ierfc}(b\sqrt{t}) + \sqrt{\alpha}\,\text{erfc}(b\sqrt{t})\right]
\end{aligned}
\tag{2.62}
$$

Eqs. (2.60), (2.62) are the complete quasi-steady solution of the governing equation of the heat transfer. Eqs. (2.60), (2.62) can be used to acquire a more accurate solution, which can be obtained by an iterative procedure. It is expected that this solution

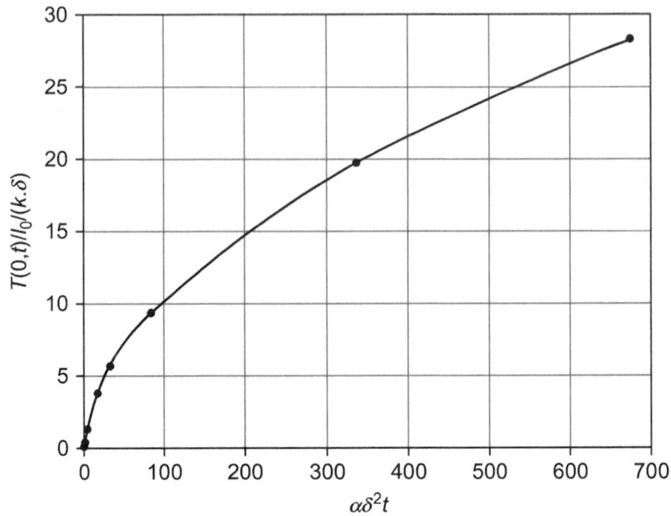

Fig. 2.2 Temporal variation of dimensionless surface temperature.

would be obtained by developing the solution from time $t=0$. In the initial stages, the evaporation rate is small and the solution approaches that of the pure conduction process. As the surface temperature rises so do the evaporation rates. The values for the velocity and the surface temperature can be obtained by using time steps which are small enough so that the change in the surface velocity between steps is also small; therefore, the velocity derived in the previous step can be used directly in Eq. (2.62). With this new value of the surface temperature, an improved estimate of the surface velocity (V_s) can be obtained and the iteration repeated to give a convergent solution.

Fig. 2.2 shows the dimensionless surface temperature ($T(0,t)k\delta/I_o$) with $\alpha\delta^2 t$ for steel. It is evident that the slope of the curve changes when the phase change occurs. Consequently, the temperature distribution with the heating time may be distinguished into two zones. In the conduction limited zone, three distinct regions may be identified. In the first region, for $0 \leq \alpha\delta^2 t \leq 0.1$, the surface temperature increases very rapidly with time. This corresponds to initial heating of the surface layers of the material, where the conduction losses from the surface are low, and follows from the consideration of the rate of change of temperature with time. In the second region, from $0.1 \leq \alpha\delta^2 t \leq 1.0$, the rate of change of temperature begins to fall, indicating that the conduction process is playing a dominant role in the energy transfer process. In the third region, for, the curve is almost linear, indicating that the relative levels of internal energy increase and the conduction effects are reaching a balance. Although the temperature in the material does not reach a steady state value at any time, it follows that in this sense an equilibrium condition is reached when $\alpha\delta^2 t = 1.0$. In this case, the surface temperature, due to the conduction-only process, reduces to:

$$T(0,t) = \frac{2}{\sqrt{\pi}} \frac{I_o}{k\delta}$$

This implies that the material has an approximate equilibrium time, $\tau = \frac{1}{\alpha\delta^2}$ (equilibrium time, which is the time required for an element at a distance $1/\delta$ from the material surface to respond to a temperature disturbance at that surface).

2.3.1 Exponential pulse heating and evaporation at the surface

The solution resented in Eq. (2.62) is valid for a step input laser pulse. However, in general laser heating, pulse varies exponentially. In this case, the closed form solution of temperature fields in the cavity changes significantly once the time-dependent laser pulse intensity is introduced. The temporal variation of the laser-pulse intensity can be simplified in the exponential form as $I_o \exp(-\beta t)$. Therefore, the Fourier heat transfer equation becomes [7]:

$$k\frac{\partial^2 T}{\partial x^2} + \rho CpV\frac{\partial T}{\partial x} + I_o(1 - r_f)\exp(-\beta t)\delta\exp(-\delta x) = \rho C_p\frac{\partial T}{\partial t} \tag{2.63}$$

with the boundary conditions,

$$\left.\frac{\partial T}{\partial x}\right|_{x=0} = \frac{\rho VL}{k} : T(\infty, t) = 0 \text{ and } T(x, 0) = 0 \tag{2.64}$$

where V is the recession velocity, L is the latent heat of evaporation, r_f is the surface reflectivity. The recession velocity of the surface is:

$$V = \frac{I_1}{\rho[CpTs + L]} \tag{2.65}$$

where $I_1 = I_o(1 - r_f)$. Eq. (2.63) can be written as:

$$\frac{\partial^2 T}{\partial x^2} + \frac{V}{\alpha}\frac{\partial T}{\partial x} + I_1\exp(-\beta t)\delta\ \exp(-\delta x) = \frac{1}{\alpha}\frac{\partial T}{\partial t} \tag{2.66}$$

The Laplace transformation of Eq. (2.66) with respect to t can be written as:

$$\frac{\partial^2 \overline{T}}{\partial x^2} + \frac{V}{\alpha}\frac{\partial \overline{T}}{\partial x} + I_1\frac{\delta}{k}\frac{1}{(p+\beta)}\exp(-\delta x) = \frac{1}{\alpha}\left[p\overline{T}\right] \tag{2.67}$$

where $\overline{T}$ is the Laplace transformation of the temperature. Eq. (2.67) yields:

$$\frac{\partial^2 \overline{T}}{\partial x^2} + \frac{V}{\alpha}\frac{\partial \overline{T}}{\partial x} - \frac{p}{\alpha}\overline{T} = -I_1\frac{\delta}{k}\frac{1}{(p+\beta)}\exp(-\delta x) \tag{2.68}$$

Eq. (2.68) has homogeneous and particular solutions. Therefore, $\overline{T}$ yields:

$$\overline{T} = \overline{Th} + \overline{Tp} \tag{2.69}$$

The characteristic equation for the homogeneous solution can be written as:

$$r^2+\frac{V}{\alpha}r-\frac{p}{\alpha}=0 \tag{2.70}$$

which yields the solution of:

$$r_{1,2}=-\frac{V}{2\alpha}\pm\frac{\sqrt{V^2+4p\alpha}}{2\alpha} \tag{2.71}$$

Therefore, the homogenous solution ($\overline{Th}$) yields:

$$\overline{Th}=c_1e^{r_1x}+c_2e^{r_2x} \tag{2.72}$$

or,

$$\overline{Th}=e^{-\frac{Vx}{2\alpha}}\left[c_1e^{-\frac{\sqrt{V^2+4p\alpha}}{2\alpha}x}+c_2e^{\frac{\sqrt{V^2+4p\alpha}}{2\alpha}x}\right] \tag{2.73}$$

For the particular solution, one can propose $\overline{Tp}=A_oe^{-\delta x}$. Consequently, substitution of $\overline{Tp}$ in Eq. (2.68) results:

$$A_o\delta^2e^{-\delta x}-\frac{V}{\alpha}\delta A_oe^{-\delta x}-\frac{p}{\alpha}A_oe^{-\delta x}=H_oe^{-\delta x} \tag{2.74}$$

where

$$H_o=I_1\frac{\delta}{k}\frac{1}{(p+\beta)} \tag{2.75}$$

and

$$A_o=-\frac{H_o\alpha}{p+V\delta-\alpha\delta^2} \tag{2.76}$$

or

$$\overline{T}=e^{-\frac{Vx}{2\alpha}}\left[c_1e^{-\frac{\sqrt{V^2+4p\alpha}}{2\alpha}x}+c_2e^{\frac{\sqrt{V^2+4p\alpha}}{2\alpha}x}\right]-\frac{H_o\alpha}{p+V_1}e^{-\delta x} \tag{2.77}$$

where $V_1=V\delta-\alpha\delta^2$. Because $T(\infty,t)=0$, then $c_2=0$. Therefore:

$$\overline{T}=c_1e^{-\frac{1}{2\alpha}\left[V+\sqrt{V^2+4\alpha p}\right]x}+\frac{I_1\delta\alpha}{k}\frac{e^{-\delta x}}{(p+\beta)(p+V_1)} \tag{2.78}$$

Let $w_1 = \frac{I_1 \delta \alpha}{k}$ and use the boundary condition $\left[\frac{\partial T}{\partial x}\right]_{x=0} = \frac{\rho V L}{k}$, one can obtain c_1, that is:

$$\frac{\partial \overline{T}}{\partial x} = \left\{ -\frac{1}{2\alpha}\left[V + \sqrt{V^2 + 4\alpha p}\right] c_1 e^{-\frac{V}{2\alpha}\left[V + \sqrt{V^2 + 4\alpha p}\right]x} - \frac{\delta w_1 e^{-\delta x}}{(p+\beta)(p+V_1)} \right\}_{x=0} = \frac{\rho V L}{k} \tag{2.79}$$

or

$$c_1 = -\frac{2\alpha \delta w_1}{(p+\beta)(p+V_1)\left[V + \sqrt{V^2 + 4\alpha p}\right]} - \frac{\frac{\rho V L}{k} 2\alpha}{V + \sqrt{V^2 + 4\alpha p}} \tag{2.80}$$

Hence:

$$\overline{T} = -\frac{2\alpha \delta w_1 e^{-\frac{1}{2\alpha}\left[V + \sqrt{V^2 + 4\alpha p}\right]x}}{(p+\beta)(p+V_1)\left[V + \sqrt{V^2 + 4\alpha p}\right]} + \frac{w_1 e^{-\delta x}}{(p+\beta)(p+V_1)} + \overline{T_{mo}} \tag{2.81}$$

where

$$\overline{T_{mo}} = -2\alpha \frac{\rho V L}{k} \frac{e^{-\frac{1}{2\alpha}\left[V + \sqrt{V^2 + 4\alpha p}\right]x}}{\left(V + \sqrt{V^2 + 4\alpha p}\right)} \tag{2.82}$$

Let H_1, H_2, and H_5 are:

$$H_1 = \frac{e^{-\frac{1}{2\alpha}\left[V + \sqrt{V^2 + 4\alpha p}\right]x}}{(p+\beta)(p+V_1)\left[V + \sqrt{V^2 + 4\alpha p}\right]} \tag{2.83}$$

and

$$H_2 = \frac{1}{(p+\beta)(p+V_1)} = \frac{1}{(V_1 - \beta)}\left[\frac{1}{p+\beta} - \frac{1}{p+V_1}\right] \tag{2.84}$$

and

$$H_5 = \frac{e^{-\frac{1}{2\alpha}\left[V + \sqrt{V^2 + 4\alpha p}\right]x}}{V + \sqrt{V^2 + 4\alpha p}} \tag{2.85}$$

Consequently,

$$T(x,t) = \pounds^{-1}\overline{T} = -2\alpha \delta w_1 \pounds^{-1} H_1 + w_1 e^{-\delta x} \pounds^{-1} H_2 - 2\alpha \frac{\rho V L}{k} \pounds^{-1} H_5 \tag{2.86}$$

To obtain the inverse transformation of functions $\pounds^{-1}H_1$ and $\pounds^{-1}H_5$, lets introduce $s = V^2 + 4\alpha p$ or $ds = 4\alpha dp$ and $p = \frac{1}{4\alpha}(s - V^2)$. Therefore:

$$\pounds^{-1}H_1 = \frac{1}{2\pi i}\int_{c-i\infty}^{c+i\infty} e^{tp} H_1(p)dp \tag{2.87}$$

or

$$\pounds^{-1}H_1 = \frac{1}{4\alpha}e^{-\left(\frac{Vx}{2\alpha}+\frac{tV^2}{4\alpha}\right)}\left\{\frac{1}{2\pi i}\int_{\overline{c}-i\infty}^{\overline{c}+i\infty}\frac{e^{-\left(\frac{x\sqrt{s}}{2\alpha}-\frac{st}{4\alpha}\right)}}{\left(\frac{s}{4\alpha}-\frac{V^2}{4\alpha}+\beta\right)\left(\frac{s}{4\alpha}-\frac{V^2}{4\alpha}+V_1\right)(V+\sqrt{s})}ds\right\} \tag{2.88}$$

where $\overline{c} = 4\alpha c + V^2$.

We use one more transformation by letting $\sigma = \frac{s}{4\alpha}$ or $ds = 4\alpha d\sigma$ and after lengthy algebra, one obtains:

$$\pounds^{-1}H_1 = \frac{1}{4\alpha}e^{-\frac{V}{2\alpha}\left(x+\frac{V}{2}t\right)}\left\{\frac{1}{2\pi i}\int_{\overline{c}-i\infty}^{\overline{c}+i\infty}\frac{e^{-\frac{x\sqrt{s}}{2\alpha}}e^{\sigma t}}{\left(\sigma-\frac{V^2}{4\alpha}+\beta\right)\left(\sigma-\frac{V^2}{4\alpha}+V_1\right)(V+2\sqrt{\alpha}\sqrt{\sigma})}d\sigma\right\} \tag{2.89}$$

where $\hat{c} = \frac{\overline{c}}{4\alpha}$.

Hence:

$$\pounds^{-1}H_1 = e^{-\frac{V}{2\alpha}\left(x+\frac{V}{2}t\right)}\pounds^{-1}H_3 \tag{2.90}$$

where:

$$H_3 = \frac{e^{-\left(\frac{x}{\sqrt{\alpha}}\sqrt{p}\right)}}{\left(p-\frac{V^2}{4\alpha}+\beta\right)\left(p-\frac{V^2}{4\alpha}+V_1\right)(V+2\sqrt{\alpha}\sqrt{p})} \tag{2.91}$$

Similarly, $\pounds^{-1}H_5$ can be obtained, that is:

$$\pounds^{-1}H_5 = e^{-\frac{V}{2\alpha}\left(x+\frac{V}{2}t\right)}\left\{\frac{1}{2\pi i}\int_{\overline{c}-i\infty}^{\overline{c}+i\infty}\frac{e^{-\frac{x\sqrt{\sigma}}{\sqrt{\alpha}}}e^{\sigma t}}{(V+2\sqrt{\alpha}\sqrt{\sigma})}d\sigma = e^{-\frac{V}{2\alpha}\left(x+\frac{V}{2}t\right)}\pounds^{-1}H_7\right\} \tag{2.92}$$

where:

$$H_7 = \frac{e^{-\left(\frac{x}{\sqrt{\alpha}}\sqrt{p}\right)}}{(V+2\sqrt{\alpha}\sqrt{p})} \tag{2.93}$$

From the Laplace table [8], it is known that:

$$£^{-1}\frac{e^{-k\sqrt{p}}}{a+\sqrt{p}}=\frac{1}{\sqrt{\pi t}}\exp\left(-\frac{k^2}{4t}\right)-ae^{ak}e^{a^2t}erfc\left(a\sqrt{t}+\frac{k}{2\sqrt{t}}\right) \tag{2.94}$$

where $k\geq 0$.

Let $w_2^2=\frac{V^2}{4\alpha}-\beta : k_1=\frac{x}{\sqrt{\alpha}} : w_3^2=\frac{V^2}{4\alpha}-V_1 : w_4=\frac{V}{2\sqrt{\alpha}}$, then, H_3 becomes:

$$H_3=\frac{1}{\left[\left(\sqrt{p}\right)^2-w_2^2\right]\left[\left(\sqrt{p}\right)^2-w_3^2\right]\left[\sqrt{p}+w_4\right]}\frac{1}{2\sqrt{\alpha}}e^{-k_1\sqrt{p}} \tag{2.95}$$

and, after using partial fraction expansion, H_3 becomes:

$$H_3=\left[\frac{D_1}{\sqrt{p}-w_2}+\frac{D_2}{\sqrt{p}+w_2}+\frac{D_3}{\sqrt{p}-w_3}+\frac{D_4}{\sqrt{p}+w_3}+\frac{D_5}{\sqrt{p}+w_4}\right]\frac{1}{2\sqrt{\alpha}}e^{-k_1\sqrt{p}} \tag{2.96}$$

where:

$$\begin{aligned}
D_1&=\frac{1}{2w_2\left(w_2^2-w_3^2\right)\left(w_2+w_4\right)}\\
D_2&=-\frac{1}{2w_2\left(w_2^2-w_3^2\right)\left(-w_2+w_4\right)}\\
D_3&=\frac{1}{2w_3\left(w_3^2-w_2^2\right)\left(w_3+w_4\right)}\\
D_4&=-\frac{1}{2w_3\left(w_3^2-w_2^2\right)\left(-w_3+w_4\right)}\\
D_5&=\frac{1}{\left(w_4^2-w_2^2\right)\left(w_4^2-w_3^2\right)}
\end{aligned} \tag{2.97}$$

After substituting Eq. (2.97) into Eq. (2.96), $£^{-1}H_3$ can be written as:

$$£^{-1}H_3=\frac{1}{2\sqrt{\alpha}}\left\{\begin{array}{l}
D_1\left[\frac{1}{\sqrt{\pi t}}e^{-\frac{k_1^2}{4t}}+w_2e^{-w_2k_1}e^{w_2^2t}\mathrm{erfc}\left(-w_2\sqrt{t}+\frac{k_1}{2\sqrt{t}}\right)\right]\\
+D_2\left[\frac{1}{\sqrt{\pi t}}e^{-\frac{k_1^2}{4t}}-w_2e^{w_2k_1}e^{w_2^2t}\mathrm{erfc}\left(w_2\sqrt{t}+\frac{k_1}{2\sqrt{t}}\right)\right]\\
D_3\left[\frac{1}{\sqrt{\pi t}}e^{-\frac{k_1^2}{4t}}+w_3e^{-w_3k_1}e^{w_3^2t}\mathrm{erfc}\left(-w_3\sqrt{t}+\frac{k_1}{2\sqrt{t}}\right)\right]\\
+D_4\left[\frac{1}{\sqrt{\pi t}}e^{-\frac{k_1^2}{4t}}-w_3e^{w_3k_1}e^{w_3^2t}\mathrm{erfc}\left(w_3\sqrt{t}+\frac{k_1}{2\sqrt{t}}\right)\right]\\
+D_5\left[\frac{1}{\sqrt{\pi t}}e^{-\frac{k_1^2}{4t}}-w_4e^{w_4k_1}e^{w_4^2t}\mathrm{erfc}\left(w_4\sqrt{t}+\frac{k_1}{2\sqrt{t}}\right)\right]
\end{array}\right\} \tag{2.98}$$

Similarly, $£^{-1}H_7$ yields:

$$£^{-1}H_7 = \frac{1}{2\sqrt{\alpha}}\left[\frac{1}{\sqrt{\pi t}}e^{-\frac{x^2}{4\alpha t}} - \frac{V}{2\sqrt{\alpha}}e^{-\frac{Vx}{2\alpha}}e^{\frac{V^2 t}{4\alpha}}\operatorname{erfc}\left(\frac{V\sqrt{t}}{2\sqrt{\alpha}} + \frac{x}{2\sqrt{\alpha t}}\right)\right] \tag{2.99}$$

Therefore,

$$\begin{aligned} T(x,t) &= £^{-1}\overline{T} \\ &= -2\alpha\delta w_1 e^{-\frac{V}{2\alpha}\left(x+\frac{V}{2}t\right)}£^{-1}H_3 + w_1 e^{-\delta x}£^{-1}H_2 - 2\alpha\frac{\rho VL}{k}£^{-1}H_5 \end{aligned} \tag{2.100}$$

Noting that:

$$£^{-1}H_2 = \frac{1}{V_1-\beta}\left(e^{-\beta t} - e^{-V_1 t}\right) \tag{2.101}$$

Substituting of Eqs. (2.101), (2.98), (2.92) into Eq. (2.101), it yields:

$$\begin{aligned} T(x,t) = &-\frac{\alpha\delta w_1}{\sqrt{\alpha}}e^{-\frac{V}{2\alpha}\left(x+\frac{V}{2}t\right)} \\ &\left\{\begin{aligned} &\frac{1}{\sqrt{\pi t}}e^{-\frac{k_1^2}{4t}}[D_1+D_2+D_3+D_4+D_5] + D_1\left[w_2 e^{-w_2k_1}e^{w_2^2 t}\operatorname{erfc}\left(-w_2\sqrt{t}+\frac{k_1}{2\sqrt{t}}\right)\right] \\ &+D_2\left[-w_2 e^{-w_2k_1}e^{w_2^2 t}\operatorname{erfc}\left(w_2\sqrt{t}+\frac{k_1}{2\sqrt{t}}\right)\right] + D_3\left[w_3 e^{-w_3k_1}e^{w_3^2 t}\operatorname{erfc}\left(-w_3\sqrt{t}+\frac{k_1}{2\sqrt{t}}\right)\right] \\ &+D_4\left[-w_3 e^{-w_3k_1}e^{w_3^2 t}\operatorname{erfc}\left(w_3\sqrt{t}+\frac{k_1}{2\sqrt{t}}\right)\right] + D_5\left[-w_4 e^{-w_4k_1}e^{w_4^2 t}\operatorname{erfc}\left(w_4\sqrt{t}+\frac{k_1}{2\sqrt{t}}\right)\right] \\ &+\frac{w_1}{V_1-\beta}e^{-\delta x}\left[e^{-\beta t} - e^{-V_1 t}\right] \end{aligned}\right\} \\ &-\frac{\alpha\rho L}{2k}\left(e^{-\frac{Vx}{\alpha}}\operatorname{erfc}\left(\frac{x-Vt}{2\sqrt{\alpha t}}\right) - \left(1+\frac{Vx}{\alpha}+\frac{V^2t}{\alpha}\right)\operatorname{erfc}\left(\frac{x+Vt}{2\sqrt{\alpha t}}\right)\right) - \frac{\rho VL}{k}\frac{\sqrt{\alpha t}}{\sqrt{\pi}}e^{-\frac{(x+Vt)^2}{4\alpha t}} \end{aligned} \tag{2.102}$$

From Eq. (2.108), $D_1+D_2+D_3+D_4+D_5=0$. Therefore, Eq. (2.102) becomes:

$$\begin{aligned} T(x,t) = &-\sqrt{\alpha}\delta w_1 e^{-\frac{V}{2\alpha}\left(x+\frac{Vt}{2}\right)} \\ &\left\{\begin{aligned} &\frac{e^{w_2^2 t}}{2\left(w_2^2-w_3^2\right)}\left[e^{-w_2k_1}\frac{\operatorname{erfc}\left(-w_2\sqrt{t}+\frac{k_1}{2\sqrt{t}}\right)}{w_2+w_4} + e^{w_2k_1}\frac{\operatorname{erfc}\left(w_2\sqrt{t}+\frac{k_1}{2\sqrt{t}}\right)}{w_4-w_2}\right] \\ &+\frac{e^{w_3^2 t}}{2\left(w_3^2-w_2^2\right)}\left[e^{-w_3k_1}\frac{\operatorname{erfc}\left(-w_3\sqrt{t}+\frac{k_1}{2\sqrt{t}}\right)}{w_3+w_4} + e^{w_3k_1}\frac{\operatorname{erfc}\left(w_3\sqrt{t}+\frac{k_1}{2\sqrt{t}}\right)}{w_4-w_3}\right] \\ &-D_5 w_4 e^{w_4^2 t}e^{w_4k_1}\operatorname{erfc}\left(w_4\sqrt{t}+\frac{k_1}{2\sqrt{t}}\right) \end{aligned}\right\} \\ &+\frac{w_1}{V_1-\beta}e^{-\delta x}\left(e^{-\beta t}-e^{-V_1 t}\right) - \frac{\alpha\rho L}{2k}\left(e^{-\frac{Vx}{\alpha}}\operatorname{erfc}\left(\frac{x-Vt}{2\sqrt{\alpha t}}\right)\right. \\ &\left.-\left(1+\frac{Vx}{\alpha}+\frac{V^2t}{\alpha}\right)\operatorname{erfc}\left(\frac{x+Vt}{2\sqrt{\alpha t}}\right)\right) - \frac{\rho VL}{k}\frac{\sqrt{\alpha t}}{\sqrt{\pi}}e^{-\frac{(x+Vt)^2}{4\alpha t}} \end{aligned} \tag{2.103}$$

Eq. (2.103) can be nondimensionalized using the following relations:

$$V^* = \frac{1}{\alpha\delta}V \; : \; \beta^* = \frac{1}{\alpha\delta^2}\beta \; : \; t^* = \alpha\delta^2 t : x^* = x\delta$$

This results in:

$$\begin{aligned}
T^*(x,t) =& \in -\frac{I_1}{k\delta} e^{-\frac{V^{*2}}{2}\left(x^* + \frac{V^*}{2}t^*\right)} \\
&\left\{ \begin{array}{l}
\frac{e^{\left(\frac{V^{*2}}{4} - (V^*-1)\right)t^*}}{(V^* - (1+\beta^*))} \left[e^{-\left(\sqrt{\frac{V^{*2}}{4} - \beta^*}\right)x^*} \frac{\operatorname{erfc}\left(-\sqrt{\left(\frac{V^{*2}}{4} - \beta^*\right)t^*} + \frac{x^*}{2\sqrt{t^*}}\right)}{\left(\sqrt{V^{*2} - \beta^*} + V^*\right)} \right. \\
\left. + e^{\left(\sqrt{\frac{V^{*2}}{4} - \beta^*}\right)x^*} \frac{\operatorname{erfc}\left(\sqrt{\left(\frac{V^{*2}}{4} - \beta^*\right)t^*} + \frac{x^*}{2\sqrt{t^*}}\right)}{\left(V^* - \sqrt{V^{*2} - \beta^*}\right)} \right] \\
+ \frac{e^{\left(\frac{V^{*2}}{4} - (V^*-1)\right)t^*}}{(\beta^* - (V^* - 1))} \left[e^{-\sqrt{\frac{V^{*2}}{4} - (V^*-1)}} \frac{\operatorname{erfc}\left(-\sqrt{\left(\frac{V^{*2}}{4} - (V^*-1)\right)t^*} + \frac{x^*}{2\sqrt{t^*}}\right)}{\left(\sqrt{V^{*2} - 4(V^*-1)} + V^*\right)} \right. \\
\left. + e^{\sqrt{\frac{V^{*2}}{4} - (V^*-1)}} \frac{\operatorname{erfc}\left(\sqrt{\left(\frac{V^{*2}}{4} - (V^*-1)\right)t^*} + \frac{x^*}{2\sqrt{t^*}}\right)}{\left(V^* - \sqrt{V^{*2} - 4(V^*-1)}\right)} \right] \\
- \frac{1}{2(\beta^*(V^*-1))} e^{\frac{V^{*2}}{4}t^*} e^{\frac{V^*}{2}x^*} \operatorname{erfc}\left(\frac{V^*}{2}t^* + \frac{x^*}{2\sqrt{t^*}}\right) +
\end{array} \right\} \\
&+ \frac{I_1}{k\delta} \frac{1}{((V^*-1) - \beta^*)} e^{-x^*} \left(e^{-\beta^* t^*} - e^{-(V^*-1)t^*}\right) \\
&- \frac{\alpha\rho L}{2k} \left[\begin{array}{l}
\left(e^{-V^* x^*} \operatorname{erfc}\left(\frac{x^* - V^* t^*}{2\sqrt{t^*}}\right) - (1 + V^* x^* + V^{*2} t^*) \operatorname{erfc}\left(\frac{x^* + V^* t^*}{2\sqrt{t^*}}\right)\right) \\
+ \frac{2V^* t^*}{\sqrt{\pi}} e^{-\left(\frac{V^{*2}}{4\tau} - \frac{V^* x^*}{2} + \frac{V^{*2} t^*}{4}\right)}
\end{array} \right]
\end{aligned} \tag{2.104}$$

Replacing $x=0$ at the surface and $k_1=0$ (because $k_1=\frac{x}{\sqrt{\alpha}}$), then Eq. (2.103) gives the surface temperature, that is

$$T(0,t)=-\sqrt{\alpha}\delta w_1 e^{-\frac{V}{2\alpha}\left(\frac{Vt}{2}\right)}\left\{\begin{array}{c}\frac{e^{w_2^2 t}}{2\left(w_2^2-w_3^2\right)}\left[\frac{erfc\left(-w_2\sqrt{t}\right)}{w_2+w_4}+\frac{erfc\left(w_2\sqrt{t}\right)}{w_4-w_2}\right]\\ +\frac{e^{w_3^2 t}}{2\left(w_3^2-w_2^2\right)}\left[\frac{erfc\left(-w_3\sqrt{t}\right)}{w_3+w_4}+\frac{erfc\left(w_3\sqrt{t}\right)}{w_4-w_3}\right]\\ -D_5 w_4 e^{w_4^2 t}\operatorname{erfc}\left(w_4\sqrt{t}\right)\end{array}\right\}$$
$$+\frac{w_1}{V_1-\beta}\left(e^{-\beta t}-e^{-V_1 t}\right)$$
$$-\frac{\alpha\rho L}{2k}\left(\operatorname{erfc}\left(-\frac{Vt}{2\sqrt{\alpha t}}\right)-\left(1+\frac{V^2 t}{\alpha}\right)\operatorname{erfc}\left(\frac{Vt}{2\sqrt{\alpha t}}\right)\right)$$
$$-\frac{\rho VL}{k}\frac{\sqrt{\alpha t}}{\sqrt{\pi}}e^{-\frac{(Vt)^2}{4\alpha t}} \tag{2.105}$$

Noting that:

$$\operatorname{erfc}(-x)=2-\operatorname{erfc}(x) \quad \text{and} \quad w_2^2-w_3^2=V_1-\beta$$

hence:

Lets recall that:

$$\alpha=\frac{k}{\rho Cp} \; : \; V_1=V\delta-\alpha\delta^2 \; : \; w_1=\frac{I_1\alpha\delta}{k} \; : \; w_2^2=\frac{V^2}{4\alpha}-\beta$$

$$w_3^2=\frac{V^2}{4\alpha}-V_1 \; : \; w_4^2=\frac{V^2}{4\alpha} \; : \; k_1=\frac{x}{\sqrt{\alpha}}$$

$$D_5=\frac{1}{\left(w_4^2-w_2^2\right)\left(w_4^2-w_3^2\right)}=\frac{1}{\beta V_1}$$

Substituting these values into Eq. (2.105) and after the lengthy algebra, it yields:

$$T(0,t)=\frac{I_1\alpha\delta}{k}\frac{1}{(V_1-\beta)}\left\{\begin{array}{c}\frac{-\delta\sqrt{\alpha}}{\beta}w_2e^{-\beta t}\left[\mathrm{erfc}\left(w_2\sqrt{t}\right)\right]\\+\frac{\delta\sqrt{\alpha}}{V_1}w_3e^{-V_1t}\left[\mathrm{erfc}\left(w_3\sqrt{t}\right)\right]\\+\frac{(V_1-\beta)}{\beta V_1}\delta\sqrt{\alpha}w_4\mathrm{erfc}\left(w_4\sqrt{t}\right)\\\left(\frac{-\delta\sqrt{\alpha}}{w_2+w_4}+1\right)e^{-\beta t}+\left(\frac{\delta\sqrt{\alpha}}{w_2+w_4}-1\right)e^{-V_1t}\end{array}\right\}+\frac{\rho VL}{k}\left[\frac{V}{2}\mathrm{erfc}\left(\frac{V\sqrt{t}}{2\sqrt{\alpha}}\right)-\frac{\sqrt{\alpha}}{\sqrt{\pi t}}e^{-\frac{V^2t}{4\alpha}}\right] \tag{2.106}$$

In order to make Eq. (2.106) nondimensionalized, the following relations are introduced:

$$V'=\frac{1}{\alpha\delta}V:\ \beta'=\frac{1}{\alpha\delta^2}\beta:\ V'=1+\frac{V_1}{\alpha\delta^2}\ \text{and}\ \tau=\alpha\delta^2t$$

$$w_2^2=\left(\frac{V'}{4}-\beta'\right)\alpha\delta^2:\ w_3^2=\left(\frac{V'-2}{2}\right)^2\alpha\delta^2:\ w_4^2=\frac{V'}{2}\sqrt{\alpha}\delta$$

$$D_5=\frac{1}{\beta V_1}=\frac{1}{\beta'(V'-1)\left(\alpha\delta^2\right)^2}$$

The surface temperature yields:

$$T(0,\tau)=\frac{I_1}{k\delta}\frac{1}{(V'-\beta'-1)}\left\{\begin{array}{l}\frac{-\sqrt{\frac{V'^2}{4}-\beta'}}{\beta'}e^{-\beta'\tau}\left[\mathrm{erfc}\left(\sqrt{\left(\frac{V'}{4}-\beta'\right)\tau}\right]+\frac{(V'-2)}{2(V'-1)}e^{-(V'-1)\tau}\left[\mathrm{erfc}\left(\left(\frac{V'-2}{2}\right)\sqrt{\tau}\right)\right]\\+\frac{(V'-\beta'-1)V'}{2\beta'(V'-1)}\mathrm{erfc}\left(\frac{V'}{2}\sqrt{\tau}\right)\\\left(\frac{-1}{\sqrt{\frac{V'^2}{4}-\beta'}+\frac{V'}{2}}+1\right)e^{-\beta'\tau}+\frac{(V'-2)}{(V'-1)}e^{-(V'-1)\tau}\end{array}\right\}-\frac{\alpha\rho L}{2k}\left[\begin{array}{l}\left(\mathrm{erfc}\left(-\frac{V^*t^*}{2\sqrt{t^*}}\right)-\left(1+V^{*2}t^*\right)\mathrm{erfc}\left(\frac{V^*t^*}{2\sqrt{t^*}}\right)\right)\\+\frac{2V^*t^*}{\sqrt{\pi}}e^{-\left(\frac{V^{*2}}{4\tau}+\frac{V^{*2}t^*}{4}\right)}\end{array}\right] \tag{2.107}$$

Noting that:

$$1-\operatorname{erfc}(\xi)=erf(\xi) \quad \text{and} \quad \operatorname{erfc}(-\xi)=2-\operatorname{erfc}(\xi)$$

Finally, the dimensionless surface temperature becomes:

$$\frac{T(0,\tau)}{\frac{I_1}{k\delta}}=\frac{1}{(V'-\beta'-1)}\left\{\begin{array}{l}\frac{\sqrt{\frac{V'^2}{4}-\beta'}}{\beta'}e^{-\beta'\tau}\left[\operatorname{erfc}\left(\sqrt{\left(\frac{V'}{4}-\beta'\right)\tau}\right]\right.\\ +\frac{(2-V')}{2(V'-1)}e^{-(V'-1)\tau}\left[\operatorname{erfc}\left(\left(\frac{2-V'}{2}\right)\sqrt{\tau}\right)\right]\\ +\frac{(V'-\beta'-1)V'}{2\beta'(V'-1)}\operatorname{erfc}\left(\frac{V'}{2}\sqrt{\tau}\right)\\ \left(\frac{-1}{\sqrt{\frac{V'^2}{4}-\beta'}+\frac{V'}{2}}+1\right)e^{-\beta'\tau}+\frac{(V'-2)}{2(V'-1)}e^{-(V'-1)\tau}\\ +\left(1-\frac{V'}{2\beta'}\right)e^{-\beta'\tau}\end{array}\right\}-\frac{\alpha\rho L}{2k}\left[\begin{array}{l}\left(\operatorname{erfc}\left(-\frac{V^*t^*}{2\sqrt{t^*}}\right)-(1+V^{*2}t^*)\operatorname{erfc}\left(\frac{V^*t^*}{2\sqrt{t^*}}\right)\right)\\ +\frac{2V^*t^*}{\sqrt{\pi}}e^{-\left(\frac{V^{*2}}{4\tau}+\frac{V^{*2}t^*}{4}\right)}\end{array}\right] \tag{2.108}$$

It can be shown that the nondimensional surface temperature becomes identical to the previously-obtained surface temperature for conduction limit only. In this case, the effect of boundary condition $\left(\left(\frac{\partial T}{\partial x}\right)_{x=0}=\frac{\rho VL}{k}\right)$ should be replaced with $\left(\left(\frac{\partial T}{\partial x}\right)_{x=0}=0\right)$ and V' should be set to zero. Therefore, it yields:

$$\frac{T(0,\tau)}{\frac{I_1}{k\delta}}=\frac{1}{-(\beta'+1)}\left\{\begin{array}{l}\frac{\sqrt{-\beta'}}{\beta'}e^{-\beta'\tau}[\operatorname{erfc}(\sqrt{(-\beta')\tau}]\\ +e^{\tau}[\operatorname{erfc}(\sqrt{\tau})]+e^{-\beta'\tau}\end{array}\right\} \tag{2.109}$$

where $\beta'>0$. Rearrangement yields:

$$\frac{T(0,\tau)}{\frac{I_1}{k\delta}}=\frac{1}{(\beta'+1)}\left\{\frac{-i}{\sqrt{\beta'}}e^{-\beta'\tau}\left[\operatorname{erfc}\left(i\sqrt{\beta'\tau}\right]+e^{\tau}\left[\operatorname{erfc}(\sqrt{\tau})\right]-e^{-\beta'\tau}\right\} \tag{2.110}$$

Noting that:

$$F\left(\sqrt{\beta'\tau}\right)=\frac{\sqrt{\pi}}{2i}e^{-\beta'\tau}\mathrm{erfc}\left(i\sqrt{\beta'\tau}\right) \tag{2.111}$$

where:

$$F(\chi)=e^{-\chi^2}\int_0^{\chi}e^{v^2}dv \tag{2.112}$$

where $F(\chi)$ is the Dawson's integral. To prove the above relation for the Dawson's integral, the following steps are adopted.

It is known that:

$$\mathrm{erf}\left(i\sqrt{\beta'\tau}\right)=\frac{2}{\sqrt{\pi}}\int_0^{i\sqrt{\beta'\tau}}e^{v^2}dv \tag{2.113}$$

Let $v=\frac{\xi}{i}$, then $dv=\frac{d\xi}{i}$. Substituting these into the above relation (Eq. 2.113), one can have:

$$\mathrm{erf}\left(i\sqrt{\beta'\tau}\right)=\frac{2}{\sqrt{\pi}}\int_0^{-i\sqrt{\beta'\tau}}e^{\xi^2}\frac{d\xi}{i}=-\frac{2i}{\sqrt{\pi}}\int_0^{-i\sqrt{\beta'\tau}}e^{\xi^2}d\xi \tag{2.114}$$

or

$$\mathrm{erf}\left(i\sqrt{\beta'\tau}\right)=\frac{2i}{\sqrt{\pi}}\int_0^{i\sqrt{\beta'\tau}}e^{\xi^2}d\xi \tag{2.115}$$

Substitution of the above, in relation into Dawson's integral, completes the proof. Therefore:

$$e^{-\beta'\tau}erf\left(i\sqrt{\beta'\tau}\right)=\frac{2i}{\sqrt{\pi}}F\left(\sqrt{\beta'\tau}\right) \tag{2.116}$$

Substituting the above relations, the surface temperature reduces to:

$$\frac{T(0,\tau)}{\frac{I_1}{k\delta}}=\frac{1}{(\beta'+1)}\left\{\frac{2}{\sqrt{\pi}\sqrt{\beta'}}F\left[\sqrt{\beta'\tau}\right]+e^{\tau}\left[\mathrm{erfc}\left(\sqrt{\tau}\right)\right]-e^{-\beta'\tau}\right\} \tag{2.117}$$

Fig. 2.3 shows the dimensionless temperature profiles inside the substrate material for pulse parameter $\beta^*=0.132$ as the dimensionless time is variable. The temperature at some depth below the surface does not reach boiling. In this case, recession velocity of the surface reduces to zero and the closed-form solution becomes identical to the solution obtained from the numerical study. As the heating progresses, the surface recession propagates toward the solid bulk. In this case, the decay of the temperature profile in the surface vicinity becomes lower than that corresponding to some depth below the surface. The temperature gradient attains minimum at some depth below the surface.

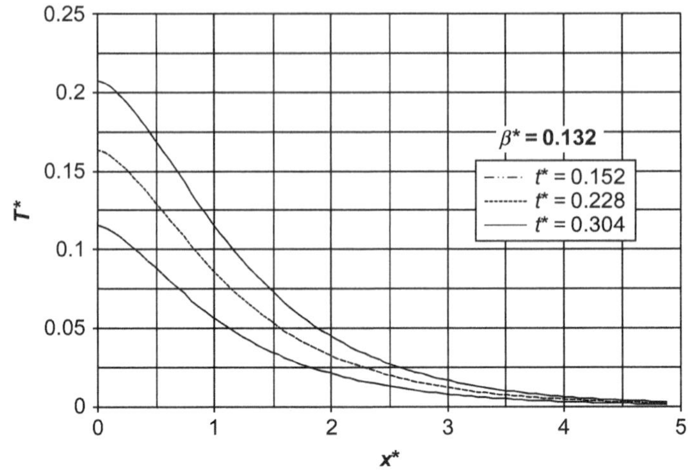

Fig. 2.3 Dimensionless temperature distribution inside the material obtained from closed form solution for $\beta^* = 0.132$ [7].

Moreover, the internal energy gain increases in the surface region of the substrate material due to the absorption of irradiated laser energy, and the diffusional energy transport from the surface vicinity to solid bulk is not as high as the internal energy gain. Although surface evaporation lowers the temperature rise in the surface vicinity due to convective boundary conditions at the surface, the high power intensity ($I_1 = 10^{13}\ \mathrm{W/m^3}$) suppresses the effect of convective losses on the temperature rise and enhances the internal energy gain through the absorption process. Consequently, this gives rise to low temperature gradient attainment in the surface vicinity of the substrate material, particularly up to the depth less than the absorption depth of the substrate material. The point of minimum $\frac{\partial T^*}{\partial x^*}$ moves toward the solid bulk as the heating period progresses. This indicates that the region of the low-temperature gradient, where internal energy gain dominates over the convectional and diffusional losses, extends further inside the substrate material.

2.3.2 Thermal stress analysis

The temporal variation of the laser-pulse intensity can be considered to decay exponentially with increasing time; therefore, this variation needs to be incorporated in the stress analysis. Because the boundary conditions are important to formulate the stress states in the irradiated material, the closed form solutions for the thermal stress generated inside the substrate material are presented in line with the previous studies [7,9,10]. Moreover, thermal stress in the key-hole formation takes place in the solid phase of the substrate material, because the liquid phase can easily be deformed without generating the high rate of thermal strain. Therefore, thermal stress analysis is limited within the solid form. Consequently, the governing heat equation reduces to the conduction limited heat situation where the phase change term at the boundary can be ignored. The Fourier heat transfer equation for the time exponentially decaying laser heating pulse can be written as:

$$\frac{\partial^2 T}{\partial x^2} + \frac{I_1 \delta}{k}\left(e^{-\beta t}\right)e^{-\delta x} = \frac{1}{\alpha}\frac{\partial T}{\partial t} \tag{2.118}$$

where $I_1 = (1 - r_f)\, I_o$.

It can be assumed that the convection losses from the surface are neglected during the heating pulse. This implies an insulated boundary at the substrate surface. In addition, the substrate material is assumed to be a semi-infinite body; therefore, temperature at a depth approaching to infinity becomes the same as the initial temperature of the substrate material. Therefore, the corresponding boundary conditions are:

$$\text{At the free surface: } x = 0 \Rightarrow \left[\frac{\partial T}{\partial x}\right]_{x=0} = 0 \tag{2.119}$$

Temperature can be considered as approaching an initial temperature at a depth infinitely below the surface, that is:

$$\text{At } x = \infty \Rightarrow T(\infty, t) = 0 \tag{2.120}$$

Due to the simplicity, the initial temperature of the substrate material is set to zero and temperature is assumed to be uniform inside the solid substrate at a time equal to zero (before heating initiates). Therefore, the initial condition becomes:

$$\text{At } t = 0 \Rightarrow T(x, 0) = 0 \tag{2.121}$$

The closed-form solution of Eq. (2.118) can be obtained through a Laplace transformation method after employing the appropriate boundary and initial conditions. In this case, the Laplace transformation of Eq. (2.118) with respect to t, results:

$$\frac{\partial^2 \overline{T}}{\partial x^2} + \frac{I_1 \delta}{k}\frac{e^{-\delta x}}{(s + \beta)} = \frac{1}{\alpha}\left[s\overline{T}(x, s) - T(x, 0)\right] \tag{2.122}$$

Introducing the initial condition and rearranging Eq. (2.118) yields:

$$\frac{\partial^2 \overline{T}}{\partial x^2} - h^2 \overline{T} = -\frac{I_1 \delta}{k}\frac{e^{-\delta x}}{(s + \beta)} \tag{2.123}$$

where $h^2 = s/\alpha$ and s is the transform variable. Eq. (2.123) has the solution:

$$T(x, s) = Ae^{hx} + Be^{-hx} - \frac{I_1 \delta e^{-\delta x}}{k(s + \beta)\left(\delta^2 - h^2\right)} \tag{2.124}$$

where A and B are constants. Introducing the boundary conditions will determine the constants A and B, that is:

$$A = 0,$$

and:

$$B = \frac{I_1 \delta^2}{kh(s+\beta)\left(\delta^2 - h^2\right)} \tag{2.125}$$

After substituting the values of A and B in Eq. (2.124), it yields:

$$\overline{T}(x, s) = -\frac{I_1 \delta}{k(s+\beta)}\left[\frac{\delta \exp(-hx)}{h\left(h^2-\delta^2\right)} - \frac{\exp(-\delta x)}{\left(h^2-\delta^2\right)}\right] \tag{2.126}$$

which gives the solution for temperature in the Laplace domain.

The inverse Laplace transformation of Eq. (2.126) gives the temperature distribution inside the substrate material in space (x) and time (t) domain as follows [9,10]:

$$T(x,t) = \frac{I_1\delta}{2k}\left(\frac{\alpha}{\beta+\alpha\delta^2}\right) \times \left\{ \begin{array}{l} i\delta\sqrt{\frac{\alpha}{\beta}}\exp(-\beta t)\cdot \left[\begin{array}{l} \exp\left(ix\sqrt{\frac{\beta}{\alpha}}\right)\cdot \mathrm{Erfc}\left(\frac{x}{2\sqrt{\alpha t}}+i\sqrt{\beta t}\right) \\ -\exp\left(-ix\sqrt{\frac{\beta}{\alpha}}\right)\cdot \mathrm{Erfc}\left(\frac{x}{2\sqrt{\alpha t}}-i\sqrt{\beta t}\right) \end{array}\right] \\ +\exp\left(\alpha\delta^2 t\right)\left[\begin{array}{l} \exp(\delta x)\cdot \mathrm{Erfc}\left(\frac{x}{2\sqrt{\alpha t}}+\delta\sqrt{\alpha t}\right) - \exp(-\delta x)\cdot \mathrm{Erfc}\left(\delta\sqrt{\alpha t}-\frac{x}{2\sqrt{\alpha t}}\right) \\ -2\exp(-(\beta t+\delta x)) \end{array}\right] \end{array}\right\} \tag{2.127}$$

where Erfc is the complementary error function. Eq. (2.127) is the closed-form solution for temperature distribution. The temperature distribution in a nondimensional form is possible by defining dimensionless quantities and substituting in Eq. (2.136). The dimensionless quantities are:

$$x^* = x\delta: \quad t^* = \alpha\delta^2 t: \quad T^* = \frac{Tk\delta}{I_1}: \quad \beta^* = \frac{\beta}{\alpha\delta^2} \tag{2.128}$$

The dimensionless temperature distribution becomes:

$$T^*(x^*,t^*) = \frac{1}{2}\left(\frac{t^*}{\beta^*+t^*}\right) \times \left\{ \begin{array}{l} i\sqrt{\frac{t^*}{\beta^*}}\exp(-\beta^*)\cdot \left[\begin{array}{l} \exp\left(ix^*\sqrt{\frac{\beta^*}{t^*}}\right)\cdot \mathrm{Erfc}\left(\frac{x^*}{2\sqrt{t^*}}+i\sqrt{\beta^*}\right) \\ -\exp\left(-ix^*\sqrt{\frac{\beta^*}{t^*}}\right)\cdot \mathrm{Erfc}\left(\frac{x^*}{2\sqrt{t^*}}-i\sqrt{\beta^*}\right) \end{array}\right] \\ +\exp(t^*)\left[\begin{array}{l} \exp(x^*)\cdot \mathrm{Erfc}\left(\frac{x^*}{2\sqrt{t^*}}+\sqrt{t^*}\right) - \exp(-x^*)\cdot \mathrm{Erfc}\left(\sqrt{t^*}-\frac{x^*}{2\sqrt{t^*}}\right) \\ -2\exp(-(\beta^*+x^*)) \end{array}\right] \end{array}\right\} \tag{2.129}$$

To obtain the closed-form solution for the stress distribution inside the semi-infinite one-dimensional substrate material, the equation governing the momentum in a one-dimensional solid for a linear elastic case can be considered. In this case, the momentum equation becomes [11]:

$$\frac{\partial^2 \sigma_x}{\partial x^2} - \frac{1}{c_1{}^2}\frac{\partial^2 \sigma_x}{\partial t^2} = c_2 \frac{\partial^2 T}{\partial t^2} \tag{2.130}$$

where c_1 is the wave speed in the solid

$$c_1 = \sqrt{\frac{E}{\rho}} \quad \text{and} \quad c_2 = \frac{1+\upsilon}{1-\upsilon}\rho\alpha_T$$

where υ is Poisson's ratio, ρ is the density of the solid, and α_T is the thermal expansion coefficient of the solid.

Because the substrate surface is free to expand during the heating process, a stress-free condition is adopted at the surface of the substrate material. Moreover, as the depth below the stress-free surface increases to infinity, the thermal stress approaches the initial stress conditions, which is the stress-free condition, that is, it is assumed that the substrate material is initially free from any stresses. Therefore, the corresponding boundary conditions are:

$$\text{At } x = 0 \Rightarrow \sigma_x = 0 \tag{2.131}$$

and

$$\text{At } x = \infty \Rightarrow \sigma_x = 0 \tag{2.132}$$

For the temperature, the boundary conditions are:

$$\text{At } x = 0 \Rightarrow \frac{\partial T}{\partial x} = 0 \tag{2.133}$$

and

$$\text{At } x = \infty \Rightarrow T = 0 \tag{2.134}$$

Because the substrate material is assumed to be initially stress free, and the thermal stress becomes the same as the initial stress condition as time approaches infinity, the initial and final conditions for the stress field yield:

$$\text{At } t = 0 \Rightarrow \sigma_x = 0 \tag{2.135}$$

and

$$\text{At } t = \infty \Rightarrow \sigma_x = 0 \tag{2.136}$$

The initial condition for temperature is:

$$\text{At } t = 0 \Rightarrow T(x, 0) = 0 \tag{2.137}$$

Because the laser pulse decays exponentially with time, the final and initial temperature of the substrate material becomes the same as the time approaches infinity, that is:

$$\text{At } t = \infty \Rightarrow T(x, \infty) = 0 \tag{2.138}$$

Taking the Laplace Transformation of Eq. (2.130), with respect to time, yields:

$$\frac{\partial^2 \overline{\sigma}_x}{\partial x^2} - \frac{1}{c_1{}^2}\left[s^2\overline{\sigma}_x(x, s) - s\sigma_x(x, 0) - \dot{\sigma}_x(x, 0)\right] = c_2\left[s^2\overline{T}(x, s) - sT(x, 0) - \dot{T}(x, 0)\right] \tag{2.139}$$

where $\overline{\sigma}_x(x, s)$ and $\overline{T}(x, s)$ are the Laplace Transformations of thermal stress and temperature, respectively, in the x and s domains.

By substituting the initial conditions, Eq. (2.139) reduces to:

$$\frac{\partial^2 \overline{\sigma}_x}{\partial x^2} - \frac{s^2}{c_1{}^2}\overline{\sigma}_x(x, s) = c_2 s^2 \overline{T}(x, s) \tag{2.140}$$

For time exponentially varying pulse, substituting Eq. (2.126) into Eq. (2.140) yields:

$$\frac{\partial^2 \overline{\sigma}_x}{\partial x^2} - \frac{s^2}{c_1{}^2}\overline{\sigma}_x(x, s) = -c_2 s^2 \frac{I_1 \delta}{k(s+\beta)}\left[\frac{\delta \exp(-hx)}{h\left(h^2 - \delta^2\right)} - \frac{\exp(-\delta x)}{\left(h^2 - \delta^2\right)}\right] \tag{2.141}$$

which can be arranged further:

$$\frac{\partial^2 \overline{\sigma}_x}{\partial x^2} - \frac{s^2}{c_1{}^2}\overline{\sigma}_x(x, s) = \frac{-I_1 \delta^2 c_2 s^2}{k(s+\beta)h\left(h^2 - \delta^2\right)} e^{-hx} + \frac{I_1 \delta c_2 s^2}{k(s+\beta)\left(h^2 - \delta^2\right)} e^{-\delta x} \tag{2.142}$$

Now let W_1 and W_2 are defined as:

$$W_1 = \frac{-I_1 \delta^2 c_2 s^2}{k(s+\beta)h\left(h^2 - \delta^2\right)} \tag{2.143}$$

and

$$W_2 = \frac{I_1 \delta c_2 s^2}{k(s+\beta)\left(h^2 - \delta^2\right)} \tag{2.144}$$

Then Eq. (2.153) becomes:

$$\frac{\partial^2 \overline{\sigma}_x}{\partial x^2} - \frac{s^2}{{c_1}^2}\overline{\sigma}_x(x, s) = W_1 e^{-hx} + W_2 e^{-\delta x} \tag{2.145}$$

The homogeneous and the particular solutions for Eq. (2.145) are:

$$(\overline{\sigma}_x)_h = C_1 e^{\frac{sx}{c_1}} + C_2 e^{\frac{-sx}{c_1}} \tag{2.146}$$

while the particular solution has two parts, the first part is:

$$(\overline{\sigma}_x)_{p_1} = Q_1 e^{-hx} \tag{2.147}$$

Substituting Eq. (2.147) in Eq. (2.145) yields:

$$Q_1 = \frac{W_1}{h^2 - \frac{s^2}{{c_1}^2}} \tag{2.148}$$

The second part of the particular solution is:

$$(\overline{\sigma}_x)_{p_2} = Q_2 e^{-\delta x} \tag{2.149}$$

Substituting Eq. (2.149) in Eq. (2.145) yields:

$$Q_2 = \frac{W_2}{\delta^2 - \frac{s^2}{{c_1}^2}} \tag{2.150}$$

Therefore, the general solution for the stress field becomes:

$$(\overline{\sigma}_x)_g = C_1 e^{\frac{sx}{c_1}} + C_2 e^{\frac{-sx}{c_1}} + Q_1 e^{-hx} + Q_2 e^{-\delta x} \tag{2.151}$$

From the boundary condition ($x = \infty \Rightarrow \sigma_x = 0$), it yields $C_1 = 0$:
Then,

$$(\overline{\sigma}_x)_g = C_2 e^{\frac{-sx}{c_1}} + \frac{W_1}{h^2 - \frac{s^2}{{c_1}^2}} e^{-hx} + \frac{W_2}{\delta^2 - \frac{s^2}{{c_1}^2}} e^{-\delta x} \tag{2.152}$$

Substituting for W_1 and W_2 in Eq. (2.152) results:

$$(\overline{\sigma}_x)_g = C_2 e^{\frac{-sx}{c_1}} + g_1(s) e^{-hx} + g_2(s) e^{-\delta x} \tag{2.153}$$

where:

$$g_1(s) = \frac{-I_1\delta^2 c_2 s^2}{k(s+\beta)h\left(h^2-\delta^2\right)\left(h^2-\dfrac{s^2}{{c_1}^2}\right)} \tag{2.154}$$

and

$$g_2(s) = \frac{I_1\delta c_2 s^2}{k(s+\beta)\left(h^2-\delta^2\right)\left(\delta^2-\dfrac{s^2}{{c_1}^2}\right)} \tag{2.155}$$

Now substituting for $h = s/\alpha$ and simplifying the expressions for $g_1(s)$ and $g_2(s)$ yields:

$$g_1(s) = \frac{I_1\delta^2 c_2\sqrt{s\alpha}\,\alpha {c_1}^2}{k(s+\beta)h\left(s-\alpha\delta^2\right)\left(s-\dfrac{{c_1}^2}{\alpha}\right)} \tag{2.156}$$

and

$$g_2(s) = \frac{-I_1\delta c_2 s^2 {c_1}^2}{k(s+\beta)\left(s-\alpha\delta^2\right)\left(s^2-{c_1}^2\delta^2\right)} \tag{2.157}$$

Rearranging the two expressions as:

$$g_1(s) = C_3\left[\frac{\sqrt{s}}{(s+\beta)h\left(s-\alpha\delta^2\right)\left(s-{c_1}^2/\alpha\right)}\right] \tag{2.158}$$

and

$$g_2(s) = C_4\left[\frac{s^2}{(s+\beta)\left(s-\alpha\delta^2\right)(s-c_1\delta)(s+c_1\delta)}\right] \tag{2.159}$$

where:

$$C_3 = \frac{I_1\delta^2 c_2\sqrt{\alpha}\,\alpha {c_1}^2}{k} \tag{2.160}$$

and

$$C_4 = \frac{-I_1 c_2 {c_1}^2\delta\alpha}{k} \tag{2.161}$$

Using partial fractions, the following relations can be obtained:

$$g_1(s) = C_3 \left[\begin{array}{l} \dfrac{\sqrt{s}}{(\alpha\delta^2+\beta)(\alpha\delta^2-c_1{}^2/\alpha)(s-\alpha\delta^2)} + \dfrac{\alpha^2\sqrt{s}}{(c_1{}^2+\alpha\beta)(c_1{}^2-\alpha^2\delta^2)(s-c_1{}^2/\alpha)} \\ + \dfrac{\sqrt{s}}{(\beta+\alpha\delta^2)(\beta+c_1{}^2/\alpha)(s+\beta)} \end{array} \right] \tag{2.162}$$

and

$$g_2(s) = C_4 \left[\begin{array}{l} \dfrac{\alpha\delta^2}{(\alpha\delta^2+\beta)(\alpha^2\delta^2-c_1{}^2)(s-\alpha\delta^2)} + \dfrac{c_1}{2(c_1\delta+\beta)(c_1-\alpha\delta)(s-c_1\delta)} \\ + \dfrac{c_1}{2(c_1\delta-\beta)(c_1+\alpha\delta)(s+c_1\delta)} + \dfrac{\beta^2}{(-\beta-\alpha\delta^2)(\beta^2-c_1{}^2\delta^2)(s+\beta)} \end{array} \right] \tag{2.163}$$

Consider the boundary condition at the surface in the stress field, where at $x=0 \Rightarrow \overline{\sigma}_x = 0$, the constant in Eq. (2.152) becomes:

$$C_2 = -g_1(s) - g_2(s) \tag{2.164}$$

Then:

$$\overline{\sigma}_x(x,s) = g_1(s)e^{-\sqrt{\frac{s}{\alpha}}x} - g_1(s)e^{\frac{-sx}{c_1}} + g_2(s)e^{-\delta x} - g_2(s)e^{\frac{-sx}{c_1}} \tag{2.165}$$

Finding the solution for σ_x in the x and t domain, we should take the inverse Laplace Transformation for each term in Eq. (2.165). To perform this, the following designations are introduced:

$$\begin{array}{ll} \text{Term1} = g_1(s)e^{-\sqrt{\frac{s}{\alpha}}x} & \text{Term2} = g_1(s)e^{\frac{-sx}{c_1}} \\ \text{Term3} = g_2(s)e^{-\delta x} & \text{Term4} = g_2(s)e^{\frac{-sx}{c_1}} \end{array} \tag{2.166}$$

Consequently, the solution for stress distribution becomes the summation of the inverse Laplace Transformation of the above terms. Therefore, the Laplace Inversion of Terms (Term1, Term2, Term3, Term4) are:

$$\pounds^{-1}[\text{Term1}] = C_3 \left\{ \begin{array}{l} \dfrac{1}{2(\alpha\delta^2+\beta)\left(\alpha\delta^2-c_1^2/\alpha\right)} \left[\dfrac{2}{\sqrt{\pi t}} \cdot e^{\frac{-x^2}{4\alpha t}} + \delta\sqrt{\alpha} \cdot e^{\delta^2 \alpha t} \left(\begin{array}{l} e^{-\delta x} \cdot \text{Erfc}\left(-\delta\sqrt{\alpha t} + \dfrac{x}{2\sqrt{\alpha t}}\right) \\ -e^{\delta x} \cdot \text{Erfc}\left(\delta\sqrt{\alpha t} + \dfrac{x}{2\sqrt{\alpha t}}\right) \end{array} \right) \right] \\ + \dfrac{\alpha^2}{2\left(c_1^2+\alpha\beta\right)\left(c_1^2-\alpha^2\delta^2\right)} \left[\dfrac{2}{\sqrt{\pi t}} \cdot e^{\frac{-x^2}{4\alpha t}} + \dfrac{c_1}{\sqrt{\alpha}} \cdot e^{\frac{c_1^2 t}{\alpha}} \left(\begin{array}{l} e^{\frac{-c_1 x}{\alpha}} \cdot \text{Erfc}\left(-c_1\sqrt{\dfrac{t}{\alpha}} + \dfrac{x}{2\sqrt{\alpha t}}\right) \\ -e^{\frac{c_1 x}{\alpha}} \cdot \text{Erfc}\left(c_1\sqrt{\dfrac{t}{\alpha}} + \dfrac{x}{2\sqrt{\alpha t}}\right) \end{array} \right) \right] \\ + \dfrac{1}{2(\beta+\alpha\delta^2)\left(\beta+c_1^2/\alpha\right)} \left[\dfrac{2}{\sqrt{\pi t}} \cdot e^{\frac{-x^2}{4\alpha t}} + i\sqrt{\beta} \cdot e^{-\beta t} \left(\begin{array}{l} e^{-xi\sqrt{\frac{\beta}{\alpha}}} \cdot \text{Erfc}\left(-i\sqrt{\beta t} + \dfrac{x}{2\sqrt{\alpha t}}\right) \\ -e^{xi\sqrt{\frac{\beta}{\alpha}}} \cdot \text{Erfc}\left(i\sqrt{\beta t} + \dfrac{x}{2\sqrt{\alpha t}}\right) \end{array} \right) \right] \end{array} \right\} \tag{2.167}$$

and

$$\pounds^{-1}[\text{Term2}] = C_3 \left\{ \begin{array}{l} \dfrac{\dfrac{1}{\sqrt{\pi(t-x/c_1)}} + \sqrt{\alpha\delta^2} e^{\alpha\delta^2(t-x/c_1)} \text{Erf}\left[\sqrt{\alpha\delta^2(t-x/c_1)}\right]}{(\alpha\delta^2 - c_1{}^2/\alpha)(\alpha\delta^2+\beta)} \\ +\alpha^2 \left(\dfrac{\dfrac{1}{\sqrt{\pi(t-x/c_1)}} + \sqrt{\dfrac{c_1{}^2}{\alpha}} e^{\frac{-c_1{}^2}{\alpha}(t-x/c_1)} \text{Erf}\left[\sqrt{\dfrac{c_1{}^2}{\alpha}(t-x/c_1)}\right]}{(c_1{}^2+\alpha\beta)(c_1{}^2-\alpha^2\delta^2)} \right) \\ + \dfrac{\dfrac{1}{\sqrt{\pi(t-x/c_1)}} + i\sqrt{\beta} e^{-\beta(t-x/c_1)} \text{Erf}\left[i\sqrt{\beta(t-x/c_1)}\right]}{(\beta + c_1{}^2/\alpha)(\alpha\delta^2+\beta)} \end{array} \right\} \cdot 1\left(t - \frac{x}{c_1}\right) \tag{2.168}$$

and

$$\pounds^{-1}[\text{Term3}] = C_4 e^{-\delta x} \left\{ \begin{array}{l} \dfrac{\alpha^2\delta^2 e^{\alpha\delta^2 t}}{(\alpha\delta^2+\beta)(\alpha^2\delta^2 - c_1{}^2)} + \dfrac{c_1 e^{c_1\delta t}}{2(c_1\delta+\beta)(c_1-\alpha\delta)} \\ + \dfrac{c_1 e^{-c_1\delta t}}{2(c_1\delta-\beta)(c_1+\alpha\delta)} + \dfrac{\beta^2 e^{-\beta t}}{(-\beta-\alpha\delta^2)(\beta^2 - c_1{}^2\delta^2)} \end{array} \right\} \tag{2.169}$$

and

$$\pounds^{-1}[\text{Term4}] = C_4 \left[\begin{array}{l} \dfrac{\alpha^2\delta^2 e^{\alpha\delta^2(t-x/c_1)}}{(\alpha\delta^2+\beta)(\alpha^2\delta^2 - c_1{}^2)} + \dfrac{c_1 e^{c_1\delta(t-x/c_1)}}{2(c_1\delta+\beta)(c_1-\alpha\delta)} \\ + \dfrac{c_1 e^{-c_1\delta(t-x/c_1)}}{2(c_1\delta-\beta)(c_1+\alpha\delta)} + \dfrac{\beta^2 e^{-\beta(t-x/c_1)}}{(-\beta-\alpha\delta^2)(\beta^2 - c_1{}^2\delta^2)} \end{array} \right] \cdot 1\left(t - \frac{x}{c_1}\right) \tag{2.170}$$

where $1\left(t-\dfrac{x}{c_1}\right)$ is a unit step function and Erf(y) is the error function of the variable y. It should be noted that the value of a unit step function takes one or zero, that is:

$$t \leq \frac{x}{c_1} \Rightarrow 1\left(t-\frac{x}{c_1}\right) = 0 \quad \text{and} \quad t > \frac{x}{c_1} \Rightarrow 1\left(t-\frac{x}{c_1}\right) = 1 \tag{2.171}$$

The closed-form solution of stress distribution can be written as:

$$\sigma_x(x,t) = \pounds^{-1}[\text{Term1}] + \pounds^{-1}[\text{Term2}] + \pounds^{-1}[\text{Term3}] + \pounds^{-1}[\text{Term4}] \tag{2.172}$$

where $\pounds^{-1}$ represents the inverse sign of the Laplace transformation.

Presenting the stress distribution in dimensionless form, the additional dimensionless quantities are defined, that is:

$$c_1^* = \frac{c_1}{\alpha\delta} \tag{2.173}$$

and

$$\sigma_x^* = \frac{k\sigma_x}{I_1\delta\alpha^2 c_2 (c_1^*)^2} \tag{2.174}$$

$U[1]$ is the unit step function, which is $U[1] = \left(t^* - \frac{x^*}{c_1^*}\right)$. The values of unit step function are:

$$t^* \le \frac{x^*}{c_1^*} \Rightarrow U[1]\left(t^* - \frac{x^*}{c_1^*}\right) = 0 \quad \text{and} \quad t^* > \frac{x^*}{c_1^*} \Rightarrow U[1]\left(t^* - \frac{x^*}{c_1^*}\right) = 1 \tag{2.175}$$

Therefore, for the dimensionless stress distribution, the following results are:

$$(\sigma_x^*)_1 = \left\{ \begin{array}{l} \dfrac{1}{(t^*+\beta^*)\left(1-(c_1^*)^2\right)}\left[\dfrac{\sqrt{t^*}e^{\frac{-(x^*)^2}{4t^*}}}{\sqrt{\pi}} + \dfrac{t^*}{2}\begin{pmatrix} e^{-x^*}\cdot \mathrm{erfc}\left(-\sqrt{t^*}+\dfrac{x^*}{2\sqrt{t^*}}\right) \\ -e^{x^*}\cdot\mathrm{erfc}\left(\sqrt{t^*}+\dfrac{x^*}{2\sqrt{t^*}}\right)\end{pmatrix}\right] \\ + \dfrac{1}{\left((c_1^*)^2-1\right)\left((c_1^*)^2 t^* + \beta^*\right)}\left[\dfrac{\sqrt{t^*}e^{\frac{-(x^*)^2}{4t^*}}}{\sqrt{\pi}} + \dfrac{c_1^* t^* e^{-(c_1^*)^2 t^*}}{2}\begin{pmatrix} e^{-c_1^*x^*}\cdot\mathrm{erfc}\left(-c_1^*\sqrt{t^*}+\dfrac{x^*}{2\sqrt{t^*}}\right) \\ -e^{c_1^*x^*}\cdot \mathrm{erfc}\left(c_1^*\sqrt{t^*}+\dfrac{x^*}{2\sqrt{t^*}}\right)\end{pmatrix}\right] \\ + \dfrac{(t^*)^{3/2}}{(\beta^*+t^*)\left(\beta^*+(c_1^*)^2 t^*\right)}\left[\dfrac{e^{\frac{-(x^*)^2}{4t^*}}}{\sqrt\pi} + \dfrac{i\sqrt{\beta^*}e^{-\beta^*}}{2}\begin{pmatrix} e^{-ix^*\sqrt{\frac{\beta^*}{t^*}}}\cdot\mathrm{erfc}\left(-i\sqrt{\beta^*}+\dfrac{x^*}{2\sqrt{t^*}}\right) \\ -e^{ix^*\sqrt{\frac{\beta^*}{t^*}}}\cdot\mathrm{erfc}\left(i\sqrt{\beta^*}+\dfrac{x^*}{2\sqrt{t^*}}\right)\end{pmatrix}\right]\end{array}\right\} \tag{2.176}$$

$$(\sigma_x^*)_2 = \left\{ \begin{array}{l} \dfrac{1}{(t^*+\beta^*)\left(1-(c_1^*)^2\right)}\left[\dfrac{t^*}{\sqrt{\pi}\sqrt{\left(t^*-x^*/c_1^*\right)}} + t^* e^{\left(t^*-x^*/c_1^*\right)}\cdot \mathrm{Erf}\left(\sqrt{\left(t^*-x^*/c_1^*\right)}\right)\right] \\ \dfrac{1}{\left((c_1^*)^2-1\right)\left((c_1^*)^2+\beta^*/t^*\right)}\left[\dfrac{1}{\sqrt{\pi}\sqrt{\left(t^*-x^*/c_1^*\right)}} + c_1^* e^{-\left((c_1^*)^2 t^*-c_1^*x^*\right)}\cdot \mathrm{Erf}\left(\sqrt{\left((c_1^*)^2 t^*-c_1^*x^*\right)}\right)\right] \\ + \dfrac{(t^*)^{3/2}}{(t^*+\beta^*)\left(\beta^*+(c_1^*)^2 t^*\right)}\left[\dfrac{\sqrt{t^*}}{\sqrt{\pi}\sqrt{\left(t^*-x^*/c_1^*\right)}} + i\sqrt{\beta^*}e^{-\beta^*\left(1-x^*/c_1^*t^*\right)}\cdot \mathrm{Erf}\left(i\sqrt{\beta^*\left(1-x^*/c_1^*t^*\right)}\right)\right] \end{array}\right\}\cdot U[1] \tag{2.177}$$

$$(\sigma_x^*)_3 = \left\{ \begin{array}{l} \dfrac{-t^* e^{t^*-x^*}}{\left(1-(c_1^*)^2\right)(t^*+\beta^*)} - \dfrac{t^* c_1^* e^{c_1^* t^*-x^*}}{2(c_1^* t^* + \beta^*)(c_1^*-1)} - \dfrac{t^* c_1^* e^{-c_1^* t^*-x^*}}{2(c_1^* t^* - \beta^*)(c_1^*+1)} \\ -\dfrac{(\beta^*)^2 t^* e^{-\beta^*-x^*}}{(-\beta^*-t^*)\left((\beta^*)^2-(c_1^* t^*)^2\right)} \end{array} \right\} \tag{2.178}$$

$$(\sigma_x^*)_4 = \left\{ \begin{array}{l} \dfrac{-t^* e^{\left(t^*-x^*/c_1^*\right)}}{\left(1-(c_1^*)^2\right)(t^*+\beta^*)} - \dfrac{t^* c_1^* e^{c_1^* t^*-x^*}}{2(c_1^* t^* + \beta^*)(c_1^*-1)} - \dfrac{t^* c_1^* e^{-\left(c_1^* t^*-x^*\right)}}{2(c_1^* t^* - \beta^*)(c_1^*+1)} \\ -\dfrac{(\beta^*)^2 t^* e^{-\beta^*\left(1-x^*/c_1^* t^*\right)}}{(-\beta^*-t^*)\left((\beta^*)^2-(c_1^* t^*)^2\right)} \end{array} \right\} \cdot U[1] \tag{2.179}$$

Consequently, the dimensionless form of stress equation (Eq. 2.171) is:

$$\sigma_x^* = (\sigma_x^*)_1 + (\sigma_x^*)_2 + (\sigma_x^*)_3 + (\sigma_x^*)_4 \tag{2.180}$$

Eqs. (2.130), (2.180) are used to compute the dimensionless temperature and stress distributions inside the substrate material.

Fig. 2.4 shows temporal variation of dimensionless temperature at different locations inside the substrate material. Temperature rises rapidly in the early heating period, and the rate of rising temperature reduces as the heating progresses. This occurs because of the energy gain and energy transport mechanisms inside the

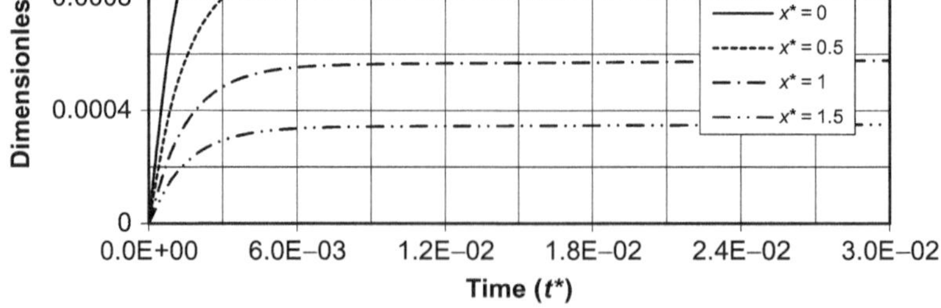

Fig. 2.4 Temporal variation of dimensionless temperature at different locations inside the substrate material [9].

substrate material. In this case the internal energy gain by the substrate material from the irradiated field dominates over the conduction losses in the early heating period. As the heating period progresses internal energy gain and temperature gradient in the surface region increase. Consequently, a stage is reached when the conduction losses from the surface region to the solid bulk becomes important due to the high-temperature gradient developed in this region with progressing time. Because the pulse intensity decreases with time (exponentially decaying intensity) temperature also decreases with further progressing heating time. Therefore, the temperature curve decays after reaching its maximum value. The rise of temperature at some depth below the surface is not as high as that corresponding to the surface. This occurs because of the energy absorbed by the substrate material at some depth below the surface, that is, energy absorbed decreases exponentially with increasing depth (Lambert's law). As the heating progresses further ($t^* > 6 \times 10^{-3}$) temperature in the surface region ($x^* < 1$) reduces, while at some depth below the surface ($x^* \geq 1$). increases gradually with the heating period. This occurs because of the diffusional heating of the substrate material at some depth below the surface due to high temperature gradient in this region. In the region limited by the absorption depth ($x^* < 1$) energy gain by the substrate material reduces considerably with further progressing of time because time exponentially decays pulse intensity. However, a high-temperature gradient enhances the energy loss from this region to solid bulk, that is, conduction losses dominates over the internal energy gain of the substrate material. This, in turn, results in temperature reduction in this region with further progressing time.

2.4 Jet impingement and flow field in the key-hole

In the laser gas-assisted cutting process, an impinging jet is used to shield the irradiated region from the oxidation reactions. In most cases, the jet emanates from a conical nozzle coaxially with the laser beam. The surface of the molten material in the irradiated region moves in the direction of gravity with the influence of the drag force generated by the assisting gas. Therefore, the heat transfer rates, as well as the skin friction, are influenced by the coupling effect of the impinging jet and the moving molten surface. This, in turn, alters the end-product quality. In this case, the moving molten surface generates a slip boundary condition for the impinging jet in the irradiated region. However, the velocity of the molten surface depends on the laser output power such that it increases with increased laser output power, that is, the velocity of the molten surface changes once the laser output power intensity is modified. Consequently, investigation into jet impingement onto the moving surface resembling the laser induced key-hole formation becomes necessary. The closed-form solution for the flow and temperature fields is very difficult due to the complex nature of the equations. However, the numerical solution for the problem is possible incorporating the appropriate boundary conditions. The mathematical arguments related to the flow and temperature field are given in relation to the previous study [12].

The geometric arrangement of the nozzle and the key-hole hole is shown in Fig. 2.5, while the nozzle and the hole configurations are given in Table 2.1. It should

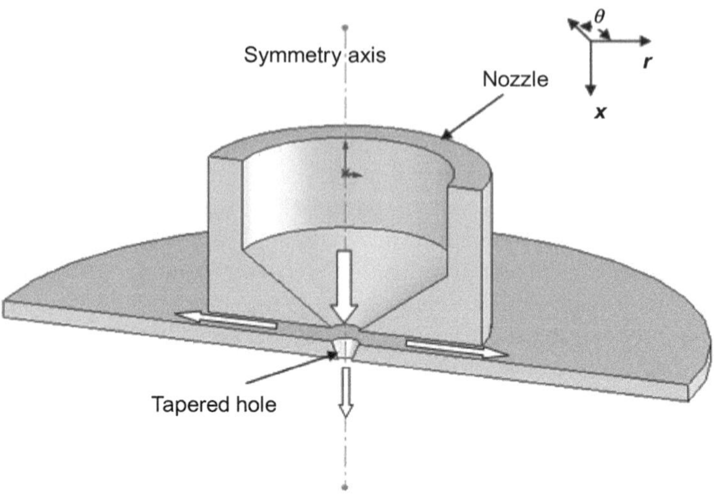

Fig. 2.5 Schematic view of the nozzle and the tapered hole resembling the key-hole. Arrows show the flow directions.

Table 2.1 Diameter and depth of the hole, and nozzle exit diameter

Hole diameter (m)	Hole thickness (depth) (m)	Hole taper angle (degree)	Nozzle exit diameter (m)
0.001	0.002	45	0.0004

be noted that the dimensions used in the simulations are similar to those used in the actual situations.

The steady and axisymmetric flow conditions are considered and the compressibility and variable properties of the working fluid are accommodated in the analysis. A constant temperature is considered at the hole wall to resemble the laser produced hole, that is, the hole wall temperature is kept at 1500 K as similar to the melting temperature of the substrate material. In addition, the hole wall is assumed to move with a velocity similar to the molten metal velocity as determined from the previous study [13]. Consequently, six-hole wall velocities are considered in the simulations to cover the large variation of hole wall velocities.

2.4.1 Flow equations

The governing flow and energy equations for the axisymmetric impinging steady jet can be written in the Cartesian tensor notation as:

(i) The continuity equation:

$$\frac{\partial}{\partial x_i}(\rho U_i) = 0; \qquad (2.181)$$

(ii) The momentum equation:

$$\frac{\partial}{\partial x_i}(\rho U_i U_j) = \frac{\partial p}{\partial x_j} + \frac{\partial}{\partial x_i}\left[\mu\left(\frac{\partial U_i}{\partial x_j} + \frac{\partial U_j}{\partial x_i}\right) - \rho R_{ij}\right]; \tag{2.182}$$

(iii) The energy equation:

$$\frac{\partial}{\partial x_i}(\rho U_i H) = \frac{\partial}{\partial x_i}\left[\frac{\mu}{\sigma}\frac{\partial H}{\partial x_i} - \rho R_{ih}\right]. \tag{2.183}$$

The flow field is turbulent; therefore, Reynolds stress turbulence model (RSM), which is based on the second-moment closure, is used in the analysis. The transport equation of the Reynolds stress (R_{ij}) is:

$$\frac{\partial}{\partial x_m}\left(U_m R_{ij}\right) = P_{ij} + \Lambda_{ij} - \varepsilon_{ij} + \Pi_{ij} + \Pi_{ij}^{w}, \tag{2.184}$$

where P, Λ, ε_{ij}, Π, and Π^{w} are the rate of production, transport by diffusion, rate of dissipation, transport due to turbulent pressure excluding strain interactions and transport due to wall reflection respectively. Eq. (2.183) consists of six partial differential equations; one for the transport of each of the six independent Reynolds stresses. The production term (P_{ij}), diffusion (Λ_{ij}), dissipation (ε_{ij}), transport due to turbulent pressure (Π_{ij}) and the modeling of the wall reflection (Π_{ij}^{w}) are defined as in Ref. [14].

2.4.2 Flow boundary conditions

Four boundary conditions are considered in accordance with the geometric arrangement of the problem, as shown in Fig. 2.6.

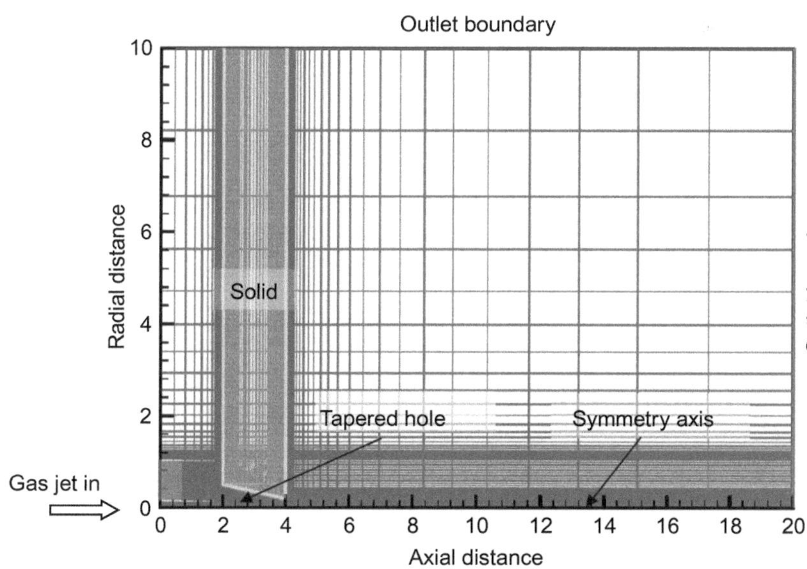

Fig. 2.6 Solution domain and grid used in the simulations [12]

2.4.2.1 Solid wall

For the solid wall, the slip condition is assumed (except $U_{wall}=0$ m/s) according to the hole velocity resembling the molten flow [13], and the boundary condition for the velocity at the solid wall, therefore is:

$$U_i = U_{wall} \tag{2.185}$$

2.4.2.2 Generalized wall functions for normal and shear turbulent stresses for the RSM model

When the flow is very near the wall, it undergoes a rapid change in direction; therefore, the wall-functions approach is not successful in reproducing the details of the flow. Consequently the turbulent stresses and fluxes at the near-wall grid points are calculated directly from their transport equations. In this case, the near-wall region lying between the wall and the near-wall computational node at x_p can be represented by two layers: the fully viscous sublayer, defined by $Re_\upsilon = x_\upsilon\sqrt{k_\upsilon}/\nu \asymp 20$, and a fully-turbulent layer. The wall shear stress near the wall is employed, that is, $\overline{\upsilon w}|_{z_\upsilon} = \tau_w/\rho$, which serves as the boundary condition for the $\overline{\upsilon w}$ transport equation.

In relation to normal stresses, the turbulence energy must decrease quadratically towards a value of zero at the wall; therefore a zero-gradient condition for the normal stresses is physically realistic. This is because of the viscous shear layer; in which the turbulence shear stress is zero within this layer. In addition, this situation is insufficient to ensure an accurate numerical representation of near-wall effects. An improved approach for internal cells is needed with respect to evaluating volume integrated production and dissipation of normal stresses (these are normally evaluated at cell centers, using linear interpolation, and then multiplied by the cell volume). Considering $\overline{\upsilon^2}$ as an example, the volume-integrated production of $\overline{\upsilon^2}$ between the wall and the P-node may be approximated by [15].

$$\int_{\Delta r}\int_0^{z_p} P_{22} d\forall \cong \int_{\Delta r}\int_{x_\upsilon}^{x_p} -2\overline{\upsilon w}\frac{\partial V}{\partial x} d\forall = 2\tau_w\left(\frac{V_p - V_\upsilon}{x_p - x_\upsilon}\right)x_p\Delta r; \tag{2.186}$$

where V_p and V_υ follow from the log-law. No contribution arises from the viscous sublayer, because $\overline{\upsilon w}=0$ in this layer. An analogous integration of the dissipation rate with the assumptions,

$$\varepsilon = \frac{2\nu k_\upsilon}{x_\upsilon^2} \qquad 0 \le x \le x_\upsilon,$$

$$\varepsilon = \frac{C_\mu^{3/4} k_p^{3/2}}{\kappa x_\upsilon} \qquad x_\upsilon \le x \le x_p,$$

leads to

$$\int_{\Delta r}\int_0^{x_p} \varepsilon d\forall \cong \left[\frac{2\nu k_p}{x_\upsilon} + \frac{C_\mu^{3/4} k_p^{3/2}}{\kappa} \ln\left(\frac{x_p}{x_\upsilon}\right)\right]\Delta r, \tag{2.187}$$

and an analogous treatment is applied to $\overline{v^2}$, while the production of $\overline{w^2}$ in the viscous and turbulent near-wall layers region is zero.

The values resulting from Eqs. (2.185), (2.186) are added, respectively, to the volume-integrated generation and dissipation computed for the upper half of the near-wall volume. It should be noted that for the wall-law approach, the near-wall dissipation (ε_p) is not determined from its differential equation applied to the near-wall cell surrounding the node. Instead, and in accordance with the log law, this value is obtained via the length scale from $\varepsilon_p = C_\mu^{3/4} k_p^{3/2} / \kappa z_p$, which serves as the boundary conditions for inner cells.

2.4.2.3 Inlet conditions

The boundary conditions for temperature (T = specified (300 K)) and the mass flow rate at the nozzle exit are introduced. The mass flow rate at the nozzle exit is changed for each case considering such that the mean jet velocities at the nozzle exit are 50 and 100 m/s. The selection of the average velocities at the nozzle exit is based on the actual laser cutting situations [13].

The values of k and ε are not known at the inlet, but can be determined from the turbulent kinetic energy, that is:

$$k = \lambda \bar{u}^2, \tag{2.188}$$

where $\bar{u}$ is the average inlet velocity and λ is a percentage.

The dissipation is calculated from $\varepsilon = C_\mu k^{\frac{3}{2}} / aD$, where D is the diameter. The values $\lambda = 0.03$ and $a = 0.005$ are commonly used and may vary slightly in the literature [16].

2.4.2.4 Outlet

The flow is considered to be extended over a long domain; therefore, the boundary condition (outflow boundaries) for any variable ϕ is

$$\frac{\partial \phi}{\partial x_i} = 0, \tag{2.189}$$

where x_i is the normal direction at outlet.

2.4.2.5 Symmetry axis

At the symmetry axis, the radial derivative of the variables is set to zero, that is:

$$\frac{\partial \phi}{\partial r} = 0, \tag{2.190}$$

except for

$$V = \overline{vw} = \overline{vh} = \overline{wh} = 0 \tag{2.191}$$

2.4.3 Solid side

2.4.3.1 Constant wall temperature boundary

Two constant temperature boundaries are considered. The first one is in the radial direction far away from the symmetry axis at a constant temperature $T=T_{\mathrm{amb}}$ (300 K). It should be noted that the constant temperature boundary condition is set at different locations in the radial directions. The boundary condition (T=constant) located in the radial direction had no significant effect on the temperature and flow field in the stagnation region. Therefore, this boundary condition is set for a radial distance of 0.0005 m from the symmetry axis. The second constant temperature boundary is set at the hole walls at T=constant (1500 K).

2.4.3.2 Solid fluid interface

The coupling of conduction within the solid and convection within the fluid is required for the present analysis. At the solid fluid interface, the appropriate boundary condition is the continuity of heat flux that is,

$$K_{W_{solid}} \frac{\partial T_{W_{solid}}}{\partial x} = K_{W_{gas}} \frac{\partial T_{W_{gas}}}{\partial x} \quad (2.192)$$

At the hole wall, a constant temperature boundary is assumed. The assumption of the constant temperature boundary is due to the high hole temperature resulted during the laser processing; in which case, the hole-wall temperature remains almost at the melting temperature of the substrate material (for mild steel, $T_w \sim 1500$ K).

No radiation loss from the solid surface is assumed, which is because of the small hole size.

The heat transfer coefficient is defined from the Fourier heat law:

$$h_s = \frac{K_f \left(\frac{\partial T}{\partial n}\right)_{wall}}{\left(T_w - T_{ref}\right)} \quad (2.193)$$

where K_f is the fluid thermal conductivity, $\left(\frac{\partial T}{\partial n}\right)_{wall}$ is the temperature gradient within the neighborhood of the wall, T_w is the wall temperature, and T_{ref} is the reference temperature. In the present case $T_{ref}=T_j$.

2.4.3.3 Solid heating

In the solid, the conduction heat transfer is considered and the Fourier heat law is accommodated. Temperature is assumed to reduce the reference temperature at a distance far away from the hole wall. Consequently, the boundary condition for the solid is at a distance far way from the wall, $T=300$ K.

2.4.3.4 Working fluid and solid

Air is used as a working fluid and it is assumed to be an ideal gas; therefore, the equation of state and related equations are adopted for the property calculations. Steel is used as solid substrate in the simulations. Table 2.2 gives the thermal properties of steel at 300 K.

2.4.4 Numerical method and simulation

A control volume approach is employed to discretize the governing equations. The discretization procedure is given in the literature [17]. The problem of determining the pressure and satisfying continuity may be overcome by adjusting the pressure field so as to satisfy continuity. A staggered grid arrangement is used in which the velocities are stored at a location midway between the grid points, that is, on the control volume faces. All other variables including pressure are calculated at the grid points. This arrangement gives a convenient way of handling the pressure linkages through the continuity equation and is known as Semi-Implicit Method for Pressure-Linked Equations (SIMPLE) algorithm. The details of this algorithm are given in the literature [17].

The computer program used for the present simulation can handle a nonuniform grid spacing. In each direction, a fine-grid spacing near the gas jet impingement point and the hole is allocated while gradually increased spacing for locations away from the hole is considered (Fig. 2.8). Elsewhere the grid spacing is adjusted to maintain a constant ratio of any of two adjacent spacing. The numbers of grid planes used, normal to the x and r directions, are 70 and 95 (70×95), respectively, in the simulations. The grid independence tests were conducted. The mesh resulting in grid independent solution is selected for the simulations (70×95 mesh size). The pressure variation along the symmetry axis for the jet velocity of 100 m/s and the wall velocity of 30 m/s.

Fig. 2.7 shows velocity vectors in the closed region of the hole for two impinging jet and six-hole wall velocities. In general, the jet bending and the streamline curvature are evident in the neighborhood of the hole inlet. This is because of the blockage effect of the hole; in which case, flow entering into the hole results in pressure increase in the region of the hole inlet because of the hole convergence (tapering). This is more pronounced in the region of the hole circumference. The streamline curvature above the hole inlet causes flow spilling in the region of the hole circumference. This situation is observed for all jet and wall velocities. Because the flow in the hole is subsonic, hole convergence results in the acceleration of flow in the hole towards the hole exit. Consequently, flow forms a jet upon leaving the hole. This situation is more pronounced for the high-jet velocity (100 m/s). Although the shear layer developed at the jet

Table 2.2 **Thermal properties of steel at 300 K**

	Specific heat capacity (J/kg K)	Thermal conductivity (W/m K)	Density (kg/m^3)	Thermal diffusivity (m^2/s)$\times 10^{-6}$
Steel	502.48	150	8030	37.175

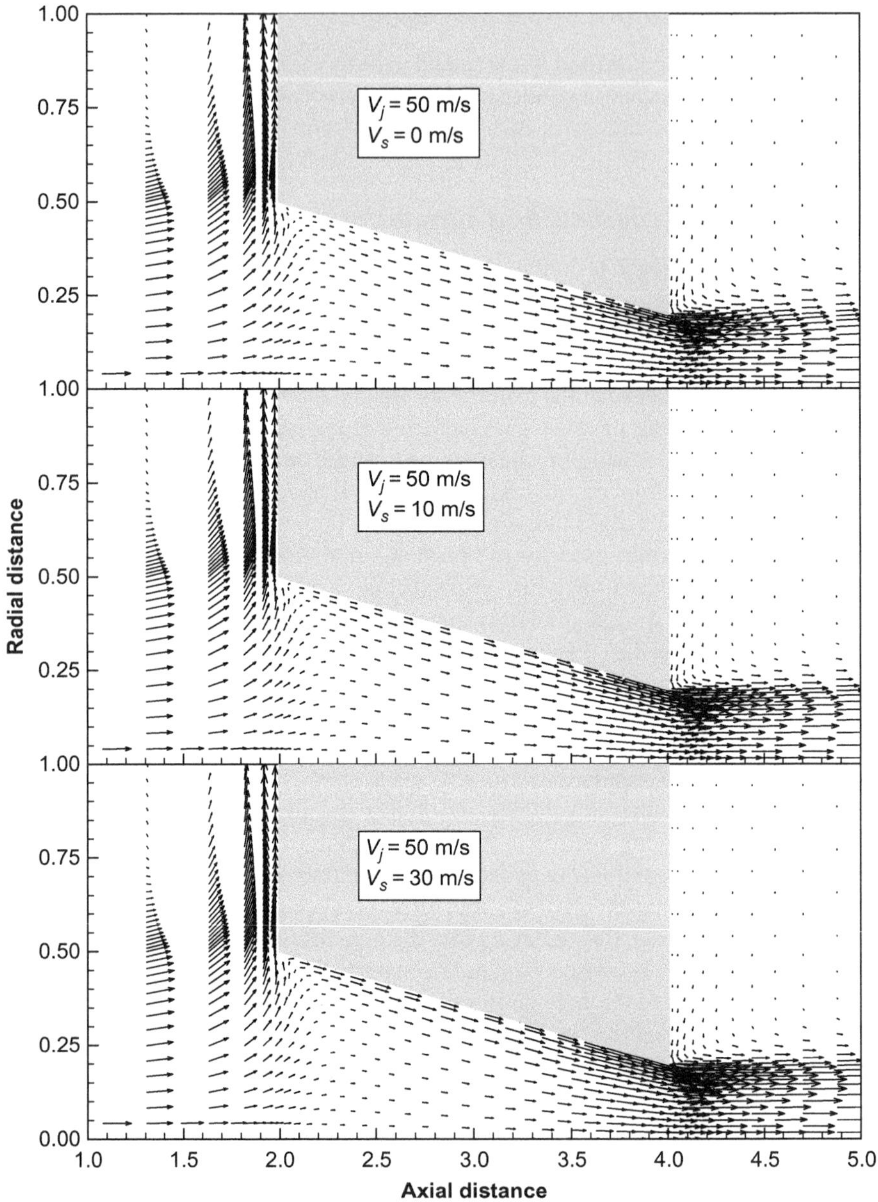

Fig. 2.7 Normalized velocity vectors (V/V_j) around the hole. Mean jet velocity at the nozzle exit is 50 m/s. The axial and radial distances are normalized by dividing the nozzle exit diameter (D). Taper angle is 45 degree [12].

boundary is evident, no circulation is formed in the neighborhood of the jet boundary leaving the hole. The influence of the hole wall velocity on the flow structure is not significant because of the confined flow situation in the hole, which suppresses the radial expansion of the flow. However, at the hole inlet radial momentum toward the hole wall generates a radial flow in this region. This decays significantly as the flow progresses in the axial direction. The effect of jet velocity on the flow field in the hole is not significant, except close to the moving wall. This is more pronounced for high-hole wall velocity (30 m/s); in which case, the axial momentum of the flow in this region increases the velocity magnitude in the near-wall region.

Fig. 2.8 shows temperature contours in the closed region of the hole for two jet and three different wall velocities. Fluid temperature increases in the region of the hole

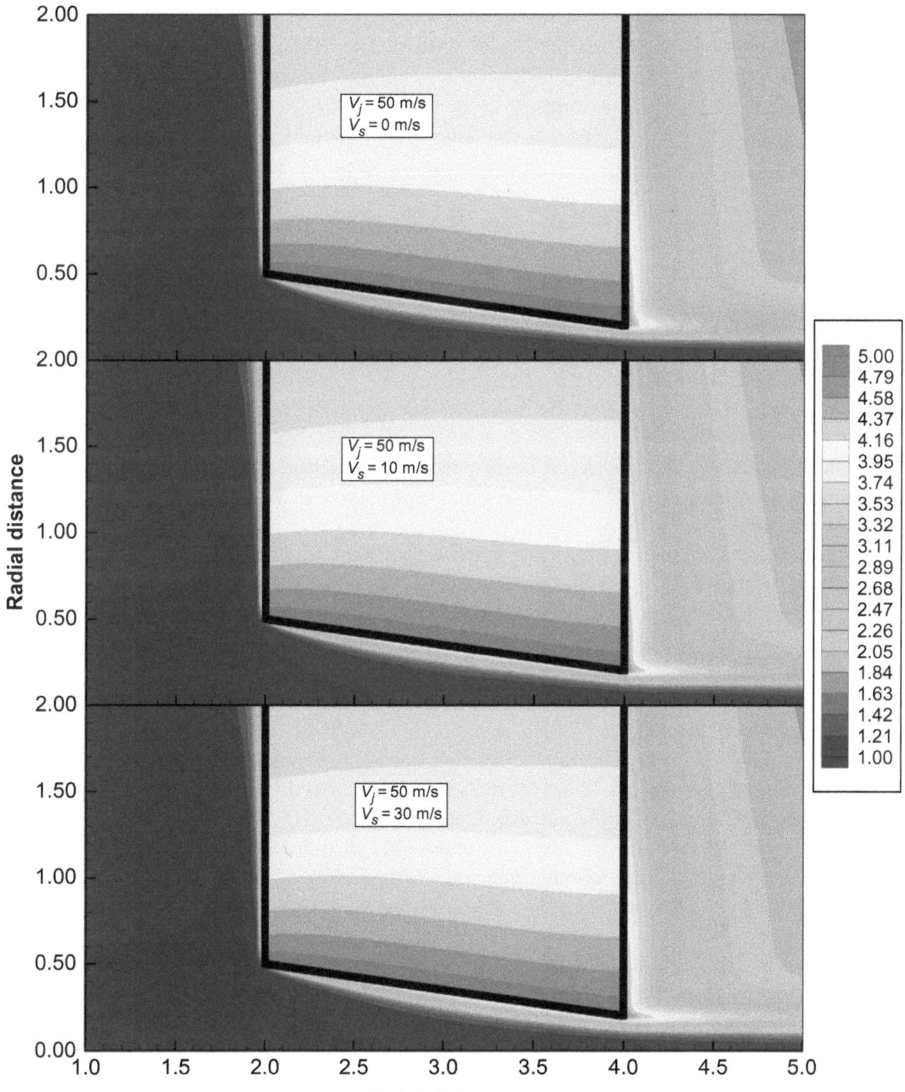

Fig. 2.8 Normalized temperature (T/T_j) distribution around the hole. Mean jet velocity at nozzle exit is 50 m/s. The axial and radial distances are normalized by dividing the nozzle exit diameter (D). Taper angle is 45 degree [12].

walls due to convective and conductive heating of the fluid from the hole wall, because the hole wall temperature is 1500 K. The stagnation point heating in the hole entry region contributes to fluid temperature enhancement. This extends in the radial direction toward the hole circumference because of the streamline curvature. The fluid heating in the region of the hole wall increases the static pressure because of the assumption of an ideal gas for the working fluid. Although the thermal boundary thickness is small, increased pressure in this region enhances the blockage effect in the hole. However, the hole wall velocity slightly influences the temperature of the jet at the hole exit. In this case, increasing the hole wall velocity increases the size of the high temperature region in the jet exiting the hole. This is more pronounced for high wall velocity (30 m/s). Moreover, increasing the jet velocity suppresses the thickness of the thermal boundary layer in the region near to the hole wall which, in turn, lowers the blockage effect in the hole, particularly at the hole exit region. This enhances the flow acceleration at the exit of the hole.

References

[1] W.D. Seka, D. Fried, J.D.B. Featherstone, R.E. Glena, Time-dependent reflection and surface temperatures during CO_2 laser irradiation of dental hard tissues with 100-μs pulses. in: Proc. SPIE 2394, Lasers in Dentistry, May 1, 1995. https://doi.org/10.1117/12.207456.

[2] R.W. Ricci, M.A. Ditzler, L.P. Nestor, Discovering the Beer-Lambert law, J. Chem. Educ. 71 (11) (1994) 983–985.

[3] S.Z. Shuja, B.S. Yilbas, Z. Ayar, Laser pulse heating and phase change process: comparison of volumetric heat source and surface heat source, Proc. IME C J. Mech. Eng. Sci. 224 (8) (2010) 1697–1706.

[4] B.S. Yilbas, K. Apalak, The basic concepts of heat transfer mechanism during laser drilling of metals, Egypt. J. Phys. 18 (1) (1987) 25–34.

[5] D. Tabor, Gasses, Liquids and Solids, second ed., Cambridge University Press, New York, 1979.

[6] M. Galeev, A. Sudan, Plasma Physics, North Holand Publ. Ltd., New York, 1983.

[7] M. Kalyon, B.S. Yilbas, Closed from solution for exponentially decaying laser pulse heating: evaporation at the surface, Jpn. J. Appl. Phys. 41 (2002) 3737–3746.

[8] M. Abramowitz, I.A. Stegun, Handbook of Mathematical Functions, Dover, New York, 1965.

[9] B.S. Yilbas, N. Al-Aqeeli, Thermal stresses due to exponentially decaying laser pulse and a convection boundary at the surface, Opt. Lasers Eng. 16 (2006) 235–265.

[10] B.S. Yilbas, N. Al-Aqeeli, Thermal stresses due to time exponentially decaying laser pulse: elasto plastic wave propagation, Int. J. Mech. Sci. 46 (2004) 57–80.

[11] A.D. Kovalenko, Thermoelasticity – Basic Theory and Applications, Wolters-Noordhoff, Groningen, 1969.

[12] S.Z. Shuja, B.S. Yilbas, S. Khan, Jet impingement onto a tapered hole: influence of jet velocity and hole wall velocities on heat transfer and skin friction, Int. J. Numer. Methods Fluids 60 (2009) 972–991.

[13] B.S. Yilbas, B.J. Abdul Aleem, Dross formation during laser cutting process, J. Phys. D Appl. Phys. 39 (2006) 1451–1461.

[14] B.E. Launder, Second-moment closure and its use in modelling turbulent industrial flows, Int. J. Numer. Methods Fluids 9 (1989) 963–985.

[15] S. Hogg, M.A. Leschziner, Second-moment-closure calculation of strongly swirling confined flow with large density gradients, Int. J. Heat Fluid Flow 10 (1989) 16–271.
[16] D. Elkaim, M. Reggio, R. Camarero, Simulating two-dimensional turbulent flow by using the k-ε model and the vorticity-stream function formulation, Int. J. Numer. Methods Fluids 14 (1992) 961–980.
[17] S.V. Patankar, Numerical Heat Transfer, McGraw-Hill, New York, 1980.

Analytical methods in laser cutting 3

3.1 Introduction

The laser cutting process is involved with multiphysical phenomena including initial solid heating, melting and evaporation toward keyhole formation, cutting initiation, kerf formation, and dross ejection. The efforts toward introducing analytical approaches that capture the engineering and multiphysics aspects of the laser cutting process need to be appreciated. Although a considerable number of assumptions are usually introduced to describe the process in a mathematical frame, the resulting equations describing the process are, in most cases, useful and predict results close to the experimental data. In general, the laser cutting process is associated with high-temperature phenomena taking place under the high-pressure assisting gas environment. When modeling the laser cutting process, treatment of the assisting gas needs to be incorporated such that the effects of cooling and the flow resistance on the cut formation should be considered. Because the assisting gas has a high velocity, it acts like an impinging jet onto the surface while forming a stagnation zone and radial flow in the cut section. In addition, in some regions of radial flow, the behavior of the flow can be expressed as laminar, and for other regions, it behaves turbulent. Therefore, flow resistance and cooling effects should be examined in the cases when the flow behavior is laminar as well as turbulent. On the other hand, oxidation reactions taking place during the cutting process at high temperatures also influences the physical processes taking place in the cutting section. This is because the high temperature oxidation reactions are exothermic and provide excessive heat while modifying the physical ambient of the cutting section. In some situations, the molten flow instability in the cut section results in the formation of striation and dross attachments in the kerf sites. The high-temperature oxidation reactions contribute to the molten flow instability and the geometric features of the striation patterns being formed. In addition, oxygen diffusion in the molten material modifies the cooling rates and the fracture toughness of the recast layer at the kerf surfaces. Therefore, in analytical approaches toward formulating the multiphysics processes taking place during laser cutting, care must be taken to tackle the problem as close to realistic as possible. In this chapter, analytical approaches for the laser cutting process, the effect of assisting gas on the physical processes, oxygen diffusion, and entropy and efficiency analyses are introduced.

3.2 Pulse heating analysis of solid surfaces in relation to initiation of laser cutting

In some laser cutting processes, laser repetitive pulses are more favorable than laser continuous wave heating (c.w.). This is because of the control of surface plasma effects on the cutting process and the efficient processing through energy savings

The Laser Cutting Process. http://dx.doi.org/10.1016/B978-0-12-812982-1.00003-5

in laser cutting. When a high-intensity beam interacts with the solid surface, solid phase heating takes place prior to key-hole formation via surface evaporation. The temperature and stress fields formed during the solid heating plays a crucial role in the key hole formation because of excessive stresses, which cause micro or large-scale cracks around the cut sections. Although the process of pulse repetition modifies the heat transfer characteristics and temperature distribution in the irradiated region, single pulse heating also provides useful information on the solid heating of surfaces prior to the initiation of evaporation. Therefore, in this section, single pulse heating and thermal-stress formation in the solid phase is presented in line with the previous studies [1–3].

The pulses delivered by a laser beam can create energies sufficient to heat surfaces rapidly, which leads to possible thermal integration at the workpiece surface. The thermal integration can be defined as the integration of the temperature at the surface during the repetitive pulse-heating process, in which case, the temperature remains almost steady during the repetitive pulses. The repetitive-pulse laser heating has several advantages over a continuous laser processing. This is because the temperature rise in the surface vicinity of the substrate can be controlled easily through varying the pulse parameters and repetition rates; consequently, heating with an almost-constant temperature at the surface can be achieved.

The heat conduction equation based on the Fourier theory of heating is employed to describe the laser pulse heating process. The heating is limited to conduction only, in which case, the phase change process is not included. In order to simplify the problem, some useful assumptions are made, which include omission of convection and radiation losses from the surface, and constant thermal properties. The governing equation of heat conduction can be written as [4]:

$$\rho Cp\frac{\partial T}{\partial t} = k\frac{\partial^2 T}{\partial x^2} + I\delta\exp(-\delta x) \tag{3.1}$$

where I is the time varying pulse intensity, which can be formulated as:

$$I = I_o[c_1\exp(-a_1t) + c_2\exp(-a_2t) + c_3\exp(-a_3t) + \cdots + c_n\exp(-a_nt)] \tag{3.2}$$

The time-varying pulse intensity described in Eq. (3.2) can simulate the actual laser output pulse because the laser output pulse intensity can be represented by several time exponentially varying functions.

The equation governing the heating process (Eq. 3.1) is linear and the complete solution of Eq. (3.1) can be obtained after solving the equation for a first term ($c_1\exp(-a_1t)$) in the time exponentially varying pulse, and then summing the solution of Eq. (3.1) for the other terms in the time varying exponential pulse. Consequently, the equation of interest reduces to:

$$\rho Cp\frac{\partial T}{\partial t} = k\frac{\partial^2 T}{\partial x^2} + I_o[c_1\exp(-a_1t)\delta\exp(-\delta x)] \tag{3.3}$$

The boundary conditions for Eq. (3.3) are:

At the surface $x=0\left(\frac{\partial T}{\partial x}\right)_{x=0}=0$ and at $x=\infty$ $T(\infty,t)=0$. The initial condition is: At $t=0$ $T(x,0)=0$.

The closed form solution of Eq. (3.3) is obtained previously using Dawson's integral method [4].

$$T(0,t)=I_o\frac{c_1}{a_1+\alpha\delta^2}\left\{e^{\alpha\delta^2 t}\operatorname{erfc}\left(\delta\sqrt{\alpha t}\right)+2\delta\frac{\sqrt{\alpha}}{\sqrt{\pi a_1}}F\left(\sqrt{a_1 t}\right)-e^{-\beta a_1 t}\right\} \tag{3.4}$$

Similarly, the temperature distribution, including all the terms in the time exponentially varying pulse (Eq. 3.2) can be written as:

$$T(0,t)=I_o\sum_{i=1}^{n}\left\{\frac{c_n}{a_n+\alpha\delta^2}\left\{\begin{array}{l}e^{\alpha\delta^2 t}\operatorname{erfc}\left(\delta\sqrt{\alpha t}\right)\\ +2\delta\frac{\sqrt{\alpha}}{\sqrt{\pi a_n}}F\left(\sqrt{a_n t}\right)-e^{-\beta a_n t}\end{array}\right\}\right\} \tag{3.5}$$

In the present formulation, n is considered as 2 in Eq. (3.5) because two of the terms in the time-exponential function (Eq. 3.2) are sufficient to represent the laser pulse. Consequently, the closed-form solution for the surface temperature (Eq. 3.5) reduces to:

$$T(0,t)=I_o\left\{\begin{array}{l}\frac{c_1}{a_1+\alpha\delta^2}\left\{e^{\alpha\delta^2 t}\operatorname{erfc}\left(\delta\sqrt{\alpha t}\right)+2\delta\frac{\sqrt{\alpha}}{\sqrt{\pi a_1}}F\left(\sqrt{a_1 t}\right)-e^{-\beta a_1 t}\right\}+\\ \frac{c_2}{a_2+\alpha\delta^2}\left\{e^{\alpha\delta^2 t}\operatorname{erfc}\left(\delta\sqrt{\alpha t}\right)+2\delta\frac{\sqrt{\alpha}}{\sqrt{\pi a_2}}F\left(\sqrt{a_2 t}\right)-e^{-\beta a_2 t}\right\}\end{array}\right\} \tag{3.6}$$

Eq. (3.6) can be non-dimensionalized introducing the following variables:
$a_1'=\frac{a_1}{\alpha\delta^2}$; $a_2'=\frac{a_2}{\alpha\delta^2}$; $b_2=\frac{c_2}{c_1}$, provided that $c_1\neq 0$; $\tau=\alpha\delta^2 t$

The non-dimensional form in Eq. (3.6) becomes:

$$T(0,\tau)=\frac{I_o}{k\delta}c_1\left\{\begin{array}{l}\left(\frac{1}{a_1'+1}+\frac{b_2}{a_2'+1}\right)e^{\tau}\operatorname{erfc}(\sqrt{\tau})+\frac{2}{\sqrt{\pi}}\left[\frac{F\left(\sqrt{a_1'\tau}\right)}{\sqrt{a_1'}\left(a_1'+1\right)}+\frac{b_2F\left(\sqrt{a_2'\tau}\right)}{\sqrt{a_2'}\left(a_2'+1\right)}\right]\\ -\frac{e^{-a_1'\tau}}{\left(a_1'+1\right)}+\frac{e^{-a_2'\tau}}{\left(a_2'+1\right)}\end{array}\right\} \tag{3.7}$$

Eq. (3.7) shows that the surface temperature rises to reach maximum before decaying to its initial value asymptotically. Therefore, in order to formulate the maximum temperature, the condition $\frac{\partial T(0,\tau)}{\partial\tau}=0$ should be satisfied. After lengthy algebra, it yields:

$$\frac{\frac{T_{max}}{c_1 I_o}}{k\delta} = \left\{ \frac{2}{\sqrt{\pi}} \left[\frac{F\left(\sqrt{a_1'\tau}\right)}{\sqrt{a_1'}} - \frac{F\left(\sqrt{a_2'\tau}\right)}{\sqrt{a_2'}} \right] - e^{-a_1'\tau} + b_2 e^{-a_2'\tau} \right\}_{\tau=\tau_{max}} \quad (3.8)$$

where τ_{max} is the dimensionless time at which the surface temperature is maximum. The solution of the following equation for τ gives the value of τ_{max}, i.e.:

$$\left(\frac{1}{a_1'+1} + \frac{b_2}{a_2'+1}\right) e^{\tau} \operatorname{erfc}\left(\sqrt{\tau}\right) = \left\{ \frac{2}{\sqrt{\pi}} \left[\frac{F\left(\sqrt{a_1'\tau}\right)}{\sqrt{a_1'}} - \frac{F\left(\sqrt{a_2'\tau}\right)}{\sqrt{a_2'}} \right] - e^{-a_1'\tau} + b_2 e^{-a_2'\tau} \right\} \quad (3.9)$$

A single time exponentially varying pulse results in a surface temperature decaying asymptotically toward the end of heating process. Therefore, steady temperature attainment at the surface is impossible with a single pulse heating process. In order to obtain a steady temperature at the surface a repetitive pulsing should be considered. In this case, the intensity profile resembling the consecutive pulses is necessary. This can be achieved introducing a shift in time in the intensity function. In this case, the heat transfer equation employing the consecutive pulses for the repetitive pulsing process can be written as:

$$\frac{\partial^2 T}{\partial x^2} - \frac{1}{\alpha}\frac{\partial T}{\partial t} = -\frac{I_1 \delta}{k} e^{-\delta x} 1(t-t_0) e^{-\beta t} \quad (3.10)$$

where $1(t-t_0)$ is a step function, which is introduced to resemble the consecutive pulses, i.e.:

$$1(t-t_0) = \begin{cases} 0, & t < t_0 \\ 1, & t \geq t_0 \end{cases} \quad (3.11)$$

It should be noted that the first term in the time exponentially varying pulse in considered as mentioned earlier for Eq. (3.10). The boundary conditions are:

$$\left.\frac{\partial T}{\partial x}\right|_{x=0} = 0, \quad T(x,0) = 0, \quad T(\infty, t) = 0 \quad (3.12)$$

Laplace transformation of Eq. (3.12) with respect to t yields:

$$\frac{\partial^2 \overline{T}}{\partial x^2} - \frac{1}{\alpha}\left[\left(p\overline{T} - T(x,0)\right)\right] = -I_o e^{-\delta x} \frac{1}{p+\beta} e^{-t_0 p} \quad (3.13)$$

where $\overline{T} = T(x,p)$ and $I_o = I\delta/k$.

If we let $\lambda^2 = \frac{p}{\alpha}$ and $H_0 = -\frac{\hat{I}}{p+\beta} e^{-t_0 p}$ and use the boundary condition $T(x,0) = 0$, Eq. (3.158) becomes:

$$\frac{\partial^2 \overline{T}}{\partial x^2} - \lambda^2 \overline{T} = H_0 e^{-\delta x} \quad (3.14)$$

Eq. (3.13) has a homogeneous solution, $\overline{T}_h$, and a particular solution, $\overline{T}_p$, as follows:

$$\overline{T}_h = c_1 e^{\lambda x} + c_2 e^{-\lambda x} \tag{3.15}$$

And $\overline{T}_p = A_0 e^{-\delta x} \Rightarrow \frac{\partial^2 \overline{T}_p}{\partial x^2} - \lambda^2 \overline{T}_p = H_0 e^{-\delta x} \Rightarrow A_0 = \frac{H_0}{\delta^2 - \lambda^2}$ or $\overline{T}_p = \frac{H_0}{\delta^2 - \lambda^2} e^{-\delta x}$.

Hence, the solution of Eq. (3.15) yields:

$$\overline{T} = \overline{T}_h + \overline{T}_p = c_1 e^{\lambda x} + c_2 e^{-\lambda x} + \frac{H_0}{\delta^2 - \lambda^2} e^{-\delta x}. \tag{3.16}$$

Using the boundary condition, $T(\infty, t) = 0$, in Eq. (3.16), one gets $c_1 = 0$. Thus, Eq. (3.16) with $\lambda^2 = \frac{p}{\alpha}$ and $H_0 = -\frac{I_o}{p+\beta} e^{-t_0 p}$, yields:

$$\overline{T} = c_2 e^{-\lambda x} - \frac{I_o}{(p+\beta)\left(\delta^2 - \frac{p}{\alpha}\right)} e^{-t_0 p} e^{-\delta x} \tag{3.17}$$

If the remaining boundary condition, $\left.\frac{\partial T}{\partial x}\right|_{x=0} = 0$, is introduced, c_2 can be obtained as follows:

$$c_2 = \frac{I_o \delta \sqrt{\alpha}}{\sqrt{p}(p+\beta)\left(\delta^2 - \frac{p}{\alpha}\right)} e^{-t_0 p} \tag{3.18}$$

Substitution of Eq. (3.18) into Eq. (3.17), yields:

$$\overline{T} = I_o e^{-t_0 p}\left(\delta\sqrt{\alpha} H_2 - I_o H_1 e^{-\delta x}\right) \tag{3.19}$$

where

$$H_1 = \frac{1}{(p+\beta)\left(\delta^2 - \frac{p}{\alpha}\right)} \quad \text{and} \quad H_2 = \frac{1}{\sqrt{p}(p+\beta)\left(\delta^2 - \frac{p}{\alpha}\right)} e^{-\frac{x}{\sqrt{\alpha}}\sqrt{p}} \tag{3.20}$$

By using the partial fraction expansion method, H_1 and H_2 can be rewritten in the following forms, respectively:

$$H_1 = \frac{\alpha}{\beta + \alpha\delta^2}\left[-\frac{1}{p - \alpha\delta^2} + \frac{1}{p+\beta}\right] \tag{3.21}$$

and

$$H_2 = \frac{\alpha}{\beta + \alpha\delta^2}\left[\frac{e^{-\frac{x}{\sqrt{\alpha}}\sqrt{p}}}{\sqrt{p}\left(p - \alpha\delta^2\right)} + \frac{e^{-\frac{x}{\sqrt{\alpha}}\sqrt{p}}}{\sqrt{p}(p+\beta)}\right] \tag{3.22}$$

The inverse Laplace transformation of H_1, which is trivial, is given by:

$$L^{-1}(H_1) = \frac{\alpha}{\beta + \alpha\delta^2}\left[-e^{\alpha\delta^2 t} + e^{-\beta t}\right] \tag{3.23}$$

where L^{-1} is the inverse Laplace transformation operator.

It should be noted from Laplace transform that the translated function can be written as:

$$\text{if } L(f(t)) = F(p), \text{ then } L(f(t-t_0)1(t-t_0)) = e^{-t_0 p}F(p) \tag{3.24}$$

where $t_0 > 0$. In this case,

$$L^{-1}(H_1 e^{-t_0 p}) = \frac{\alpha}{\beta + \alpha\delta^2}\left[-e^{\alpha\delta^2(t-t_0)} + e^{-\beta(t-t_0)}\right]1(t-t_0) \tag{3.25}$$

On the other hand, the inverse Laplace transformation of H_2 requires a special treatment. The approach of obtaining $L^{-1}(H_2)$ is summarized as follow:

From the inverse Laplace relation [5], one can have:

$$L^{-1}\left[\frac{e^{-\sigma\sqrt{p}}}{\sqrt{p}(\sqrt{p}+a)}\right] = e^{a\sigma}e^{a^2 t}\operatorname{erfc}\left(a\sqrt{t} + \frac{\sigma}{2\sqrt{t}}\right) \tag{3.26}$$

where $\sigma > 0$.

One can note that:

$$\frac{1}{p - \alpha\delta^2} = \frac{1}{(\sqrt{p} - \sqrt{\alpha}\delta)(\sqrt{p} + \sqrt{\alpha}\delta)} = \frac{1}{2\sqrt{\alpha}\delta}\left[\frac{1}{\sqrt{p} - \sqrt{\alpha}\delta} - \frac{1}{\sqrt{p} + \sqrt{\alpha}\delta}\right] \tag{3.27}$$

and

$$\frac{1}{p + \beta} = \frac{1}{(\sqrt{p} - i\sqrt{\beta})(\sqrt{p} + i\sqrt{\beta})} = \frac{1}{2\sqrt{\beta}i}\left[\frac{1}{\sqrt{p} - i\sqrt{\beta}} - \frac{1}{\sqrt{p} + i\sqrt{\beta}}\right] \tag{3.28}$$

Substituting these identities into Eq. (3.22), yields:

$$H_2 = \frac{\alpha}{\beta + \alpha\delta^2}\left\{-\frac{1}{2\sqrt{\alpha}\delta}[z_1 - z_2] + \frac{1}{2\sqrt{\beta}i}[z_3 - z_4]\right\} \tag{3.29}$$

where

$$z_1 = \frac{e^{-\frac{x}{\sqrt{\alpha}}\sqrt{p}}}{\sqrt{p}(\sqrt{p} - \sqrt{\alpha}\delta)}, \quad z_2 = \frac{e^{-\frac{x}{\sqrt{\alpha}}\sqrt{p}}}{\sqrt{p}(\sqrt{p} + \sqrt{\alpha}\delta)},$$

and

$$z_3 = \frac{e^{-\frac{x}{\sqrt{\alpha}}\sqrt{p}}}{\sqrt{p}\left(\sqrt{p} - i\sqrt{\beta}\right)}, \quad z_4 = \frac{e^{-\frac{x}{\sqrt{\alpha}}\sqrt{p}}}{\sqrt{p}\left(\sqrt{p} + i\sqrt{\beta}\right)} \tag{3.30}$$

If we take the inverse Laplace transformation of Eq. (3.29), we obtain:

$$L^{-1}(H_2) = \frac{\alpha}{\beta + \alpha\delta^2}\left\{ -\frac{1}{2\sqrt{\alpha}\delta}\left[L^{-1}(z_1) - L^{-1}(z_2)\right] + \frac{1}{2\sqrt{\beta}i}\left[L^{-1}(z_3) - L^{-1}(z_4)\right] \right\} \tag{3.31}$$

where $L^{-1}(z_1)$, $L^{-1}(z_2)$, $L^{-1}(z_3)$, and $L^{-1}(z_4)$ are given below:

$$L^{-1}(z_1) = e^{-\delta x} e^{\alpha\delta^2 t} \operatorname{erfc}\left(-\sqrt{\alpha}\delta\sqrt{t} + \frac{x}{2\sqrt{\alpha}\sqrt{t}} \right) \tag{3.32}$$

and

$$L^{-1}(z_2) = e^{\delta x} e^{\alpha\delta^2 t} \operatorname{erfc}\left(\sqrt{\alpha}\delta\sqrt{t} + \frac{x}{2\sqrt{\alpha}\sqrt{t}} \right) \tag{3.33}$$

and

$$L^{-1}(z_3) = e^{-i\frac{\sqrt{\beta}}{\sqrt{\alpha}}x} e^{-\beta t} \operatorname{erfc}\left(-i\sqrt{\beta}\sqrt{t} + \frac{x}{2\sqrt{\alpha}\sqrt{t}} \right) \tag{3.34}$$

and

$$L^{-1}(z_4) = e^{i\frac{\sqrt{\beta}}{\sqrt{\alpha}}x} \, e^{-\beta t} \, erfc\left(i\sqrt{\beta}\sqrt{t} + \frac{x}{2\sqrt{\alpha}\sqrt{t}} \right) \tag{3.35}$$

Because $L^{-1}(H_2)$ is already achieved, one can obtain $L^{-1}(H_2 e^{-t_0 p})$ as follows:

$$L^{-1}(H_2 e^{-t_0 p}) = \frac{\alpha}{\beta + \alpha\delta^2}$$
$$\left\{ -\frac{1}{2\sqrt{\alpha}\delta}\left[L^{-1}(z_1) - L^{-1}(z_2)\right] + \frac{1}{2\sqrt{\beta}i}\left[L^{-1}(z_3) - L^{-1}(z_4)\right] \right\} 1(t - t_0) \tag{3.36}$$

where $L^{-1}(z_1)$, $L^{-1}(z_2)$, $L^{-1}(z_3)$, and $L^{-1}(z_4)$ are given below:

$$L^{-1}(z_1) = e^{-\delta x} e^{\alpha\delta^2 (t - t_0)} \operatorname{erfc}\left(-\delta\sqrt{\alpha(t - t_0)} + \frac{x}{2\sqrt{\alpha(t - t_0)}} \right) \tag{3.37}$$

and

$$L^{-1}(z_2) = e^{\delta x}\, e^{\alpha\delta^2(t-t_0)}\, erfc\left(\delta\sqrt{\alpha(t-t_0)} + \frac{x}{2\sqrt{\alpha(t-t_0)}}\right) \tag{3.38}$$

and

$$L^{-1}(z_3) = e^{-i\frac{\sqrt{\beta}}{\sqrt{\alpha}}x}\, e^{-\beta(t-t_0)}\,\mathrm{erfc}\left(-i\sqrt{\beta(t-t_0)} + \frac{x}{2\sqrt{\alpha(t-t_0)}}\right) \tag{3.39}$$

and

$$L^{-1}(z_4) = e^{i\frac{\sqrt{\beta}}{\sqrt{\alpha}}x}\, e^{-\beta(t-t_0)}\,\mathrm{erfc}\left(i\sqrt{\beta(t-t_0)} + \frac{x}{2\sqrt{\alpha(t-t_0)}}\right) \tag{3.40}$$

If the inverse Laplace transformation of Eq. (3.19) is taken, one can obtain:

$$L^{-1}(\overline{T}) = \hat{I}\delta\sqrt{\alpha}L^{-1}(H_2 e^{-t_0 p}) - \hat{I}L^{-1}(H_1 e^{-t_0 p})e^{-\delta x} \tag{3.41}$$

Substituting inverse of H_2 and H_1 into Eq. (3.41) and setting $x=0$, yields:

$$\begin{aligned} T(0,t) = I_o\frac{\alpha/2}{\beta+\alpha\delta^2}\left\{-\left[L^{-1}(z_1) - L^{-1}(z_2)\right]\big|_{x=0} + \frac{\delta\sqrt{\alpha}}{\sqrt{\beta}i}\left[L^{-1}(z_3) - L^{-1}(z_4)\right]\big|_{x=0}\right\} \\ -\hat{I}\frac{\alpha}{\beta+\alpha\delta^2}\left[-e^{\alpha\delta^2 t} + e^{-\beta t}\right] \end{aligned} \tag{3.42}$$

Noting that:

$$\begin{aligned} \left[L^{-1}(z_1) - L^{-1}(z_2)\right]\big|_{x=0} = e^{\alpha\delta^2(t-t_0)}\,\mathrm{erfc}\left(-\delta\sqrt{\alpha(t-t_0)}\right) \\ - e^{\alpha\delta^2(t-t_0)}\,\mathrm{erfc}\left(\delta\sqrt{\alpha(t-t_0)}\right) \end{aligned} \tag{3.43}$$

and

$$\begin{aligned} \left[L^{-1}(z_3) - L^{-1}(z_4)\right]\big|_{x=0} = e^{-\beta(t-t_0)}\,\mathrm{erfc}\left(-i\sqrt{\beta(t-t_0)}\right) \\ - e^{-\beta(t-t_0)}\,\mathrm{erfc}\left(i\sqrt{\beta(t-t_0)}\right) \end{aligned} \tag{3.44}$$

Because $\mathrm{erfc}(-\xi) = 2 - \mathrm{erfc}(\xi)$, Eqs. (3.43) and (3.44) become:

$$\left[L^{-1}(Z_1) - L^{-1}(Z_2)\right]\big|_{x=0} = 2e^{\alpha\delta^2(t-t_0)}\left(1 - \mathrm{erfc}\left(\delta\sqrt{\alpha(t-t_0)}\right)\right) \tag{3.45}$$

and

$$\left[L^{-1}(Z_3)-L^{-1}(Z_4)\right]\Big|_{x=0}=2e^{-\beta(t-t_0)}\left(1-\operatorname{erfc}\left(i\sqrt{\beta(t-t_0)}\right)\right) \tag{3.46}$$

Substituting Eqs. (3.45) and (3.46) for Eq. (3.42), noting that $1-\operatorname{erfc}(\xi)=\operatorname{erf}(\xi)$, and rearranging the resulting equation, yields:

$$T(0,t)=I_o\frac{\alpha}{\beta+\alpha\delta^2}\left\{\begin{array}{l}e^{\alpha\delta^2(t-t_0)}\operatorname{erfc}\left(\delta\sqrt{\alpha(t-t_0)}\right)\\+\dfrac{\delta\sqrt{\alpha}}{i\sqrt{\beta}}e^{-\beta(t-t_0)}\operatorname{erf}\left(i\sqrt{\beta(t-t_0)}\right)-e^{-\beta(t-t_0)}\end{array}\right\} \tag{3.47}$$

Noting that:

$$\frac{\sqrt{\pi}}{2i}e^{-\beta(t-t_0)}\operatorname{erf}\left(i\sqrt{\beta(t-t_0)}\right)=F\left(\sqrt{\beta(t-t_0)}\right) \tag{3.48}$$

where

$$F(u)=e^{-u^2}\int_0^u e^{v^2}dv \tag{3.49}$$

is Dawson's integral.

Eq. (3.48) can be proved as follows:

$$\operatorname{erf}\left(i\sqrt{\beta(t-t_0)}\right)=\frac{2}{\sqrt{\pi}}\int_0^{i\sqrt{\beta(t-t_0)}}e^{-v^2}dv \tag{3.50}$$

Let $v=\frac{\sigma}{i}$, then $dv=\frac{d\sigma}{i}$. Substitute these for Eq. (3.50), one can have:

$$\begin{aligned}\operatorname{erf}\left(i\sqrt{\beta(t-t_0)}\right)&=\frac{2}{\sqrt{\pi}}\int_0^{-\sqrt{\beta(t-t_0)}}e^{\sigma^2}\frac{d\sigma}{i}=\frac{2}{i\sqrt{\pi}}\int_0^{-\sqrt{\beta(t-t_0)}}e^{\sigma^2}d\sigma\\&=-\frac{2i}{\sqrt{\pi}}\int_0^{-\sqrt{\beta(t-t_0)}}e^{\sigma^2}d\sigma=\frac{2i}{\sqrt{\pi}}\int_0^{\sqrt{\beta(t-t_0)}}e^{\sigma^2}d\sigma\end{aligned} \tag{3.51}$$

Substituting Eq. (3.51) for Eq. (3.49) completes the proof.

Consequently, the closed form solution obtained earlier in Eq. (3.7) is the same as Eq. (3.47), but it is written in terms of the error functions.

Substituting Eq. (3.51) for Eq. (3.47), it results the following surface temperature equation as derived previously [4]:

$$T(0,t) = I_o \frac{\alpha 1(t-t_0)}{\beta+\alpha\delta^2}\left\{e^{\alpha\delta^2 t}\mathrm{erfc}\left(\delta\sqrt{\alpha(t-t_0)}\right)+2\delta\frac{\sqrt{\alpha}}{\sqrt{\pi\beta}}F\left(\sqrt{\beta(t-t_0)}\right)-e^{-\beta(t-t_0)}\right\} \quad (3.52)$$

The resulting equation for the surface temperature rise for the first term ($c_1$exp $(-a_1(t-t_0))$) in the time varying exponentially decaying pulse is:

$$T(0,t) = I_o 1(t-t_0)\left\{\frac{c_1}{a_1+\alpha\delta^2}\left\{\begin{array}{l} e^{\alpha\delta^2 t}\mathrm{erfc}\left(\delta\sqrt{\alpha(t-t_0)}\right) \\ +2\delta\frac{\sqrt{\alpha}}{\sqrt{\pi a_1}}F\left(\sqrt{a_1(t-t_0)}\right)-e^{-\beta a_1(t-t_0)}\end{array}\right\}\right\} \quad (3.53)$$

Because the governing equation of heat transfer is linear, the temperature profile included in the second term ($c_2$exp$(-a_2(t-t_0))$) in the time varying pulse can be written in a similar way.

Consider the consecutive laser pulse composes of the following time varying exponential pulses:

$$I_o\left[c_1 e^{-a_1(t-t_0)}+c_2 e^{-a_2(t-t_0)}+c_3 e^{-a_3(t-t_0)}\right]1(t-t_0) \quad (3.54)$$

Now, we replace the right-hand side of the Eq. (3.10) by:

$$\frac{I_1\delta}{k}e^{-\delta x}\left[c_1 e^{-a_1(t-t_0)}+c_2 e^{-a_2(t-t_0)}+c_3 e^{-a_3(t-t_0)}\right]1(t-t_0), \quad (3.55)$$

where c_1, c_2, c_3 are all constants, and let us search for the solution of Eq. (3.10). One can claim, by the superposition principle that the solution is given by:

$$T = c_1 T_1 + c_2 T_2 + c_3 T_3 \quad (3.56)$$

where T_1, T_2, T_3 are all satisfying the differential form of the Fourier equation of heat transfer (Eq. 3.10) with the boundary conditions.

Namely:

$$\frac{\partial^2 T_1}{\partial x^2}-\frac{1}{\alpha}\frac{\partial T_1}{\partial t}=\frac{I_1\delta}{k}e^{-a_1(t-t_0)}e^{-\delta x};\ \left.\frac{\partial T_1}{\partial x}\right|_{x=0}=0;\ T_1(x,0)=0;\ T_1(\infty,t)=0$$

and

$$\frac{\partial^2 T_2}{\partial x^2}-\frac{1}{\alpha}\frac{\partial T_2}{\partial t}=\frac{I_1\delta}{k}e^{-a_2(t-t_0)}e^{-\delta x};\ \left.\frac{\partial T_2}{\partial x}\right|_{x=0}=0;\ T_2(x,0)=0;\ T_2(\infty,t)=0$$

and

$$\frac{\partial^2 T_3}{\partial x^2}-\frac{1}{\alpha}\frac{\partial T_3}{\partial t}=\frac{I_1\delta}{k}e^{-a_3(t-t_0)}e^{-\delta x};\quad \left.\frac{\partial T_3}{\partial x}\right|_{x=0}=0;\ T_3(x,0)=0;\ T_3(\infty,t)=0 \tag{3.57}$$

The closed-form solution of sets of equations (Eq. 3.57) can be obtained using the similar method introduced for the solution of Eq. (3.10).

The surface temperature profile for the repetitive pulse, therefore, can be written as:

$$T(0,t)=I_o 1(t-t_0)\left\{\begin{array}{l}\dfrac{c_1}{a_1+\alpha\delta^2}\left\{\begin{array}{l}e^{\alpha\delta^2 t}\operatorname{erfc}\left(\delta\sqrt{\alpha(t-t_0)}\right)\\ +2\delta\dfrac{\sqrt{\alpha}}{\sqrt{\pi a_1}}F\left(\sqrt{a_1(t-t_0)}\right)-e^{-\beta a_1(t-t_0)}\end{array}\right\}\\ +\dfrac{c_2}{a_2+\alpha\delta^2}\left\{\begin{array}{l}e^{\alpha\delta^2(t-t_0)}\operatorname{erfc}\left(\delta\sqrt{\alpha(t-t_0)}\right)\\ +2\delta\dfrac{\sqrt{\alpha}}{\sqrt{\pi a_2}}F\left(\sqrt{a_2(t-t_0)}\right)-e^{-\beta a_2(t-t_0)}\end{array}\right\}\end{array}\right\} \tag{3.58}$$

The non-dimensional form of Eq. (3.58) can be obtained in a similarly way, i.e.:

$$T(0,\tau)=\frac{I_o}{k\delta}c_1 1(\tau-\tau_0)\left\{\begin{array}{l}\left(\dfrac{1}{a_1'+1}+\dfrac{b_2}{a_2'+1}\right)e^{(\tau-\tau_0)}\operatorname{erfc}\left(\sqrt{(\tau-\tau_0)}\right)\\ +\dfrac{2}{\sqrt{\pi}}\left[\dfrac{F\left(\sqrt{a_1'(\tau-\tau_0)}\right)}{\sqrt{a_1'}(a_1'+1)}+\dfrac{b_2F\left(\sqrt{a_2'(\tau-\tau_0)}\right)}{\sqrt{a_2'}(a_2'+1)}\right]\\ -\dfrac{e^{-a_1'(\tau-\tau_0)}}{(a_1'+1)}+\dfrac{e^{-a_2'(\tau-\tau_0)}}{(a_2'+1)}\end{array}\right\} \tag{3.59}$$

where

$$1(\tau-\tau_0)=\begin{cases}0, & \tau<\tau_0\\ 1, & \tau\geq\tau_0\end{cases}$$

A computer program was developed to predict the non-dimensional temperature profiles due to repetitive pulses.

The repetitive pulse heating results in the integration of conductive heating in the irradiated region of the substrate material. However, depending on the pulse shape and the repetition rate, the temperature at the surface rises continuously and the attainment of a steady temperature at the irradiated region becomes impossible.

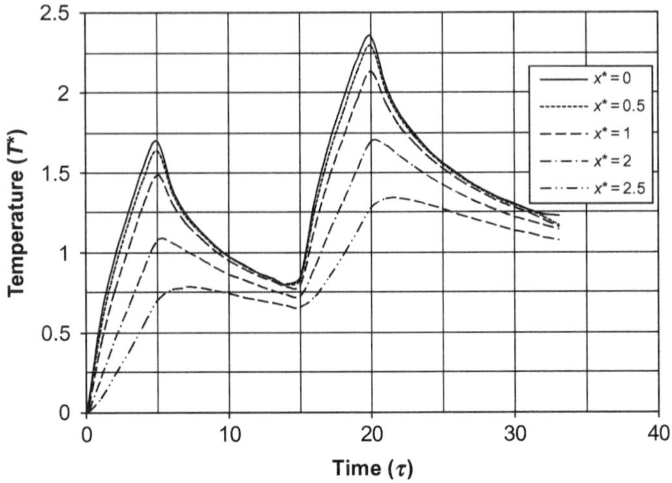

Fig. 3.1 Temporal variation of dimensionless temperature at different x-axis locations [3].

The influence of consecutive pulses on temporal behavior of temperature is demonstrated in Fig. 3.1, in which the temporal variation of non-dimensional temperature distribution at different locations inside the substrate material is shown. The temperature rises rapidly in the beginning of the heating pulse, and as the heating pulse progresses, the rate of temperature profile rise decreases slightly. This indicates that in the early heating period, the internal energy gain of the substrate material, due to absorption of irradiated energy, dominates the conduction energy transport from the surface vicinity to the solid bulk. As the heating period progresses, the temperature gradient increases and heat conduction from the surface vicinity to the solid bulk enhances. The material response to a heating pulse at some depth below the surface differs. In this case, the temperature profiles become almost similar in depths, which are less than the absorption depth of the substrate material ($x^* < 1$, where $x^* = x\delta$). As the depth increases further from $x^* = 1$, the energy gain due to absorption becomes almost negligible, and energy transport by conduction is almost the sole mechanism governing the temperature rise in this region. Consequently, the material response to a heating pulse slows down and the rate of rising temperature reduces in this region. The influence of the second pulse on the temperature profiles is similar at the surface as well as inside the substrate material, i.e., the magnitude of the temperature rise in the surface region is similar to the depth below the surface, provided that $x^* < 1$. Moreover, the domination of internal energy gain over the conduction energy transport in the surface region during the early heating period is not observed at the depth below the surface. This is because of the energy transfer mechanism, in which case, absorption replaces conduction heating in this region.

3.3 Laser solid-phase heating and convection cooling at the surface resembling assisting gas effect

Consider the case where the laser pulse intensity variation can be formulated through the exponential form ($I_1\left(e^{-\beta t}\right)$). The presence of the assisting gas generates cooling effects at the surface while modifying the thermal field in the irradiated region. This can be formulated using the Fourier heating model after incorporating convection boundary condition at the irradiated surface in line with the previous study [6]. Therefore, the Fourier heat transfer equation for a laser pulse, decaying exponentially with time, can be written as:

$$\frac{\partial^2 T}{\partial x^2}+\frac{I_1\delta}{k}\left(e^{-\beta t}\right)e^{-\delta x}=\frac{1}{\alpha}\frac{\partial T}{\partial t} \tag{3.60}$$

where:

$$I_1=\left(1-r_f\right)I_0$$

The substrate material is considered as a semi-infinite body and is heated by a laser beam on the surface. The convective boundary condition is assumed be on the substrate surface. In addition, as the depth is considered to extend to infinity, the temperature goes down to zero. Heating occurs in the surface region during the laser pulse. Therefore, the corresponding boundary conditions are:

$$\text{At } x=0 \Rightarrow \left[\frac{\partial T}{\partial x}\right]_{x=0}=\frac{h}{k}\left(T(0,t)-T_0\right) \tag{3.61}$$

$$\text{At } x=\infty \Rightarrow T(\infty,t)=0 \tag{3.62}$$

Initially, the substrate material is assumed to be at uniform temperature. Therefore, the initial condition is:

$$\text{At } t=0 \Rightarrow T(x,0)=0 \tag{3.63}$$

The Laplace transformation of Eq. (3.60) with respect to t, results in:

$$\frac{\partial^2 \overline{T}}{\partial x^2}+\frac{I_1\delta}{k}\frac{e^{-\delta x}}{(s+\beta)}=\frac{1}{\alpha}\left[s\overline{T}(x,s)-T(x,0)\right] \tag{3.64}$$

Introducing the initial condition and rearranging Eq. (3.64) yields:

$$\frac{\partial^2 \overline{T}}{\partial x^2}-g^2\overline{T}=-\frac{I_1\delta}{k}\frac{e^{-\delta x}}{(s+\beta)} \tag{3.65}$$

where $g^2 = s/\alpha$ and s is the Laplace transform variable. Eq. (3.65) has the solution:

$$\overline{T}(x,s) = A_1 e^{gx} + A_2 e^{-gx} - \frac{I_1 \delta e^{-\delta x}}{k(s+\beta)(\delta^2 - g^2)} \tag{3.66}$$

where A and B are constants. Introducing the boundary conditions will enable the calculation of the constants A_1 and A_2, i.e.:

$$A_1 = 0,$$

and:

$$A_2 = \frac{I_1 \delta (h + \delta k)}{k(s+\beta)(\delta^2 - g^2)(h + kg)} + \frac{hT_0}{s(h+kg)}$$

After substituting the values of A_1 and A_2 in Eq. (3.66), it yields:

$$\overline{T}(x,s) = \frac{I_1 \delta (h + \delta k) e^{-gx}}{k(s+\beta)(\delta^2 - g^2)(h+kg)} + \frac{hT_0 e^{-gx}}{s(h+kg)} - \frac{I_1 \delta e^{-\delta x}}{k(\delta^2 - g^2)(s+\beta)} \tag{3.67}$$

which gives the solution for the temperature distribution in the Laplace domain.

The inverse Laplace transformation of Eq. (3.67) provides the temperature distribution within the substrate material in space (x) and time (t). The mathematical arrangements of the Laplace inversion of Eq. (3.67) are given in the previous study [7]. Therefore, the equation after the Laplace inversion is given below as:

$$\begin{aligned} T(x,t) = {} & \frac{I_1 \delta \alpha^{3/2} (h + k\delta)}{k^2} \\ & \left\{ -\frac{e^{(\alpha\delta^2 t - \delta x)} \operatorname{Erfc}\left(\frac{x}{2\sqrt{\alpha t}} - \delta\sqrt{\alpha t}\right)}{2(\beta + \alpha\delta^2)\left(\frac{h\sqrt{\alpha}}{k} + \delta\sqrt{\alpha}\right)} - \frac{e^{(\alpha\delta^2 t + \delta x)} \operatorname{Erfc}\left(\frac{x}{2\sqrt{\alpha t}} - \delta\sqrt{\alpha t}\right)}{2(\beta + \alpha\delta^2)\left(\frac{h\sqrt{\alpha}}{k} + \delta\sqrt{\alpha}\right)} \right. \\ & + \frac{e^{-\beta t} e^{\sqrt{(\beta/\alpha)} xi} \operatorname{Erfc}\left(\frac{x}{2\sqrt{\alpha t}} + \sqrt{\beta t i}\right)}{2(\beta + \alpha\delta^2)\left(\frac{h\sqrt{\alpha}}{k} - \sqrt{\beta i}\right)} + \frac{e^{-\beta t} e^{-\sqrt{(\beta/\alpha)} xi} \operatorname{Erfc}\left(\frac{x}{2\sqrt{\alpha t}} - \sqrt{\beta t i}\right)}{2(\beta + \alpha\delta^2)\left(\frac{h\sqrt{\alpha}}{k} + \sqrt{\beta i}\right)} \\ & \left. + \frac{h\sqrt{\alpha}}{k} \frac{e^{hx/k} e^{(h^2/k^2)\alpha t} \operatorname{Erfc}\left(\frac{x}{2\sqrt{\alpha t}} + \frac{h}{k}\sqrt{\alpha t}\right)}{\left(\frac{h^2 \alpha}{k^2} + \beta\right)\left(\frac{h^2 \alpha}{k^2} - \alpha\delta^2\right)} - \frac{k e^{-\delta x}\left(e^{-\alpha\delta^2 t} - e^{-\beta t}\right)}{\sqrt{\alpha}(\beta + \alpha\delta^2)(h + k\delta)} \right\} \\ & + T_0 \left[\operatorname{Erfc}\left(\frac{x}{2\sqrt{\alpha t}}\right) - e^{hx/k} e^{(h^2/k^2)\alpha t} \operatorname{Erfc}\left(\frac{x}{2\sqrt{\alpha t}} + \frac{h}{k}\sqrt{\alpha t}\right) \right] \end{aligned} \tag{3.68}$$

where Erfc is the complementary error function. Eq. (3.68) is the closed-form solution for temperature distribution. The temperature distribution can be expressed in a non-dimensional form by introducing dimensionless quantities and substituting in Eq. (3.68). The dimensionless quantities are:

$$x^* = x\delta \ : \ t^* = \alpha\delta^2 t \ : \ T^* = \frac{Tk\delta}{I_1} \ : \ h^* = \frac{h}{\delta k} \ : \ \beta^* = \beta t$$

The dimensionless temperature distribution for a full pulse is then:

$$\begin{aligned}
T^*(x^*,t^*) = (h^*+1)&\\
\Bigg\{ \Bigg[&-\frac{e^{-x^*} e^{t^*} \operatorname{Erfc}\left(\frac{x^*}{2\sqrt{t^*}} - \sqrt{t^*}\right)}{2(\beta^*+1)(h^*+1)} - \frac{e^{-x^*} e^{t^*} \operatorname{Erfc}\left(\frac{x^*}{2\sqrt{t^*}} - \sqrt{t^*}\right)}{2(\beta^*+1)(h^*-1)} + \frac{e^{-\beta^*} e^{\sqrt{\beta^*} x^* i} \operatorname{Erfc}\left(\frac{x^*}{2\sqrt{t^*}} - \sqrt{\beta^* t^* i}\right)}{2(\beta^*+1)\left(h^* - \sqrt{\beta^*} i\right)} \\
&+ \frac{e^{-\beta^* t^*} e^{-\sqrt{\beta^*} x^* i} \operatorname{Erfc}\left(\frac{x^*}{2\sqrt{t^*}} - \sqrt{\beta^* t^* i}\right)}{2(\beta^*+1)\left(h^* + \sqrt{\beta^*} i\right)} + \frac{h^* e^{h^* x^*} e^{(h^*)^2 t^*} \operatorname{Erfc}\left(\frac{x^*}{2\sqrt{t^*}} - h^*\sqrt{t^*}\right)}{2\left((h^*)^2 + \beta^*\right)\left((h^*)^2 - 1\right)} - \frac{e^{-x^*}\left(e^{-t^*} - e^{-\beta^* t^*}\right)}{(\beta^*+1)(h^*+1)} \Bigg] \\
- \Bigg[&-\frac{e^{-x^*} e^{t^*} \operatorname{Erfc}\left(\frac{x^*}{2\sqrt{t^*}} - \sqrt{t^*}\right)}{2(\gamma^*+1)(h^*+1)} - \frac{e^{-x^*} e^{t^*} \operatorname{Erfc}\left(\frac{x^*}{2\sqrt{t^*}} - \sqrt{t^*}\right)}{2(\gamma^*+1)(h^*-1)} + \frac{e^{-\gamma^*} e^{\sqrt{\gamma^*} x^* i} \operatorname{Erfc}\left(\frac{x^*}{2\sqrt{t^*}} - \sqrt{\gamma^* t^* i}\right)}{2(\gamma^*+1)\left(h^* - \sqrt{\gamma^*} i\right)} \\
&+ \frac{e^{-\gamma^* t^*} e^{-\sqrt{\gamma^*} x^* i} \operatorname{Erfc}\left(\frac{x^*}{2\sqrt{t^*}} - \sqrt{\gamma^* t^* i}\right)}{2(\gamma^*+1)\left(h^* + \sqrt{\gamma^*} i\right)} + \frac{h^* e^{h^* x^*} e^{(h^*)^2 t^*} \operatorname{Erfc}\left(\frac{x^*}{2\sqrt{t^*}} - h^*\sqrt{t^*}\right)}{2\left((h^*)^2 + \gamma^*\right)\left((h^*)^2 - 1\right)} - \frac{e^{-x^*}\left(e^{-t^*} - e^{-\gamma^* t^*}\right)}{(\gamma^*+1)(h^*+1)} \Bigg] \Bigg\}
\end{aligned} \tag{3.69}$$

The solution of the above equation provides the temperature field in the irradiated region. Therefore, the findings are briskly discussed below in the line with the previous studies [6,7] and the data used in the simulations are given in Table 3.1.

Fig. 3.2 shows the temperature profiles predicted numerically and obtained from the closed-form solution. It can be observed that both profiles are almost identical. Consequently, the closed-form solution presented in this study is sufficiently accurate to calculate the temperature profiles for the exponentially decaying laser pulse with time, with the convective boundary condition at the surface.

Fig. 3.3 shows the dimensionless temperature distribution within the substrate material for various dimensionless heating periods. The influence of the heat-transfer coefficient on the temperature distribution becomes significant when the

Table 3.1 Material properties and the laser pulse parameter used in the simulations

ρ (kg/m^3)	δ (1/m)	α (m^2/s)	α_T (1/K)	ν	E (Pa)	Cp (J/kg K)	k (W/mK)	β (1/s)
7930	6.16×10^6	3.7×10^{-5}	1.6×10^{-5}	0.29	2.10×10^{11}	510	150	1.53×10^{11}

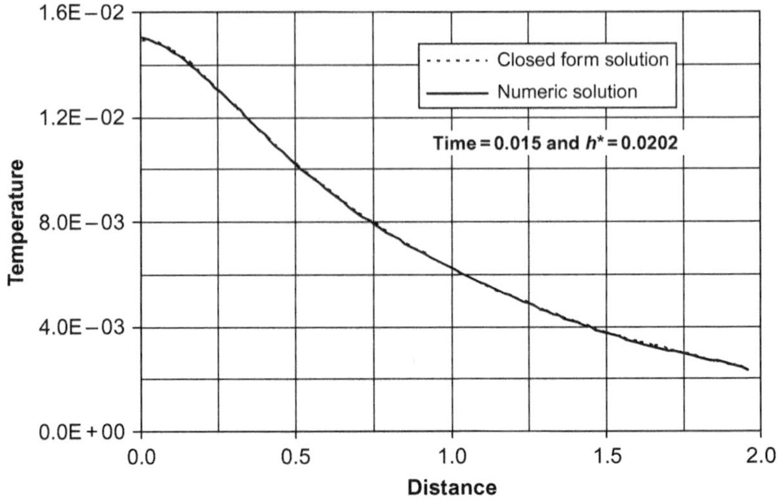

Fig. 3.2 Comparison of temperature profiles obtained from the closed-form solution and numerical predictions [7].

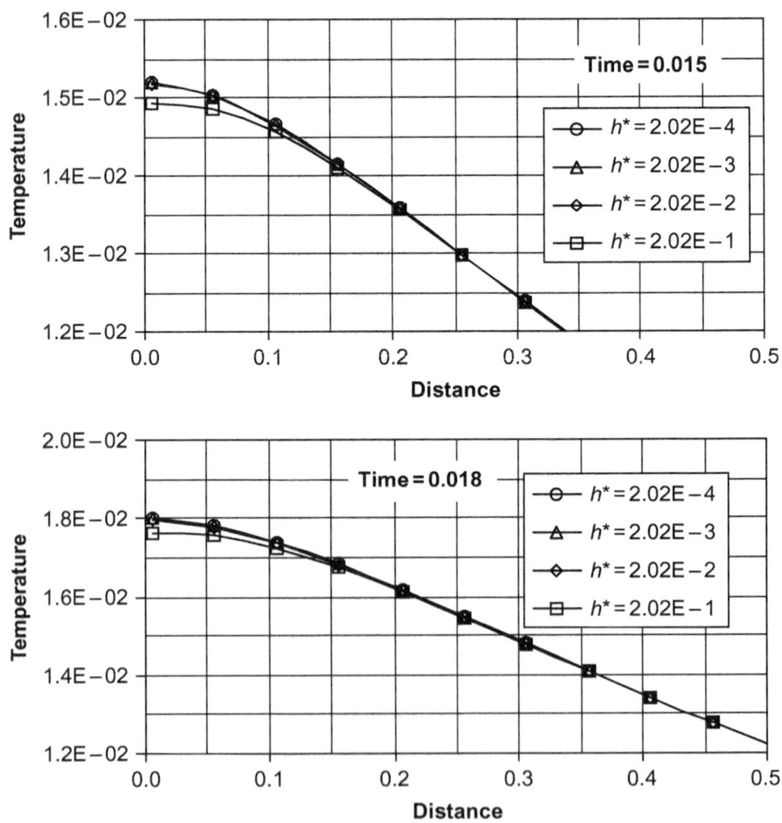

Fig. 3.3 Dimensionless temperature distributions within the substrate material [7].

dimensionless heat transfer coefficient reaches $h^{\bullet} \geq 2.02 \times 10^{-2} (\asymp 10^{8}\,\mathrm{W/m^2K})$. In this case, the temperature and its gradient in the surface region are reduced. The temperature gradient in the surface region is reduced to its minimum. At the point of a minimum temperature gradient, the internal energy gain by the substrate from the irradiated area is balanced by the diffusional energy transport from the substrate to the solid bulk. In this case, the depth beyond the point of a minimum temperature gradient diffusional energy transport dominates over the internal energy gain of the substrate material due to absorption of the irradiated field. The point of minimum temperature gradient changes with the heat transfer coefficient, which is more pronounced for the heating period of 0.018. Moreover, the sharp decay in the temperature gradient in the surface region ($x^{\bullet} \leq 0.1$) is because of: (i) the absorption process, i.e., the absorbed energy decreases exponentially with increasing depth (Lambert's law), and (ii) the internal energy gain in the surface region is high and diffusional energy transport due to the temperature gradient from this region to the solid bulk is low, i.e., the increase in temperature due to diffusional energy transport in the neighboring region is low; therefore, the temperature profile is governed by the internal energy gain in this region as consistent with the early work.

3.4 The thermal stress field generated due to the convection boundary at the surface

The cooling of the workpiece surface because of the assisting gas impingement alters the thermal stress states at the irradiated surface. Therefore, thermal stress analysis incorporating the convection cooling of the surface becomes necessary. The analytical formulation of the stress field is presented below in line with the previous study [6,7]. Therefore, in order to solve for the stress distribution within the substrate it is possible to consider the equation governing the momentum in a one-dimensional solid for the linear elastic case, i.e.:

$$\frac{\partial^2 \sigma_x}{\partial x^2} - \frac{1}{c_1{}^2}\frac{\partial^2 \sigma_x}{\partial t^2} = c_2 \frac{\partial^2 T}{\partial t^2} \tag{3.70}$$

where c_1 is the wave speed in the solid

$$c_1 = \sqrt{\frac{E}{\rho}}$$

and

$$c_2 = \frac{1+\upsilon}{1-\upsilon}\rho\alpha_T$$

where υ is Poisson's ratio, ρ is the density of the solid, and α_T is the thermal expansion coefficient of the solid.

In order to solve the momentum equation (Eq. 3.70), it is necessary to establish the initial conditions for stress and temperature fields. In this case, the substrate material is assumed to be free from stresses initially (at time = 0), and as time extends to infinity, the stress-free state must apply in the substrate. The same initial condition for the temperature is applied as in Eq. (3.63), provided that as time approaches infinity, the temperature in the substrate material reduces to zero. This is due to the fact that the laser pulse decays exponentially with time; therefore, as time approaches infinity, the laser pulse intensity becomes zero. Thus the initial and boundary conditions for the stress field are:

$$\text{At } t=0 \Rightarrow \sigma_x = 0 \tag{3.71}$$

$$\text{At } t=\infty \Rightarrow \sigma_x = 0 \tag{3.72}$$

$$\text{At } x=0 \Rightarrow \sigma_x = 0 \tag{3.73}$$

$$\text{At } x=\infty \Rightarrow \sigma_x = 0 \tag{3.74}$$

Taking the Laplace transformation of Eq. (3.70) with respect to time yields:

$$\begin{aligned}\frac{\partial^2 \overline{\sigma}_x}{\partial x^2} - \frac{1}{{c_1}^2}\left[s^2\overline{\sigma}_x(x,s) - s\sigma_x(x,0) - \dot{\sigma}_x(x,0)\right] \\ = c_2\left[s^2\overline{T}(x,s) - sT(x,0) - \dot{T}(x,0)\right]\end{aligned} \tag{3.75}$$

where $\overline{\sigma}_x(x,s)$ and $\overline{T}(x,s)$ are the Laplace transformations of thermal stress and temperature, respectively, in the x and s domains.

By substituting the initial conditions, Eq. (3.75) reduces to:

$$\frac{\partial^2 \overline{\sigma}_x}{\partial x^2} - \frac{s^2}{{c_1}^2}\overline{\sigma}_x(x,s) = c_2 s^2 \overline{T}(x,s) \tag{3.76}$$

Considering the temperature distribution in a Laplace domain for an exponentially decaying pulse with time, Eq. (3.67), and substituting it with Eq. (3.76) and solving for the stress field, yields:

$$\begin{aligned}&\frac{\partial^2 \overline{\sigma}_x}{\partial x^2} - \frac{s^2}{{c_1}^2}\overline{\sigma}_x(x,s) = \\ &-c_2 s^2\left[\frac{I_1\delta(h+\delta k)e^{-gx}}{k(s+\beta)(\delta^2-g^2)(h+kg)} + \frac{hT_0 e^{-gx}}{s(h+kg)} - \frac{I_1\delta e^{-\delta x}}{k(\delta^2-g^2)(s+\beta)}\right]\end{aligned} \tag{3.77}$$

Now let M_1 and M_2 be defined as:

$$M_1 = \frac{I_1\delta(h+\delta k)c_2 s^2}{k(s+\beta)(\delta^2 - g^2)(h+kg)} + \frac{hT_0 c_2 s}{(h+kg)} \tag{3.78}$$

and

$$M_2 = \frac{I_1 \delta c_2 s^2}{k(\delta^2 - g^2)(s+\beta)} \tag{3.79}$$

Then Eq. (3.77) becomes:

$$\frac{\partial^2 \overline{\sigma}_x}{\partial x^2} - \frac{s^2}{{c_1}^2}\overline{\sigma}_x(x, s) = M_1 e^{-gx} + M_2 e^{-\delta x} \tag{3.80}$$

The complementary and the particular solutions of Eq. (3.80) are:

$$(\overline{\sigma}_x)_h = A_3 e^{\frac{sx}{c_1}} + A_4 e^{\frac{-sx}{c_1}} \tag{3.81}$$

While the particular solution has two parts, the first part is:

$$(\overline{\sigma}_x)_{p_1} = G_1 e^{-\sqrt{\frac{s}{\alpha}}x} \tag{3.82}$$

Substituting Eq. (3.82) in Eq. (3.79) yields:

$$G_1 = \frac{M_1}{g^2 - \dfrac{s^2}{{c_1}^2}}$$

The second part of the particular solution is:

$$(\overline{\sigma}_x)_{p_2} = G_2 e^{-\delta x} \tag{3.83}$$

Substituting Eq. (3.83) in Eq. (3.79) yields:

$$G_2 = \frac{M_2}{\delta^2 - \dfrac{s^2}{{c_1}^2}}$$

So, the general solution for the stress field becomes:

$$(\overline{\sigma}_x)_g = A_3 e^{\frac{sx}{c_1}} + A_4 e^{\frac{-sx}{c_1}} + G_1 e^{-gx} + G_2 e^{-\delta x} \tag{3.84}$$

From the boundary condition ($x=\infty \Rightarrow \sigma_x=0$), this yields $A_3=0$.

Then, Eq. (3.86) reduces to:

$$(\overline{\sigma}_x)_g = A_4 e^{\frac{-sx}{c_1}} + G_1 e^{-\sqrt{\frac{s}{\alpha}}x} + G_2 e^{-\delta x} \tag{3.85}$$

Consider the boundary condition of the stress field at the surface, where at $x=0 \Rightarrow \partial\overline{\sigma}_x=0$, the constant in Eq. (3.85) becomes:

$$A_4 = -\frac{c_1}{s}\left[\sqrt{\frac{s}{\alpha}}G_1(s) + \delta G_2(s)\right] \tag{3.86}$$

Therefore, Eq. (3.86) becomes:

$$\overline{\sigma}_x(x,s) = G_1(s)e^{-\sqrt{\frac{s}{\alpha}}x} - G_1(s)\frac{c_1}{\sqrt{s\alpha}}e^{\frac{-sx}{c_1}} + G_2(s)e^{-\delta x} - G_2(s)\frac{c_1\delta}{s}e^{\frac{-sx}{c_1}} \tag{3.87}$$

Finding the solution for σ_x in the x and t domain, one should take the inverse Laplace transformation for each term in Eq. (3.87). To do this, the following terms are introduced:

$$\begin{aligned} &\text{Term1} = G_1(s)e^{-\sqrt{\frac{s}{\alpha}}x} \quad \text{Term2} = -G_1(s)\frac{c_1}{\sqrt{s\alpha}}e^{\frac{-sx}{c_1}} \\ &\text{Term3} = G_2(s)e^{-\delta x} \qquad \text{Term4} = -G_2(s)\frac{c_1\delta}{s}e^{\frac{-sx}{c_1}} \end{aligned} \tag{3.88}$$

Consequently, the solution for the stress distribution is the summation of the inverse Laplace transformations of the above terms.

Therefore, the Laplace inversion of terms (Term1, Term2, Term3, Term4) can be stated as follows.

Term1 is composed of the terms:

$$\text{Term1} = \text{Term11} + \text{Term21}$$

where

$$\text{Term11} = \frac{I_1\delta(h+\delta k)c_2}{k}\left[\frac{se^{-\sqrt{\frac{s}{\alpha}}x}}{(\delta^2 - s/\alpha)(s+\beta)\left(h+k\sqrt{s/\alpha}\right)(1/\alpha - s/c_1^2)}\right] \tag{3.89}$$

and

$$\text{Term21} = hT_0c_2\left[\frac{e^{-\sqrt{\frac{s}{\alpha}}x}}{\left(h+k\sqrt{s/\alpha}\right)(1/\alpha - s/c_1^2)}\right] \tag{3.90}$$

Let $C_{10} = \dfrac{I_1 \delta (h + \delta k) c_2}{k}$

and

$$C_{20} = h T_0 c_2$$

Therefore, the Laplace transformation of Term1 can be written as:

$$£^{-1}[\text{Term1}] = £^{-1}[\text{Term11}] + £^{-1}[\text{Term21}] \tag{3.91}$$

where

$$\begin{aligned} £^{-1}[\text{Term11}] \; = \; & £^{-1}[\text{Term111}] + £^{-1}[\text{Term211}] \\ & + £^{-1}[\text{Term311}] + £^{-1}[\text{Term411}] + £^{-1}[\text{Term511}] \\ & + £^{-1}[\text{Term611}] + £^{-1}[\text{Term711}] + £^{-1}[\text{Term811}] \end{aligned}$$

The Laplace transformations of the terms are:

$$£^{-1}[\text{Term111}] = \\ -\frac{c_1^2 h^2 k^3 \alpha^2 \sqrt{\alpha} C_{10}}{2(\alpha^2 h^2 - c_1^2 k^2)(h^2 \alpha + k^2 \beta)(h^2 - k^2 \delta^2)} \left[\begin{array}{l} \dfrac{h\sqrt{\alpha}}{k} e^{\frac{h^2 \alpha t}{k^2} - \frac{hx}{k}} \text{Erfc}\left(-\dfrac{h\sqrt{\alpha t}}{k} + \dfrac{x}{2\sqrt{\alpha t}} \right) \\ + \dfrac{h\sqrt{\alpha}}{k} e^{\frac{h^2 \alpha t}{k^2} + \frac{hx}{k}} \text{Erfc}\left(\dfrac{h\sqrt{\alpha t}}{k} + \dfrac{x}{2\sqrt{\alpha t}} \right) \end{array} \right] \tag{3.92}$$

and

$$£^{-1}[\text{Term211}] = \\ -\frac{c_1^2 h \alpha^3 \sqrt{\beta} C_{10}}{2(\beta + \alpha \delta^2)(h^2 \alpha + k^2 \beta)(c_1^2 + \alpha \beta)} \left[\begin{array}{l} \sqrt{\beta} e^{\beta t - \sqrt{\frac{\beta}{\alpha}} x} \text{Erfc}\left(-\sqrt{\beta t} + \dfrac{x}{2\sqrt{\alpha t}} \right) \\ + \sqrt{\beta} e^{\sqrt{\frac{\beta}{\alpha}} x + \beta t} \text{Erfc}\left(\sqrt{\beta t} + \dfrac{x}{2\sqrt{\alpha t}} \right) \end{array} \right] \tag{3.93}$$

and

$$£^{-1}[\text{Term311}] = \\ -\frac{c_1^3 h \alpha^4 \sqrt{\alpha} C_{10}}{2(\alpha^2 h^2 - c_1^2 k^2)(c_1^2 + \alpha \beta)(\delta^2 \alpha^2 - c_1^2)} \left[\begin{array}{l} \dfrac{c_1}{\sqrt{\alpha}} e^{\frac{c_1^2 t}{\alpha} - \frac{c_1 x}{\alpha}} \text{Erfc}\left(-c_1 \sqrt{\dfrac{t}{\alpha}} + \dfrac{x}{2\sqrt{\alpha t}} \right) \\ + \dfrac{c_1}{\sqrt{\alpha}} e^{\frac{c_1^2 t}{\alpha} + \frac{c_1 x}{\alpha}} \text{Erfc}\left(c_1 \sqrt{\dfrac{t}{\alpha}} + \dfrac{x}{2\sqrt{\alpha t}} \right) \end{array} \right] \tag{3.94}$$

and

$$£^{-1}[\text{Term411}] = \frac{c_1^2 h\alpha^2\sqrt{\alpha}\delta C_{10}}{2(\beta+\alpha\delta^2)(\delta^2\alpha^2-c_1^2)(h^2-k^2\delta^2)}\left[\begin{array}{l}\sqrt{\alpha}\delta e^{\alpha\delta^2 t-\delta x}\text{Erfc}\left(-\delta\sqrt{\alpha t}+\dfrac{x}{2\sqrt{\alpha t}}\right)\\ +\sqrt{\alpha}\delta e^{\alpha\delta^2 t+\delta x}\text{Erfc}\left(\delta\sqrt{\alpha t}+\dfrac{x}{2\sqrt{\alpha t}}\right)\end{array}\right] \tag{3.95}$$

and

$$£^{-1}[\text{Term511}] = \frac{c_1^2 h^2 k^3\alpha^2\sqrt{\alpha}C_{10}}{2\left(\alpha^2h^2-c_1^2k^2\right)(h^2\alpha+k^2\beta)(h^2-k^2\delta^2)}\left[\begin{array}{l}\dfrac{2}{\sqrt{\pi t}}\exp\left(\dfrac{-x^2}{4\alpha t}\right)+\dfrac{h\sqrt{\alpha}}{k}e^{\frac{h^2\alpha t}{k^2}-\frac{hx}{k}}\text{Erfc}\left(-\dfrac{h\sqrt{\alpha t}}{k}+\dfrac{x}{2\sqrt{\alpha t}}\right)\\ -\dfrac{h\sqrt{\alpha}}{k}e^{\frac{h^2\alpha t}{k^2}+\frac{hx}{k}}\text{Erfc}\left(\dfrac{h\sqrt{\alpha t}}{k}+\dfrac{x}{2\sqrt{\alpha t}}\right)\end{array}\right] \tag{3.96}$$

and

$$£^{-1}[\text{Term611}] = \frac{c_1^2 k\alpha^2\sqrt{\alpha}\beta C_{10}}{2\left(c_1^2+\alpha\beta\right)(h^2\alpha+k^2\beta)(\beta+\alpha\delta^2)}\left[\begin{array}{l}\dfrac{2}{\sqrt{\pi t}}\exp\left(\dfrac{-x^2}{4\alpha t}\right)+\sqrt{\beta}e^{\beta t-x\sqrt{\frac{\beta}{\alpha}}}\text{Erfc}\left(-\sqrt{\beta t}+\dfrac{x}{2\sqrt{\alpha t}}\right)\\ -\sqrt{\beta}e^{\beta t+x\sqrt{\frac{\beta}{\alpha}}}\text{Erfc}\left(\sqrt{\beta t}+\dfrac{x}{2\sqrt{\alpha t}}\right)\end{array}\right] \tag{3.97}$$

and

$$£^{-1}[\text{Term711}] = \frac{c_1^2 k\alpha^3\sqrt{\alpha}C_{10}}{2\left(\alpha^2h^2-c_1^2k^2\right)\left(c_1^2+\alpha\beta\right)\left(\delta^2\alpha^2-c_1^2\right)}\left[\begin{array}{l}\dfrac{2}{\sqrt{\pi t}}\exp\left(\dfrac{-x^2}{4\alpha t}\right)+\dfrac{c_1}{\sqrt{\alpha}}e^{\frac{c_1^2 t}{\alpha}-\frac{c_1 x}{\alpha}}\text{Erfc}\left(-c_1\sqrt{\dfrac{t}{\alpha}}+\dfrac{x}{2\sqrt{\alpha t}}\right)\\ -\dfrac{c_1}{\sqrt{\alpha}}e^{\frac{c_1^2 t}{\alpha}+\frac{c_1 x}{\alpha}}\text{Erfc}\left(c_1\sqrt{\dfrac{t}{\alpha}}+\dfrac{x}{2\sqrt{\alpha t}}\right)\end{array}\right] \tag{3.98}$$

$$£^{-1}[\text{Term811}] = \frac{c_1^2 k\alpha^2\sqrt{\alpha}\delta^2 C_{10}}{2(\beta+\alpha\delta^2)\left(\delta^2\alpha^2-c_1^2\right)(k^2\delta^2-h^2)}\left[\begin{array}{l}\dfrac{2}{\sqrt{\pi t}}\exp\left(\dfrac{-x^2}{4\alpha t}\right)+\sqrt{\alpha}\delta e^{\alpha\delta^2 t-\delta x}\text{Erfc}\left(-\delta\sqrt{\alpha t}+\dfrac{x}{2\sqrt{\alpha t}}\right)\\ -\sqrt{\alpha}\delta e^{\alpha\delta^2 t+\delta x}\text{Erfc}\left(\delta\sqrt{\alpha t}+\dfrac{x}{2\sqrt{\alpha t}}\right)\end{array}\right] \tag{3.99}$$

The Laplace inversion of Term21 can be written as:

$$£^{-1}[\text{Term21}] = £^{-1}[\text{Term121}] + £^{-1}[\text{Term221}] + £^{-1}[\text{Term321}] + £^{-1}[\text{Term421}] \tag{3.100}$$

Therefore, the Laplace inversions of composing terms are:

$$£^{-1}[\text{Term121}] = \frac{c_1^2 hk\alpha\sqrt{\alpha}C_{20}}{2h(\alpha^2h^2 - c_1^2k^2)}\left[\begin{array}{l}\frac{h\sqrt{\alpha}}{k}e^{\frac{h^2\alpha t}{k^2}-\frac{hx}{k}}\text{Erfc}\left(-\frac{h\sqrt{\alpha t}}{k}+\frac{x}{2\sqrt{\alpha t}}\right)\\ +\frac{h\sqrt{\alpha}}{k}e^{\frac{h^2\alpha t}{k^2}+\frac{hx}{k}}\text{Erfc}\left(\frac{h\sqrt{\alpha t}}{k}+\frac{x}{2\sqrt{\alpha t}}\right)\end{array}\right] \tag{3.101}$$

and

$$£^{-1}[\text{Term221}] = -\frac{c_1 h\alpha^2\sqrt{\alpha}C_{20}}{2(\alpha^2h^2 - c_1^2k^2)}\left[\begin{array}{l}\frac{c_1}{\sqrt{\alpha}}e^{\frac{c_1^2 t}{\alpha}-\frac{c_1 x}{\alpha}}\text{Erfc}\left(-c_1\sqrt{\frac{t}{\alpha}}+\frac{x}{2\sqrt{\alpha t}}\right)\\ +\frac{c_1}{\sqrt{\alpha}}e^{\frac{c_1^2 t}{\alpha}+\frac{c_1 x}{\alpha}}\text{Erfc}\left(c_1\sqrt{\frac{t}{\alpha}}+\frac{x}{2\sqrt{\alpha t}}\right)\end{array}\right] \tag{3.102}$$

and

$$£^{-1}[\text{Term321}] = -\frac{c_1^2 k\alpha\sqrt{\alpha}C_{20}}{2(\alpha^2h^2 - c_1^2k^2)}\left[\begin{array}{l}\frac{2}{\sqrt{\pi t}}\exp\left(\frac{-x^2}{4\alpha t}\right)+\frac{h\sqrt{\alpha}}{k}e^{\frac{h^2\alpha t}{k^2}-\frac{hx}{k}}\text{Erfc}\left(-\frac{h\sqrt{\alpha t}}{k}+\frac{x}{2\sqrt{\alpha t}}\right)\\ -\frac{h\sqrt{\alpha}}{k}e^{\frac{h^2\alpha t}{k^2}+\frac{hx}{k}}\text{Erfc}\left(\frac{h\sqrt{\alpha t}}{k}+\frac{x}{2\sqrt{\alpha t}}\right)\end{array}\right] \tag{3.103}$$

and

$$£^{-1}[\text{Term421}] = \frac{c_1^2 k\alpha\sqrt{\alpha}C_{20}}{2(\alpha^2h^2 - c_1^2k^2)}\left[\begin{array}{l}\frac{2}{\sqrt{\pi t}}\exp\left(\frac{-x^2}{4\alpha t}\right)+\frac{c_1}{\sqrt{\alpha}}e^{\frac{c_1^2 t}{\alpha}-\frac{c_1 x}{\alpha}}\text{Erfc}\left(-c_1\sqrt{\frac{t}{\alpha}}+\frac{x}{2\sqrt{\alpha t}}\right)\\ -\frac{c_1}{\sqrt{\alpha}}e^{\frac{c_1^2 t}{\alpha}+\frac{c_1 x}{\alpha}}\text{Erfc}\left(c_1\sqrt{\frac{t}{\alpha}}+\frac{x}{2\sqrt{\alpha t}}\right)\end{array}\right] \tag{3.104}$$

The Laplace transformation of Term2 can be written as:

$$£^{-1}[\text{Term2}] = £^{-1}[\text{Term12}] + £^{-1}[\text{Term22}] \tag{3.105}$$

Therefore, the Laplace inversions of composing terms are:

$$\begin{aligned}£^{-1}[\text{Term}12] &= £^{-1}[\text{Term}112]+£^{-1}[\text{Term}212]\\ &+£^{-1}[\text{Term}312]+£^{-1}[\text{Term}412]+£^{-1}[\text{Term}512]\\ &+£^{-1}[\text{Term}612]+£^{-1}[\text{Term}712]+£^{-1}[\text{Term}812]\end{aligned}$$

Knowing that:

$$C_{30}=-\frac{I_1\delta(h+\delta k)c_2c_1}{\sqrt{\alpha}k} \tag{3.106}$$

and

$$C_{40}=-\frac{hT_0c_2c_1}{\sqrt{\alpha}} \tag{3.107}$$

Hence:

$$£^{-1}[\text{Term}112]=$$
$$-\frac{c_1^2hk^6\alpha^2C_{30}}{k^2\left(c_1^2k^2-h^2\alpha^2\right)(h^2\alpha+k^2\beta)(h^2-k^2\delta^2)}\left[\begin{array}{l}\dfrac{1}{\sqrt{\pi(t-x/c_1)}}+\\ \dfrac{h\sqrt{\alpha}}{k}e^{\frac{h^2\alpha t}{k^2}}\text{Erf}\left(\dfrac{h\sqrt{\alpha}}{k}\sqrt{t-x/c_1}\right)\end{array}\right]\cdot 1\left(t-\frac{x}{c_1}\right) \tag{3.108}$$

and

$$£^{-1}[\text{Term}212]=$$
$$\frac{c_1^2h\alpha^3C_{30}}{(\beta+\alpha\delta^2)(h^2\alpha+k^2\beta)(c_1^2+\alpha\beta)}\left[\begin{array}{l}\dfrac{1}{\sqrt{\pi(t-x/c_1)}}+\\ \sqrt{-\beta}e^{-\beta t}\text{Erf}\left(\sqrt{-\beta(t-x/c_1)}\right)\end{array}\right]\cdot 1\left(t-\frac{x}{c_1}\right) \tag{3.109}$$

and

$$£^{-1}[\text{Term}312]=$$
$$-\frac{c_1^2h\alpha^5C_{30}}{(\alpha^2h^2-c_1^2k^2)(c_1^2+\alpha\beta)(\delta^2\alpha^2-c_1^2)}\left[\begin{array}{l}\dfrac{1}{\sqrt{\pi(t-x/c_1)}}+\\ \dfrac{c_1}{\sqrt{\alpha}}e^{\frac{c_1^2}{\alpha}t}\text{Erf}\left(\dfrac{c_1}{\sqrt{\alpha}}\sqrt{t-x/c_1}\right)\end{array}\right]\cdot 1\left(t-\frac{x}{c_1}\right) \tag{3.110}$$

and

$$£^{-1}[\text{Term}412] = \frac{c_1^2 h\alpha^2 C_{30}}{(\beta+\alpha\delta^2)(\delta^2\alpha^2-c_1^2)(h^2-k^2\delta^2)}\left[\frac{1}{\sqrt{\pi(t-x/c_1)}}+\sqrt{\alpha}\delta e^{\alpha\delta^2 t}\text{Erf}\left(\sqrt{\alpha}\delta\sqrt{t-x/c_1}\right)\right]\cdot 1\left(t-\frac{x}{c_1}\right) \tag{3.111}$$

and

$$£^{-1}[\text{Term}512] = \frac{c_1^2 k^5\alpha^2 C_{30}}{(\alpha^2h^2-c_1^2k^2)(h^2\alpha+k^2\beta)(h^2-k^2\delta^2)}\left[D(t-x/c_1)+\frac{h^2\alpha}{k^2}e^{\frac{h^2\alpha}{k^2}(t-x/c_1)}\right]\cdot 1\left(t-\frac{x}{c_1}\right) \tag{3.112}$$

and

$$£^{-1}[\text{Term}612] = -\frac{c_1^2 k\alpha^2\sqrt{\alpha}C_{30}}{(c_1^2+\alpha\beta)(h^2\alpha+k^2\beta)(\beta+\alpha\delta^2)}\left[D(t-x/c_1)-\beta e^{-\beta(t-x/c_1)}\right]\cdot 1\left(t-\frac{x}{c_1}\right) \tag{3.113}$$

and

$$£^{-1}[\text{Term}712] = -\frac{c_1^2 k\alpha^4\sqrt{\alpha}C_{30}}{(\alpha^2h^2-c_1^2k^2)(c_1^2+\alpha\beta)(\delta^2\alpha^2-c_1^2)}\left[D(t-x/c_1)+\frac{c_1^2}{\alpha}e^{\frac{c_1^2}{\alpha}(t-x/c_1)}\right]\cdot 1\left(t-\frac{x}{c_1}\right) \tag{3.114}$$

and

$$£^{-1}[\text{Term}812] = \frac{c_1^2 k\alpha\sqrt{\alpha}C_{30}}{(\beta+\alpha\delta^2)(\delta^2\alpha^2-c_1^2)(k^2\delta^2-h^2)}\left[D(t-x/c_1)+\alpha\delta^2 e^{\alpha\delta^2(t-x/c_1)}\right]\cdot 1\left(t-\frac{x}{c_1}\right) \tag{3.115}$$

The Laplace transformation of Term22 can be written as:

$$£^{-1}[\text{Term}22] = £^{-1}[\text{Term}122] + £^{-1}[\text{Term}222] + £^{-1}[\text{Term}322] + £^{-1}[\text{Term}422] \tag{3.116}$$

Therefore, the Laplace transformations of the composing terms are:

$$£^{-1}[\text{Term122}] = \frac{\alpha C_{40}}{h\sqrt{\pi(t-x/c_1)}} \cdot 1\left(t - \frac{x}{c_1}\right) \tag{3.117}$$

and

$$£^{-1}[\text{Term222}] = \frac{c_1^2 k^2 \alpha C_{40}}{h\left(\alpha^2 h^2 - c_1^2 k^2\right)}\left[\frac{1}{\sqrt{\pi(t-x/c_1)}} + \frac{h\sqrt{\alpha}}{k} e^{\frac{h^2\alpha t}{k^2}} \text{Erf}\left(\frac{h\sqrt{\alpha}}{k}\sqrt{t-\frac{x}{c_1}}\right)\right] \cdot 1\left(t-\frac{x}{c_1}\right) \tag{3.118}$$

and

$$£^{-1}[\text{Term322}] = -\frac{h\alpha^3 C_{40}}{\left(\alpha^2 h^2 - c_1^2 k^2\right)}\left[\frac{1}{\sqrt{\pi(t-x/c_1)}} + \frac{c_1}{\sqrt{\alpha}} e^{\frac{c_1^2}{\alpha}t} \text{Erf}\left(\frac{c_1}{\sqrt{\alpha}}\sqrt{t-\frac{x}{c_1}}\right)\right] \cdot 1\left(t-\frac{x}{c_1}\right) \tag{3.119}$$

and

$$£^{-1}[\text{Term422}] = -\frac{k\sqrt{\alpha}C_{40}}{h^2} D(t-x/c_1) \cdot 1\left(t-\frac{x}{c_1}\right) \tag{3.120}$$

and

$$£^{-1}[\text{Term522}] = \frac{c_1^2 k^3 \sqrt{\alpha} C_{40}}{h^2\left(c_1^2 k^2 - h^2\alpha^2\right)}\left[D(t-x/c_1) + \frac{h^2\alpha}{k^2} e^{\frac{h^2\alpha}{k^2}(t-x/c_1)}\right] \cdot 1\left(t-\frac{x}{c_1}\right) \tag{3.121}$$

and

$$£^{-1}[\text{Term622}] = -\frac{k\alpha^2\sqrt{\alpha}C_{40}}{\left(c_1^2 k^2 - h^2\alpha^2\right)}\left[D(t-x/c_1) + \frac{c_1^2}{\alpha} e^{\frac{c_1^2}{\alpha}(t-x/c_1)}\right] \cdot 1\left(t-\frac{x}{c_1}\right) \tag{3.122}$$

and

$$\text{Term3} = \frac{-I_1\delta c_2}{k}\left[\frac{s^2 e^{-\delta x}}{\left(\delta^2 - s/\alpha\right)(s+\beta)\left(\delta^2 - s^2/c_1^2\right)}\right] \tag{3.123}$$

and

$$£^{-1}[\text{Term3}] = \frac{-I_1\delta c_2}{k}\left\{\begin{array}{l} \dfrac{c_1^2\alpha e^{c_1\delta t-\delta x}}{2(\beta+c_1\delta)(c_1-\alpha\delta)} - \dfrac{c_1^3\alpha e^{-(c_1\delta t+\delta x)}}{2(c_1\delta-\beta)(c_1+\alpha\delta)} - \dfrac{c_1^2\alpha\beta^2 e^{-(\beta t+\delta x)}}{(\beta^2-c_1^2\delta^2)(\beta+\alpha\delta^2)} \\ +\dfrac{c_1^2\alpha^3\delta^2 e^{\alpha\delta^2 t-\delta x}}{(\alpha^2\delta^2-c_1^2)(\beta+\alpha\delta^2)} \end{array}\right\} \tag{3.124}$$

and

$$\text{Term4} = \frac{I_1\delta^2 c_2 c_1}{k}\left[\frac{se^{-\frac{s}{c_1}x}}{(\delta^2-s/\alpha)(s+\beta)(\delta^2-s^2/c_1^2)}\right] \tag{3.125}$$

and

$$£^{-1}[\text{Term4}] = \frac{I_1\delta^2 c_2 c_1}{k}\left\{\begin{array}{l} \dfrac{c_1^2\alpha e^{c_1\delta(t-x/c_1)}}{2\delta(\beta+c_1\delta)(c_1-\alpha\delta)} + \dfrac{c_1^2\alpha e^{-c_1\delta(t-x/c_1)}}{2\delta(c_1\delta-\beta)(c_1+\alpha\delta)} \\ -\dfrac{c_1^2\alpha\beta e^{-\beta(t-x/c_1)}}{(c_1^2\delta^2-\beta^2)(\beta+\alpha\delta^2)} - \dfrac{c_1^2\alpha^2 e^{\alpha\delta^2(t-x/c_1)}}{(c_1^2-\alpha^2\delta^2)(\beta+\alpha\delta^2)} \end{array}\right\}. \quad 1\left(t-\frac{x}{c_1}\right) \tag{3.126}$$

where $1\left(t-\frac{x}{c_1}\right)$ is a unit step function, Erf (y) is the error function of the variable y and $D(t-x/c_1)$ is the Dirac delta function. The unit step function has the values of 0 for $t \le \frac{x}{c_1}$ and 1 for $t > \frac{x}{c_1}$.

The closed-form solution of the stress distribution can be written as:

$$\sigma_x(x,t) = £^{-1}[\text{Term1}] + £^{-1}[\text{Term2}] + £^{-1}[\text{Term3}] + £^{-1}[\text{Term4}] \tag{3.127}$$

where $£^{-1}$ represents the inverse sign of the Laplace transformation.

The additional dimensionless quantities are defined to present the stress distribution in the dimensionless form, i.e.:

$$c_1^* = \frac{c_1}{\alpha\delta} : \sigma_x^* = \frac{k\sigma_x}{I_1\delta\alpha^2 c_2} \quad \text{and} \quad U[1] = \left(t^* - \frac{x^*}{c_1{}^*}\right)$$

where U [1] is the dimensionless unit step function.

Therefore, for the dimensionless stress distribution, the following results:

$$\left(\sigma_x^{\bullet}\right)_1 = \left(\sigma_x^{\bullet}\right)_{11} + \left(\sigma_x^{\bullet}\right)_{21} \tag{3.128}$$

where

$$\begin{aligned}\left(\sigma_x^{\bullet}\right)_{11} = {} & \left(\sigma_x^{\bullet}\right)_{111} + \left(\sigma_x^{\bullet}\right)_{211} + \left(\sigma_x^{\bullet}\right)_{311} + \left(\sigma_x^{\bullet}\right)_{411} \\ & + \left(\sigma_x^{\bullet}\right)_{511} + \left(\sigma_x^{\bullet}\right)_{611} + \left(\sigma_x^{\bullet}\right)_{711} + \left(\sigma_x^{\bullet}\right)_{811}\end{aligned} \tag{3.129}$$

In the dimensionless form:

$$\left(\sigma_x^{\bullet}\right)_{111} = -\frac{(h^{\bullet}+1)h^{\bullet 2}c_1^{\bullet 2}}{2(h^{\bullet 2}-1)\left(h^{\bullet 2}-c_1^{\bullet 2}\right)\left(h^{\bullet 2}+\beta^{\bullet}/t^{\bullet}\right)}\left[\begin{array}{l} h^{\bullet}e^{h^{\bullet 2}t^{\bullet}-h^{\bullet}x^{\bullet}}\operatorname{Erfc}\left(-h^{\bullet}\sqrt{t^{\bullet}}+\dfrac{x^{\bullet}}{2\sqrt{t^{\bullet}}}\right) \\ +h^{\bullet}e^{h^{\bullet 2}t^{\bullet}+h^{\bullet}x^{\bullet}}\operatorname{Erfc}\left(h^{\bullet}\sqrt{t^{\bullet}}+\dfrac{x^{\bullet}}{2\sqrt{t^{\bullet}}}\right)\end{array}\right] \tag{3.130}$$

and

$$\left(\sigma_x^{\bullet}\right)_{211} = -\frac{(h^{\bullet}+1)h^{\bullet}c_1^{\bullet 2}\beta^{\bullet}}{2(\beta^{\bullet}/t^{\bullet}+1)\left(h^{\bullet 2}+\beta^{\bullet}/t^{\bullet}\right)\left(c_1^{\bullet 2}+\beta^{\bullet}/t^{\bullet}\right)}\left[\begin{array}{l} e^{\beta^{\bullet}-\sqrt{\frac{\beta^{\bullet}}{t^{\bullet}}}x^{\bullet}}\operatorname{Erfc}\left(-\sqrt{\beta^{\bullet}}+\dfrac{x^{\bullet}}{2\sqrt{t^{\bullet}}}\right) \\ +e^{\beta^{\bullet}+\sqrt{\frac{\beta^{\bullet}}{t^{\bullet}}}x^{\bullet}}\operatorname{Erfc}\left(\sqrt{\beta^{\bullet}}+\dfrac{x^{\bullet}}{2\sqrt{t^{\bullet}}}\right)\end{array}\right] \tag{3.131}$$

and

$$\left(\sigma_x^{\bullet}\right)_{311} = -\frac{(h^{\bullet}+1)h^{\bullet}c_1^{\bullet 3}}{2\left(h^{\bullet 2}-c_1^{\bullet 2}\right)\left(1-c_1^{\bullet 2}\right)\left(c_1^{\bullet 2}+\beta^{\bullet}/t^{\bullet}\right)}\left[\begin{array}{l} c_1^{\bullet 2}e^{c_1^{\bullet 2}t^{\bullet}-c_1^{\bullet}x^{\bullet}}\operatorname{Erfc}\left(-c_1^{\bullet}\sqrt{t^{\bullet}}+\dfrac{x^{\bullet}}{2\sqrt{t^{\bullet}}}\right) \\ +c_1^{\bullet 2}e^{c_1^{\bullet 2}t^{\bullet}+c_1^{\bullet}x^{\bullet}}\operatorname{Erfc}\left(c_1^{\bullet}\sqrt{t^{\bullet}}+\dfrac{x^{\bullet}}{2\sqrt{t^{\bullet}}}\right)\end{array}\right] \tag{3.132}$$

and

$$\left(\sigma_x^{\bullet}\right)_{411} = \frac{(h^{\bullet}+1)h^{\bullet}c_1^{\bullet 2}}{2(h^{\bullet 2}-1)\left(1-c_1^{\bullet 2}\right)\left(1+\beta^{\bullet}/t^{\bullet}\right)}\left[\begin{array}{l} e^{t^{\bullet}-x^{\bullet}}\operatorname{Erfc}\left(-\sqrt{t^{\bullet}}+\dfrac{x^{\bullet}}{2\sqrt{t^{\bullet}}}\right) \\ +e^{t^{\bullet}+x^{\bullet}}\operatorname{Erfc}\left(\sqrt{t^{\bullet}}+\dfrac{x^{\bullet}}{2\sqrt{t^{\bullet}}}\right)\end{array}\right] \tag{3.133}$$

and

$$(\sigma_x^\bullet)_{511} = \frac{(h^\bullet+1)h^{\bullet 2}c_1^{\bullet 2}}{2\left(h^{\bullet 2}-1\right)\left(h^{\bullet 2}-c_1^{\bullet 2}\right)\left(h^{\bullet 2}+\beta^\bullet/t^\bullet\right)}\left[\begin{array}{l}\frac{2}{\sqrt{\pi t^\bullet}}\exp\left(\frac{-x^{\bullet 2}}{4t^\bullet}\right)+h^\bullet e^{h^{\bullet 2}t^\bullet-h^\bullet x^\bullet}\mathrm{Erfc}\left(-h^\bullet\sqrt{t^\bullet}+\frac{x^\bullet}{2\sqrt{t^\bullet}}\right)\\ -h^\bullet e^{h^{\bullet 2}t^\bullet+h^\bullet x^\bullet}\mathrm{Erfc}\left(h^\bullet\sqrt{t^\bullet}+\frac{x^\bullet}{2\sqrt{t^\bullet}}\right)\end{array}\right] \tag{3.134}$$

and

$$(\sigma_x^\bullet)_{611} = \frac{(h^\bullet+1)c_1^{\bullet 2}\beta^\bullet}{2t^\bullet(\beta^\bullet/t^\bullet+1)\left(h^{\bullet 2}+\beta^\bullet/t^\bullet\right)\left(c_1^{\bullet 2}+\beta^\bullet/t^\bullet\right)}\left[\begin{array}{l}\frac{2}{\sqrt{\pi t^\bullet}}\exp\left(\frac{-x^{\bullet 2}}{4t^\bullet}\right)+\sqrt{\frac{\beta^\bullet}{t^\bullet}}e^{\beta^\bullet-\sqrt{\frac{\beta^\bullet}{t^\bullet}}x^\bullet}\mathrm{Erfc}\left(-\sqrt{\beta^\bullet}+\frac{x^\bullet}{2\sqrt{t^\bullet}}\right)\\ -\sqrt{\frac{\beta^\bullet}{t^\bullet}}e^{\beta^\bullet+\sqrt{\frac{\beta^\bullet}{t^\bullet}}x^\bullet}\mathrm{Erfc}\left(\sqrt{\beta^\bullet}+\frac{x^\bullet}{2\sqrt{t^\bullet}}\right)\end{array}\right] \tag{3.135}$$

and

$$(\sigma_x^\bullet)_{711} = \frac{(h^\bullet+1)c_1^{\bullet 4}}{2\left(h^{\bullet 2}-c_1^{\bullet 2}\right)\left(1-c_1^{\bullet 2}\right)\left(c_1^{\bullet 2}+\beta^\bullet/t^\bullet\right)}\left[\begin{array}{l}\frac{2}{\sqrt{\pi t^\bullet}}\exp\left(\frac{-x^{\bullet 2}}{4t^\bullet}\right)+c_1^\bullet e^{c_1^{\bullet 2}t^\bullet-c_1^\bullet x^\bullet}\\ \mathrm{Erfc}\left(-c_1^\bullet\sqrt{t^\bullet}+\frac{x^\bullet}{2\sqrt{t^\bullet}}\right)\\ -c_1^\bullet e^{c_1^{\bullet 2}t^\bullet+c_1^\bullet x^\bullet}\mathrm{Erfc}\left(c_1^\bullet\sqrt{t^\bullet}+\frac{x^\bullet}{2\sqrt{t^\bullet}}\right)\end{array}\right] \tag{3.136}$$

and

$$(\sigma_x^\bullet)_{811} = \frac{(h^\bullet+1)c_1^{\bullet 2}}{2\left(1-h^{\bullet 2}\right)\left(1-c_1^{\bullet 2}\right)\left(1+\beta^\bullet/t^\bullet\right)}\left[\begin{array}{l}\frac{2}{\sqrt{\pi t^\bullet}}\exp\left(\frac{-x^{\bullet 2}}{4t^\bullet}\right)+e^{t^\bullet-x^\bullet}\mathrm{Erfc}\left(-\sqrt{t^\bullet}+\frac{x^\bullet}{2\sqrt{t^\bullet}}\right)\\ -e^{t^\bullet+x^\bullet}\mathrm{Erfc}\left(\sqrt{t^\bullet}+\frac{x^\bullet}{2\sqrt{t^\bullet}}\right)\end{array}\right] \tag{3.137}$$

The dimensionless form of $(\sigma_x{}^*)_{21}$ is:

$$(\sigma_x^\bullet)_{21} = (\sigma_x^\bullet)_{121} + (\sigma_x^\bullet)_{221} + (\sigma_x^\bullet)_{321} + (\sigma_x^\bullet)_{421} \tag{3.138}$$

where

$$(\sigma_x^\bullet)_{121} = \frac{T_0^\bullet h^\bullet c_1^{\bullet 2}}{2\left(h^{\bullet 2}-c_1^{\bullet 2}\right)}\left[\begin{array}{l}h^\bullet e^{h^{\bullet 2}t^\bullet-h^\bullet x^\bullet}\mathrm{Erfc}\left(-h^\bullet\sqrt{t^\bullet}+\frac{x^\bullet}{2\sqrt{t^\bullet}}\right)\\ +h^\bullet e^{h^{\bullet 2}t^\bullet+h^\bullet x^\bullet}\mathrm{Erfc}\left(h^\bullet\sqrt{t^\bullet}+\frac{x^\bullet}{2\sqrt{t^\bullet}}\right)\end{array}\right] \tag{3.139}$$

and

$$\left(\sigma_x^\bullet\right)_{221} = -\frac{T_0^\bullet h^{\bullet 2} c_1^\bullet}{2\left(h^{\bullet 2} - c_1^{\bullet 2}\right)} \begin{bmatrix} c_1^\bullet e^{c_1^{\bullet 2} t^\bullet - c_1^\bullet x^\bullet} \mathrm{Erfc}\left(-c_1^\bullet \sqrt{t^\bullet} + \dfrac{x^\bullet}{2\sqrt{t^\bullet}}\right) \\ + c_1^\bullet e^{c_1^{\bullet 2} t^\bullet + c_1^\bullet x^\bullet} \mathrm{Erfc}\left(c_1^\bullet \sqrt{t^\bullet} + \dfrac{x^\bullet}{2\sqrt{t^\bullet}}\right) \end{bmatrix} \tag{3.140}$$

and

$$\left(\sigma_x^\bullet\right)_{321} = -\frac{T_0^\bullet h^\bullet c_1^{\bullet 2}}{2\left(h^{\bullet 2} - c_1^{\bullet 2}\right)} \begin{bmatrix} \dfrac{2}{\sqrt{\pi t^\bullet}} \exp\left(\dfrac{-x^{\bullet 2}}{4t^\bullet}\right) + h^\bullet e^{h^{\bullet 2} t^\bullet - h^\bullet x^\bullet} \mathrm{Erfc}\left(-h^\bullet \sqrt{t^\bullet} + \dfrac{x^\bullet}{2\sqrt{t^\bullet}}\right) \\ - h^\bullet e^{h^{\bullet 2} t^\bullet + h^\bullet x^\bullet} \mathrm{Erfc}\left(h^\bullet \sqrt{t^\bullet} + \dfrac{x^\bullet}{2\sqrt{t^\bullet}}\right) \end{bmatrix} \tag{3.141}$$

and

$$\left(\sigma_x^\bullet\right)_{421} = \frac{T_0^\bullet c_1^{\bullet 2} h^\bullet}{2\left(h^{\bullet 2} - c_1^{\bullet 2}\right)} \begin{bmatrix} \dfrac{2}{\sqrt{\pi t^\bullet}} \exp\left(\dfrac{-x^{\bullet 2}}{4t^\bullet}\right) + c_1^\bullet e^{c_1^{\bullet 2} t^\bullet - c_1^\bullet x^\bullet} \mathrm{Erfc}\left(-c_1^\bullet \sqrt{t^\bullet} + \dfrac{x^\bullet}{2\sqrt{t^\bullet}}\right) \\ - c_1^\bullet e^{c_1^{\bullet 2} t^\bullet + c_1^\bullet x^\bullet} \mathrm{Erfc}\left(c_1^\bullet \sqrt{t^\bullet} + \dfrac{x^\bullet}{2\sqrt{t^\bullet}}\right) \end{bmatrix} \tag{3.142}$$

The dimensionless form of $(\sigma_x^*)_2$ is:

$$\left(\sigma_x^\bullet\right)_2 = \left(\sigma_x^\bullet\right)_{12} + \left(\sigma_x^\bullet\right)_{22} \tag{3.143}$$

where

$$\begin{aligned} \left(\sigma_x^\bullet\right)_{12} = & \left(\sigma_x^\bullet\right)_{112} + \left(\sigma_x^\bullet\right)_{212} + \left(\sigma_x^\bullet\right)_{312} + \left(\sigma_x^\bullet\right)_{412} \\ & + \left(\sigma_x^\bullet\right)_{512} + \left(\sigma_x^\bullet\right)_{612} + \left(\sigma_x^\bullet\right)_{712} + \left(\sigma_x^\bullet\right)_{812} \end{aligned} \tag{3.144}$$

and

$$\left(\sigma_x^\bullet\right)_{112} = \frac{\left(h^\bullet + 1\right) h^\bullet c_1^{\bullet 3}}{\left(h^{\bullet 2} - 1\right)\left(h^{\bullet 2} - c_1^{\bullet 2}\right)\left(h^{\bullet 2} + \beta^\bullet / t^\bullet\right)} \left[\frac{1}{\sqrt{\pi\left(t^\bullet - x^\bullet / c_1^\bullet\right)}} + h^\bullet e^{h^{\bullet 2} t^\bullet} \mathrm{Erf}\left(h^\bullet \sqrt{t^\bullet - \frac{x^\bullet}{c_1^\bullet}}\right)\right] \cdot U[1] \tag{3.145}$$

and

$$\left(\sigma_x^\bullet\right)_{212} = -\frac{\left(h^\bullet + 1\right) c_1^{\bullet 3} h^\bullet}{\left(\beta^\bullet / t^\bullet + 1\right)\left(h^{\bullet 2} + \beta^\bullet / t^\bullet\right)\left(c_1^{\bullet 2} + \beta^\bullet / t^\bullet\right)} \left[\frac{1}{\sqrt{\pi\left(t^\bullet - x^\bullet / c_1^\bullet\right)}} + \sqrt{-\frac{\beta^\bullet}{t^\bullet}} e^{-\beta^\bullet} \mathrm{Erf}\left(\sqrt{-\beta^\bullet + \frac{\beta^\bullet x^\bullet}{c_1^\bullet t^\bullet}}\right)\right] \cdot U[1] \tag{3.146}$$

and

$$(\sigma_x^\bullet)_{312} = \frac{(h^\bullet+1)c_1^{\bullet 3}h^\bullet}{(h^{\bullet 2}-c_1^{\bullet 2})(1-c_1^{\bullet 2})(c_1^{\bullet 2}+\beta^\bullet/t^\bullet)}\left[\frac{1}{\sqrt{\pi(t^\bullet - x^\bullet/c_1^\bullet)}} + c_1^\bullet e^{c_1^{\bullet 2}t^\bullet}\mathrm{Erf}\left(c_1^\bullet\sqrt{t^\bullet - \frac{x^\bullet}{c_1^\bullet}}\right)\right]\cdot U[1] \tag{3.147}$$

and

$$(\sigma_x^\bullet)_{412} = \frac{(h^\bullet+1)c_1^{\bullet 3}h^\bullet}{(1-h^{\bullet 2})(1-c_1^{\bullet 2})(1+\beta^\bullet/t^\bullet)}\left[\frac{1}{\sqrt{\pi(t^\bullet - x^\bullet/c_1^\bullet)}} + e^{t^\bullet}\mathrm{Erf}\left(\sqrt{t^\bullet - \frac{x^\bullet}{c_1^\bullet}}\right)\right]\cdot U[1] \tag{3.148}$$

and

$$(\sigma_x^\bullet)_{512} = -\frac{(h^\bullet+1)c_1^{\bullet 3}}{(h^{\bullet 2}-1)(h^{\bullet 2}-c_1^{\bullet 2})(h^{\bullet 2}+\beta^\bullet/t^\bullet)}\left[D(t^\bullet - x^\bullet/c_1^\bullet) + h^{\bullet 2}e^{h^{\bullet 2}t^\bullet - \frac{h^{\bullet 2}x^\bullet}{c_1^\bullet}}\right]\cdot U[1] \tag{3.149}$$

and

$$(\sigma_x^\bullet)_{612} = \frac{(h^\bullet+1)c_1^{\bullet 3}}{(\beta^\bullet/t^\bullet+1)(h^{\bullet 2}+\beta^\bullet/t^\bullet)(c_1^{\bullet 2}+\beta^\bullet/t^\bullet)}\left[D(t^\bullet - x^\bullet/c_1^\bullet) - \frac{\beta^\bullet}{t^\bullet}e^{-\beta^\bullet + \frac{\beta^\bullet x^\bullet}{c_1^\bullet}}\right]\cdot U[1] \tag{3.150}$$

and

$$(\sigma_x^\bullet)_{712} = \frac{(h^\bullet+1)c_1^{\bullet 3}}{(h^{\bullet 2}-c_1^{\bullet 2})(1-c_1^{\bullet 2})(c_1^{\bullet 2}+\beta^\bullet/t^\bullet)}\left[D(t^\bullet - x^\bullet/c_1^\bullet) + c_1^{\bullet 2}e^{c_1^{\bullet 2}t^\bullet - c_1^\bullet x^\bullet}\right]\cdot U[1] \tag{3.151}$$

and

$$(\sigma_x^\bullet)_{812} = -\frac{(h^\bullet+1)c_1^{\bullet 3}}{(1-h^{\bullet 2})(1-c_1^{\bullet 2})(1+\beta^\bullet/t^\bullet)}\left[D(t^\bullet - x^\bullet/c_1^\bullet) + e^{t^\bullet - \frac{x^\bullet}{c_1^\bullet}}\right]\cdot U[1] \tag{3.152}$$

The dimensionless for of $(\sigma_x^*)_{22}$ is:

$$(\sigma_x^\bullet)_{22} = (\sigma_x^\bullet)_{122} + (\sigma_x^\bullet)_{222} + (\sigma_x^\bullet)_{322} + (\sigma_x^\bullet)_{422} \tag{3.153}$$

where

$$\left(\sigma_x^\bullet\right)_{122} = -\frac{T_0^\bullet c_1^\bullet}{\sqrt{\pi\left(t^\bullet - x^\bullet/c_1^\bullet\right)}} \cdot U[1] \tag{3.154}$$

and

$$\left(\sigma_x^\bullet\right)_{222} = -\frac{T_0^\bullet c_1^{\bullet 3}}{\left(h^{\bullet 2} - c_1^{\bullet 2}\right)} \left[\frac{1}{\sqrt{\pi\left(t^\bullet - x^\bullet/c_1^\bullet\right)}} + h^\bullet e^{h^{\bullet 2} t^\bullet} \mathrm{Erf}\left(h^\bullet \sqrt{t^\bullet - \frac{x^\bullet}{c_1^\bullet}}\right)\right] \cdot U[1] \tag{3.155}$$

and

$$\left(\sigma_x^\bullet\right)_{322} = \frac{T_0^\bullet c_1^\bullet h^{\bullet 2}}{\left(h^{\bullet 2} - c_1^{\bullet 2}\right)} \left[\frac{1}{\sqrt{\pi\left(t^\bullet - x^\bullet/c_1^\bullet\right)}} + c_1^\bullet e^{c_1^{\bullet 2} t^\bullet} \mathrm{Erf}\left(c_1^\bullet \sqrt{t^\bullet - \frac{x^\bullet}{c_1^\bullet}}\right)\right] \cdot U[1] \tag{3.156}$$

and

$$\left(\sigma_x^\bullet\right)_{422} = \frac{T_0^\bullet c_1^\bullet}{h^\bullet} D\left(t^\bullet - x^\bullet/c_1^\bullet\right) \cdot U[1] \tag{3.157}$$

and

$$\left(\sigma_x^\bullet\right)_{522} = \frac{T_0^\bullet c_1^{\bullet 3}}{h^\bullet\left(h^{\bullet 2} - c_1^{\bullet 2}\right)} \left[D\left(t^\bullet - x^\bullet/c_1^\bullet\right) + h^{\bullet 2} e^{h^{\bullet 2} t^\bullet - \frac{h^{\bullet 2} x^\bullet}{c_1^\bullet}}\right] \cdot U[1] \tag{3.158}$$

and

$$\left(\sigma_x^\bullet\right)_{622} = -\frac{T_0^\bullet c_1^\bullet h^\bullet}{\left(h^{\bullet 2} - c_1^{\bullet 2}\right)} \left[D\left(t^\bullet - x^\bullet/c_1^\bullet\right) + c_1^{\bullet 2} e^{c_1^{\bullet 2} t^\bullet - c_1^\bullet x^\bullet}\right] \cdot U[1] \tag{3.159}$$

and

$$\left(\sigma_x^\bullet\right)_3 = -\frac{c_1^{\bullet 3} t^\bullet e^{c_1^\bullet t^\bullet - x^\bullet}}{2\left(\beta^\bullet + c_1^\bullet t^\bullet\right)\left(c_1^\bullet - 1\right)} + \frac{c_1^{\bullet 3} t^\bullet e^{-c_1^\bullet t^\bullet - x^\bullet}}{2\left(c_1^\bullet t^\bullet - \beta^\bullet\right)\left(c_1^\bullet + 1\right)} + \frac{c_1^{\bullet 2} t^\bullet \beta^{\bullet 2} e^{-\beta^\bullet - x^\bullet}}{\left(\beta^\bullet + t^\bullet\right)\left(\beta^{\bullet 2} - c_1^{\bullet 2} t^{\bullet 2}\right)} - \frac{c_1^{\bullet 2} t^\bullet e^{t^\bullet - x^\bullet}}{\left(\beta^\bullet + t^\bullet\right)\left(1 - c_1^{\bullet 2}\right)} \tag{3.160}$$

and

$$(\sigma_x^\bullet)_4 = \left[\frac{c_1^{\bullet 3} t^\bullet e^{c_1^\bullet t^\bullet - x^\bullet}}{2(\beta^\bullet + c_1^\bullet t^\bullet)(c_1^\bullet - 1)} + \frac{c_1^{\bullet 3} t^\bullet e^{-c_1^\bullet t^\bullet + x^\bullet}}{2(c_1^\bullet t^\bullet - \beta^\bullet)(c_1^\bullet + 1)} - \frac{c_1^{\bullet 3} t^{\bullet 2} \beta^\bullet e^{-\beta^\bullet + \frac{\beta^\bullet x^\bullet}{c_1^\bullet t^\bullet}}}{(\beta^\bullet + t^\bullet)(c_1^{\bullet 2} t^{\bullet 2} - \beta^{\bullet 2})} - \frac{c_1^{\bullet 3} t^\bullet e^{t^\bullet - \frac{x^\bullet}{c_1^\bullet}}}{(\beta^\bullet + t^\bullet)(c_1^{\bullet 2} - 1)}\right] \cdot U[1] \tag{3.161}$$

Consequently, the dimensionless form of the stress equation is:

$$\sigma_x^* = (\sigma_x^*)_1 + (\sigma_x^*)_2 + (\sigma_x^*)_3 + (\sigma_x^*)_4 \tag{3.162}$$

The dimensionless temperature (Eq. 3.68) and stress distributions (Eq. 3.162) are computed during the heating pulse.

The above equation for the thermal stress field is used to predict the stress distribution in the irradiated region. The findings are presented below in line with the previous study [6,7].

Fig. 3.4 shows the dimensionless stress distribution within the substrate material for different dimensionless heat transfer coefficients and times. The thermal stress

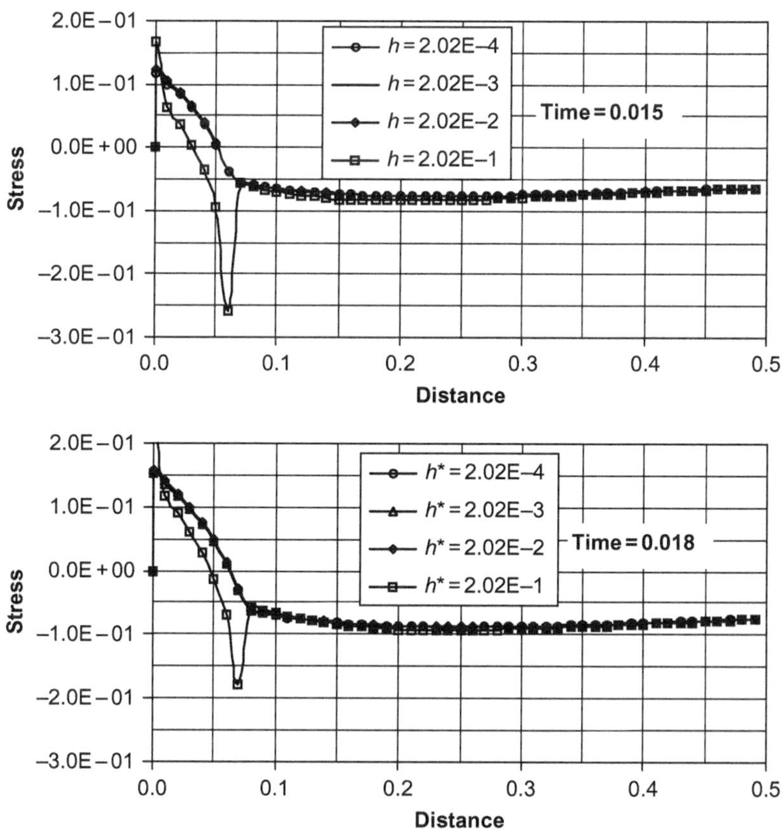

Fig. 3.4 Dimensionless stress distribution within the substrate material [7].

is zero at the surface as a result of the surface boundary condition used in the analysis, and it increases sharply close to the surface. The thermal stress is tensile in this region due to expansion of the surface. As the depth increases ($x^{\bullet} > 0.06$), the stress becomes compressive as a result of the thermal strain developed in this region, i.e., at this depth and beyond, the material contracts, resulting in a compressive thermal stress field. The influence of the heat transfer coefficient on the stress development is considerable, as illustrated by $h^{\bullet} = 0.0202\left(10^9\,\mathrm{W/m^2K}\right)$. In this case, the stress developed is compressive and with a high magnitude in the vicinity of the surface, and decays sharply as the depth increases. However, the compressive stress wave is developed at some point below the surface. The magnitude of the stress wave is lower at this point as time progresses. In addition, the magnitude of the thermal stress levels, corresponding to a heat transfer coefficient other than $h^{\bullet} = 0.0202$, increases with an increase in time, provided that this increase is less than 10%.

Fig. 3.5 shows the change in thermal stress for different dimensionless heat transfer coefficients with time. Because the substrate material is considered initially to be stress free, i.e., the magnitude of the stress level is zero at time $t = 0$. The stress level is tensile in the in the vicinity of the surface ($x^{\bullet} = 0.001$). This is because of the free expansion of the surface, in which case the strain becomes positive. However, the thermal stress becomes compressive as the depth increases beyond $x^{\bullet} = 0.01$. This occurs

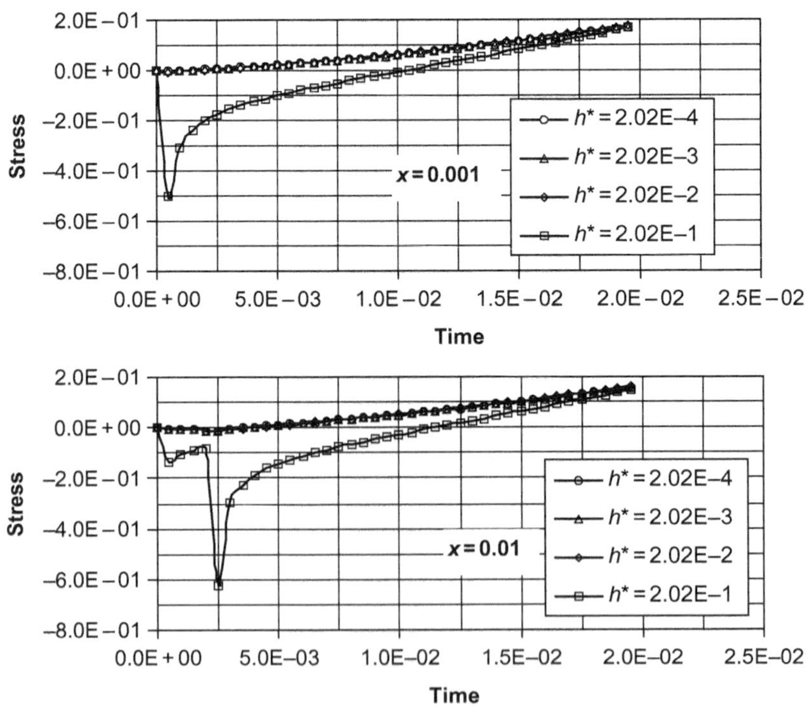

Fig. 3.5 Changes in dimensionless stress distribution with time [7].

as a result of the compression of the substrate material. The effect of the heat transfer coefficient on the stress distribution behavior with time is significant for a heat transfer coefficient of $h^{\bullet} = 0.0202$. In this case, the temperature gradient in the surface region differs significantly from those corresponding to the other heat transfer coefficients. This, in turn, generates a compressive stress wave propagating into the material. The magnitude of the stress wave is reduced as it propagates into the substrate material. Moreover, the stress wave behavior disappears at a depth of 0.1, i.e., the stress behavior beyond this depth changes significantly. This shows that the stress wave does not attain a depth of $x^{\bullet} = 0.1$ during the time interval considered in the present study.

3.5 Two-dimensional heating

Although the laser irradiated spot radius is small and the temperature distribution along the laser beam axis is more critical than along the radius of the irradiated spot, the one-dimensional analysis presented in the previous chapter needs to be extended to include the two-dimensional effect. On the other hand, the closed-form solution for a two-dimensional heating situation, including the source term, is difficult to obtain. However, the approximate solutions provide useful information associated with the laser machining process. The transient evaporation at the surface in two-dimensional space may be associated with the laser drilling process.

Two-dimensional heating of metallic substrates, with a constant velocity of a moving laser beam, results in nonsymmetric isothermal lines in the irradiated region of the substrate material. The isothermal lines are shown in Fig. 3.6 at $y = 0$ plane for the

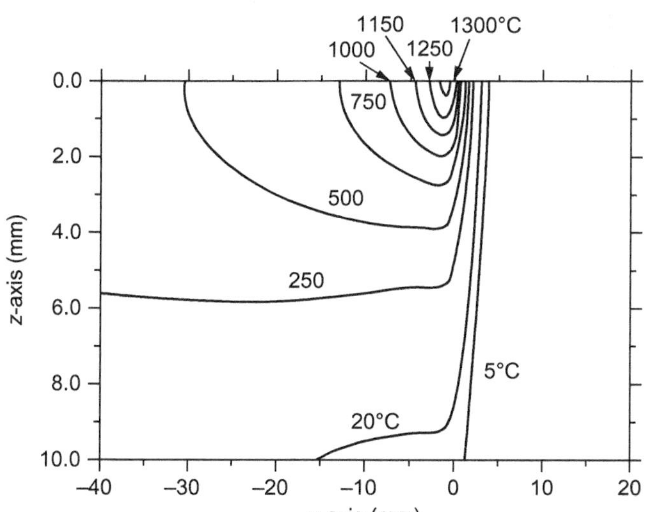

Fig. 3.6 Isothermals along z and x-axes for laser output power 350 W and scanning speed (V) 0.3 m/s.

laser power intensity of 350 W and $V = 0.3$ m/s. The isotherm at $z \cong 3.5$mm is 250°C at $y = 0$ plane and peak temperature well above melting of the substrate occurs.

3.6 Transient evaporation

As numerical solution is required for the two-dimensional case, in which a radially dependent intensity distribution that is constant in time, can be introduced. The heat transfer equation in the two-dimensional space that includes the volumetric source term can be written as:

$$k\left(\frac{\partial^2 T}{\partial z^2} + \frac{1}{r}\frac{\partial^2 T}{\partial r^2}\right) + I(r)\delta \exp\left[-\delta(z - z_s)\right] = \rho C_p \frac{\partial T}{\partial t} \tag{3.163}$$

Boundary conditions,

$$\text{At the surface}(z = 0) \text{ then } \frac{\partial T(r, 0, t)}{\partial z} = \frac{I(r)\delta}{k} \tag{3.164}$$

$$\text{At infinity}\,(\mathrm{z} = \infty) \text{ then } T(r, \infty, t) = 0 \tag{3.165}$$

$$\text{At the symmetry axis}(r = 0) \text{ then } \frac{\partial T(0, z, t)}{\partial r} = 0 \tag{3.166}$$

$$\text{At infinity}\,(r = \infty) \text{ then } T(\infty, z, t) = 0 \tag{3.167}$$

Initial condition,

$$\text{At } t = 0 \text{ then } T(r, z, 0) = 0 \tag{3.168}$$

However, it is impossible to obtain boundary conditions that are independent of temperature. A solution, therefore, is not very true. If equilibrium is not reached, a solution for the temperature is not analytic and so the depth of weld is not predictable. For non-steady state drilling, a solution can be formed by assuming that the evaporation rate can follow the changes in surface temperature.

Assuming the surface temperature is in the form of:

$$T = Q\sqrt{t} \tag{3.169}$$

and the numerical analysis of transient evaporation during laser heating results in the temperature-dependent recession velocity curve, as shown in Fig. 3.7. The curve-fitting technique enables the development of the relationship between recession velocity of the surface and temperature, which is:

$$\log(V) = C_1 T - C_2 \tag{3.170}$$

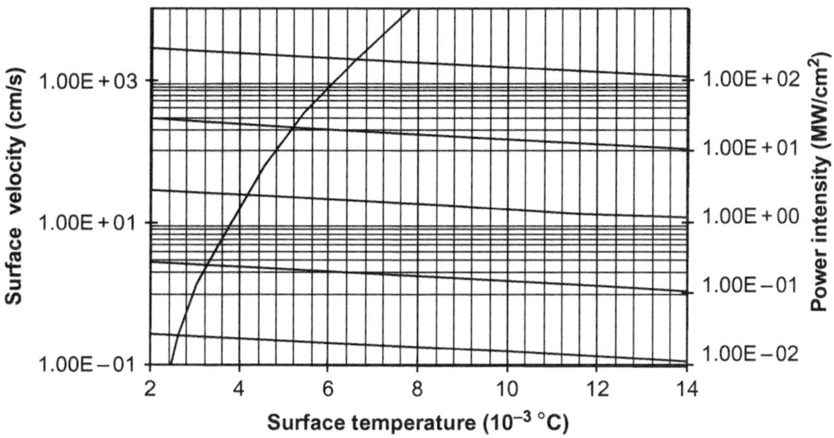

Fig. 3.7 Surface recession velocity with surface temperature and laser power intensity for aluminum.

where C_1 is the average gradient of the curve, and C_2 is the intercept at $T=0$. Writing equation as:

$$V=\exp(C_1T-C_2) \tag{3.171}$$

and substituting Eq. (3.152),

$$V=\exp\left(C_1Q\sqrt{t}-C_2\right) \tag{3.172}$$

If D is the depth of the laser-produced cavity,

$$D=\int_0^{\tau'}\exp\left(C_1Q\sqrt{t}-C_2\right)dt+\int_{t_1}^{\tau}V_s dt \tag{3.173}$$

where τ is the pulse length. If τ' is very short compared to then the expression reduces to the second integral,

$$D=\frac{1}{\exp(C_2)}\int_0^{\tau'}\exp C_1Q\sqrt{t}dt+\int_{\tau'}^{\tau}V_s dt \tag{3.174}$$

let $x=C_1Q\sqrt{t}$ and $dx/dt=1/2C_1Q\sqrt{t}$ or

$$dt=\frac{2\sqrt{t}}{C_1Q}dx \tag{3.175}$$

Because

$$\int xe^x dx = xe^x - \int e^x dx = e^x(x-1) \tag{3.176}$$

then Eq. (3.174) becomes

$$D = \frac{2}{(C_1Q)^2}\frac{\exp\left(C_1Q\sqrt{t}\right)}{\exp(C_2)}\left(C_1Q\sqrt{t}-1\right)\Bigg|_0^{\tau'} + V_s(\tau-\tau') \tag{3.177}$$

$$D = \frac{2}{(C_1Q)^2}V_s(C_1T_s-1) - \frac{2}{\exp(C_1Q)^2(C_2)} + V_s(\tau-\tau') \tag{3.178}$$

or

$$D = \frac{2}{(C_1Q)^2}V_s(TC_1T_s-1) + V_s(\tau-\tau') \tag{3.179}$$

Evaporation of the surface during the key-hole formation introduces a complicated physical process. However, the simplification of the evaporation process results in the above analysis. The power intensity required increases significantly when the evaporation temperature increases. This, in turn, increases the vapor velocity (surface velocity) significantly. Consequently, the momentum exchange across the vapor front and the liquid behind the vapor front generates extremely high recoil pressure, which causes ejection of metallic liquid from the key-hole. In this case, the above analysis fails to predict the correct surface evaporation and vapor velocity.

3.7 Three-dimensional heating and stationary source at the surface

The differential equation of heat conduction is:

$$\frac{\partial^2 T}{\partial x^2} + \frac{\partial^2 T}{\partial y^2} + \frac{\partial^2 T}{\partial z^2} = \frac{1}{\alpha}\frac{\partial T}{\partial t} \tag{3.180}$$

which is satisfied by:

$$T = \frac{Q}{8(\pi\alpha t)}\exp\left[\frac{(x-x')^2+(y-y')^2+(z-z')^2}{4Kt}\right] \tag{3.181}$$

As time approaches zero the expression tends to be zero at all points except (x', y', z'), where it becomes infinite. The solution may be interpreted as the temperature in an infinite solid due to a quantity of heat, Q, instantaneously generated at $t=0$ at the point (x', y', z'). Such a solution is inappropriate to laser heating because the time over which the heat transfer occurs is of the same order of magnitude as the pulse duration. Consider that the laser heating intensity is a continuous point source in preference to an instantaneous one, based on the following assumptions:

(i) The metal is semi-infinite, homogeneous, and isotropic with thermal properties that are independent of temperature.
(ii) There is no heat loss from the surface out of the metal.
(iii) There is no heat loss due to change in states.
(iv) No material is lost during heating.

The temperature at a location (x, y, z), due to a continuous stationary source, Q, at the surface. is found by integrating:

$$T = \frac{2}{(8\pi\alpha)^{9/2}} \int_0^t Q \exp\left[-\frac{r^2}{4\alpha(t-t')}\right] \frac{dt'}{(t-t')^{9/2}} \tag{3.182}$$

where $r^2 = (x-x')^2 + (y-y')^2 + (z-z')^2$ and the factor 2 introduced to make the model semi-infinite. Putting $t'' = (t-t')^{-1/2}$, it yields;

$$T = \frac{2Q}{4(\pi K)^{9/2}} \int_{1/t}^{\infty} \exp\left[-\frac{r^2 t''^2}{4\alpha}\right] dt'' \tag{3.183}$$

or

$$T = \frac{Q}{2\pi\alpha r} \text{erfc}\left[\frac{r}{\sqrt{4\alpha t}}\right] \tag{3.184}$$

Assumption (iv) renders the solution invalid the moment the surface reaches the normal boiling temperature. This may be after one spike output from the laser; in which case, the heat flow is one-dimensional and produced from an exponential heat source, not by constant flux absorption. In fact, there is no period, either at the beginning or toward the end of the laser pulse, in which Eq. (3.184) holds.

3.8 Laser cutting process and striation formation

The analytical approach formulating the laser beam heating of solid substrate is given in line with the previous study [8]. The possible formulation of striation patterns is also introduced in this section [8].

In the laser beam cutting process, when the beam is focused onto a plane surface, a fraction of the beam is reflected and the rest is absorbed by the substance. The energy balance of the substance at the surface region changes as a result of this absorption process. If the energy input is supplied along the z-axis and the

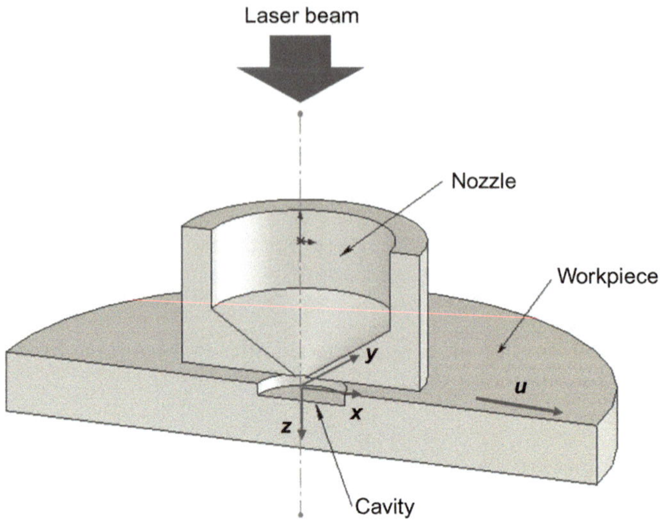

Fig. 3.8 A schematic view of the coordinate system, laser beam, and the workpiece.

source travels along the x-axis at a velocity u (Fig. 3.8), the energy equation can be written as:

$$U\frac{\partial T}{\partial x}+\frac{\partial^2 T}{\partial x^2}+\frac{\partial^2 T}{\partial y^2}+\frac{\partial^2 T}{\partial z^2}=2\lambda\frac{\partial T}{\partial t} \tag{3.185}$$

After assuming no heat convection at the surface, then the boundary conditions yield:

$$\begin{aligned}&\frac{\partial T}{\partial x}=0 \quad \text{as } x=\pm\infty\\&\frac{\partial T}{\partial y}=0 \quad \text{as } y=\pm\infty\end{aligned} \tag{3.186}$$

This may be re-arranged with the source at the origin by making the transformation: $\xi=x-Ut$ for the wavefront travel in the positive x direction and $2\lambda=1/\alpha$ is used for convenience. On substitution gives:

$$\frac{\partial^2 T}{\partial \xi^2}+\frac{\partial^2 T}{\partial y^2}+\frac{\partial^2 T}{\partial z^2}=-2\lambda u\frac{\partial T}{\partial \xi}+2\lambda\frac{\partial T}{\partial t} \tag{3.187}$$

If the solid is of sufficient length, the temperature distribution around the heat source becomes constant, i.e., $\partial T/\partial t=0$. Therefore:

$$\frac{\partial^2 T}{\partial \xi^2}+\frac{\partial^2 T}{\partial y^2}+\frac{\partial^2 T}{\partial z^2}=-2\lambda U\frac{\partial T}{\partial \xi} \tag{3.188}$$

The solution of Eq. (3.178) is straightforward. Let, $T = e^{-(\lambda u \xi)} F(\xi, y, z)$ then substituting with Eq. (3.178) gives:

$$\nabla^2 F - (\lambda u)^2 F = 0 \tag{3.189}$$

After assuming no heat convection at the surface, the boundary conditions yield:

$$\begin{aligned} \frac{\partial T}{\partial \xi} &= 0 \text{ as } \xi = \pm\infty \\ \frac{\partial T}{\partial y} &= 0 \text{ as } y = \pm\infty \end{aligned} \tag{3.190}$$

and considering a cylinder of radius r drawn around the heat source:

$$-k\frac{\partial T}{\partial r} 2\pi r \to Q \text{ as } r \to 0 \tag{3.191}$$

where Q is the rate of heat release per length of source and

$$r = \sqrt{(\xi^2 + y^2)} \tag{3.192}$$

In cylindrical co-ordinates, Eq. (3.189) becomes:

$$\frac{\partial^2 F}{\partial r^2} + \frac{1}{r}\frac{\partial F}{\partial r} - (\lambda u)^2 F = 0 \tag{3.193}$$

which is satisfied by $K_o[\lambda u r]$, i.e., the Modified Bessel function of the second kind of zero order. When

$$r \to 0 \text{ then } \frac{\partial K_o[\lambda u r]}{\partial r} = \frac{1}{r} \tag{3.194}$$

Hence the solution with the condition:

$$-k\frac{\partial T}{\partial r} 2\pi r \to q \text{ as } r \to 0 \tag{3.195}$$

is satisfied for:

$$T = \frac{q}{2\pi k} e^{-\lambda \xi} K_o[\lambda u r] \tag{3.196}$$

Re-arrangement of Eq. (3.196) gives the solution of temperature distribution in an infinite solid due to continuous line source of q (energy per unit length), which is:

$$T = \frac{q}{2\pi kD} e^{\frac{ux}{2\alpha}} K_o \left(\frac{Ur}{2\alpha} \right) \tag{3.197}$$

where r is the radial distance to the point being considered (satisfying $r^2 = x^2 + y^2$), k is the thermal conductivity, D is the thickness of the workpiece, and K_o is the modified Bessel function of the second kind of zero order. When laser heating is considered the heat flux per unit of time becomes:

$$q_{laser} = (1 - R_f) I_o \tag{3.198}$$

where R_f is the surface reflectivity and I_o is the laser incident power intensity.

In the case of a gas-assisted laser cutting process, the contribution of exothermic chemical reactions to the heat transfer taking place at the liquid surface may be written as:

$$q_{chem} = \rho_e u_e C_{H_c} \Delta I \left\{ 1 + \left[\frac{C_{H_d}}{C_{H_c}} - 1 \right] \frac{h_c}{\Delta I} - \frac{B_3}{\Delta I} \right\} \tag{3.199}$$

where

$$\Delta I = (I_f)_e - (I_f)_g + h_c, \tag{3.200}$$

and $\left[(I_f)_e - (I_f)_g \right]$ is the enthalpy change at the interface due to the assisting gas and h_c is the total chemical reaction enthalpy which is defined as:

$$h_c = \sum_{i \neq E} (h_i)_g \left[(C_i)_g - (1 + B_2)(C_i)_g \right] + (h_E)_g \left[B_2 - (1 + B_2)(C_E)_g \right] \tag{3.201}$$

where B_2 and B_3 are:

$$B_2 = B_3 \frac{C_{H_c}}{C_{H_d}} \tag{3.202}$$

and

$$B_3 = \frac{(\rho u)_g}{\rho_e u_e C_{H_c}} \tag{3.203}$$

Eq. (3.198) gives the heat transfer to the liquid metal due to chemical reactions and mass transfer occurring at the assisted gas-liquid interface when a reacting turbulent flow occurs over the liquid molten metal. The total heat flux available at the surface becomes:

$$q_{tot} = q_{laser} + q_{chem} \tag{3.204}$$

When the temperature of the surface reaches the melting temperature, melting isotherm develops at the surface, then the temperature of the melting isotherm can be obtained by rearranging Eq. (3.197), i.e.:

$$T_m = \frac{q_{tot}}{4\pi kD} e^{\frac{ux}{2\alpha}} K_o\left(\frac{ur}{2\alpha}\right) \tag{3.205}$$

Introducing dimensionless coordinates as:

$$X = \frac{ux}{2\alpha}; \; R = \frac{ur}{2\alpha} \tag{3.206}$$

Eq. (3.205) yields:

$$T_m = \frac{q_{tot}}{2\pi kD} e^{X} K_o(R) \tag{3.207}$$

To determine the area ratio of the total melted area to focused spot area, a geometric description of the focused spot and melted area should be considered (Fig. 3.9). Assume that the maximum width of the melting isotherm lies outside of the focused spot. The total area of the focused spot removed is the shaded zone in Fig. 3.9, which may be expressed as:

$$S_{tot} = S_1 + S_2 = r^2\varphi + \frac{r^2 Sin\varphi}{2} + 2\int_{X_1}^{X_2} Y dx \tag{3.208}$$

where

$$X_2 = \ln\left\{\frac{2\pi kDT_m}{q_{tot}} K_o[R]\right\}$$

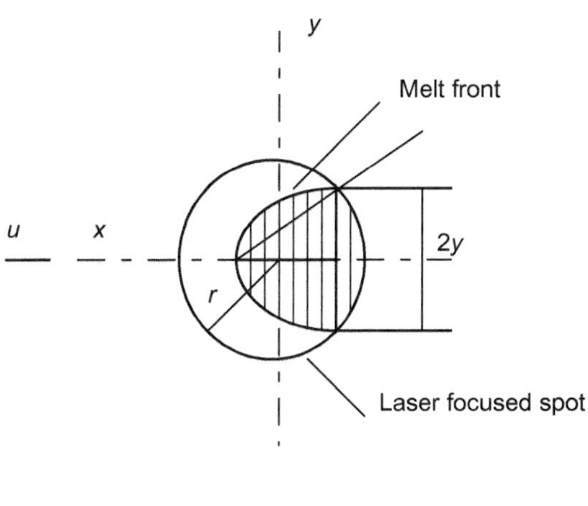

Fig. 3.9 A schematic view of the laser irradiated spot and the melt front at the irradiated surface [8].

and

$$\varphi = \mathrm{Cos}^{-1}\left(\frac{X_2}{r}\right) \tag{3.209}$$

Because

$$Y = \sqrt{R^2 - X^2} = \left\{R^2 - \ln\left[\frac{2\pi kDT_m}{q_{tot}}K_o[R]\right]^2\right\}^{1/2} \tag{3.210}$$

and

$$dX = \frac{K_1[R]}{K_o[R]}dR \tag{3.211}$$

After rearranging Eqs. (3.208)–(3.211), it yields:

$$S = \frac{\varphi}{\pi} - \frac{\mathrm{Sin}\,2\varphi}{2\pi} + \frac{2}{\pi r^2}\int_{X_1}^{r}\left\{R^2 - \left[\frac{2\pi kDT_m}{q_{tot}}{}_{K_o[R]}\right]^2\right\}^{1/2}\frac{K_1[R]}{K_o[R]}dR \tag{3.212}$$

Consider Fig. 3.9, the melted area may be approximated as an elliptic shape. Therefore, the striation width (w_k) can be calculated from:

$$w_k = \frac{S_{tot}}{2\pi y_{\max}} \tag{3.213}$$

Consider the speed of melting isotherms penetrating into the workpiece; in this case, assuming attainment of a steady state at the surface and omitting the surface evaporation, the penetration speed can be written as:

$$v = \frac{q_{tot}}{\rho(L + CpT_m)} \tag{3.214}$$

A computer program is developed to compute the relevant cutting parameters including non-dimensional maximum cut width, area ratio, striation width, and penetration speed.

To determine the maximum melt width, re-arrangement of Eq. (3.205) is necessary, i.e.,

$$T_m = \frac{(1 - R_f)I_o}{2\pi kD}e^{uX/2\alpha}K_o\left(\frac{ur}{2\alpha}\right) \tag{3.215}$$

where R_f is the surface reflectivity and I_o is peak power intensity of the laser beam. Re-arrangement of Eq. (3.215) yields:

$$\exp(X) = \frac{2\pi kDT_m}{(1-R_f)I_o}\frac{1}{K_o[R]} \tag{3.216}$$

or

$$K_o\left[\left(X^2+Y^2\right)^{1/2}\right] = \frac{2\pi kDT_m}{(1-R_f)I_o}\exp(-X) \tag{3.217}$$

where

$$R = \sqrt{(X^2+Y^2)}$$

Differentiating with respect to X gives:

$$-K_1\left\{\left(X^2+Y^2\right)^{1/2}\right]\left(X+Y\frac{dY}{dX}\right)\left(X^2+Y^2\right)^{-1/2} = -\frac{2\pi kDT_m}{(1-R_f)I_o}\exp(-X) \tag{3.218}$$

When $dY/dX = 0$ then:

$$\frac{[XK_1(R)]}{R} = \frac{2\pi kDT_m}{(1-R_f)I_o}\exp(-X) \tag{3.219}$$

where $K_1[R]$ is the modified Bessel function of of the second kind of first order. Using Eq. (3.217), it results:

$$\frac{[XK_1(R)]}{R} = K_o(R) \tag{3.220}$$

Therefore, for maximum Y:

$$X = \frac{RK_o[R]}{K_1[R]} \tag{3.221}$$

and

$$K_o[R] = \frac{2\pi kDT_m}{(1-R_f)I_o}e^{\left[-\frac{RK_o(R)}{K_1(R)}\right]} \tag{3.222}$$

The solution of Eq. (3.222) gives the values of R for melt at the point of maximum width. Y can be found from:

$$Y = R\mathrm{Sin}\varphi \tag{3.223}$$

where

$$\varphi = \text{Cos}^{-1}\left[\frac{K_o(R)}{K_1(R)}\right] \tag{3.224}$$

The above equations are limited with $D < (2\alpha r/u)$, in this case, the melt depth should be less than the characteristic distance. Some differences occur between the theoretical predictions and experimental results. This may be due to the assumptions made in the model and/or error related to the experiment. In some cases, both results are in agreement within acceptable limits. Increasing workpiece velocity increases the laser power requirements for a given focus spot diameter. Melted zone profile, normal to the surface, follows the similar profile of laser output power across the spot.

3.9 Lump parameter analysis for kerf width formulation

The overall energy balance in the laser cutting process was simplified through introducing a lumped parameter technique by Kar et al. [9]. The simple scaling law states that a relationship exists among the laser energy absorbed unit depth, convection, and conduction contributions of heat transfer and phase change. This yields the relation [9]:

$$\frac{P}{d} = \frac{vw_k + A_3\sqrt{vw_k}}{A_o} \tag{3.225}$$

where:

$$A_o = \frac{A}{a_o};\ A_3 = \frac{1}{a_o}\frac{(w_k + 2wk(T_m T_o))}{2\sqrt{\alpha w w_k}} \quad \text{and} \quad a_a$$
$$= \rho[Cp(T_m - T_o) + L_m + \beta L_b] \tag{3.226}$$

The term A_o is related to energy transfer rate to the workpiece at the surface during the laser-workpiece interaction. A is the effective energy-coupling factor at the workpiece surface, and A_3 is related to energy loss due to conduction in the solid. β is the fraction of evaporation contribution at the workpiece surface.

Using the scale law, the kerf width can be written in a simplified form for a circular laser beam with a beam waist size of w as [9]:

$$w_k = \frac{1}{v}\frac{\left[2.51\sqrt{\frac{\alpha}{w}\frac{2A\eta_u}{k(T_m - T_o)}}P\sqrt{v}\right]}{\left[1 + 3.08\times 10^8\frac{A_o}{A_3\sqrt{f}}\frac{\rho\rho_g\sigma^2\alpha}{M_w P_g f\sqrt{w}}P\sqrt{v}\right]} \tag{3.227}$$

where f is the fraction of the pressure drop in the kerf, σ is the molecular diameter (Angstrom unit), M_w is the molecular weight (gr/mol unit) of the assisting gas, P_g

Table 3.2 **Properties of steel and possible simulation conditions for Eq. (3.227)**

Properties	Values	Units
Boiling temperature	3133	K
Density of assisting gas	1.97	kg/m^3
Density of workpiece	7880	kg/m^3
Fraction of evaporation contribution ($0<\beta<1$)	0.1	–
Fraction of pressure drop ($0<f<1$)	0.1	–
Latent heat of melting	2.72×10^5	J/kg
Latent heat of boiling	6.10×10^6	J/kg
Melting temperature	1809	K
Molecular diameter of oxygen	6	Angstrom
Specific heat capacity	460	J/kg K
Super heating factor	0.02	–
Thermal conductivity	80.3	W/mK
Thermal diffusivity	2.21×10^{-5}	m^2/s

and ρ_g are the pressure and density of the assisting gas, and η_u is the superheating factor in the melt pool.

Eq. (3.231) can be used to determine the kerf width while the energy-coupling factor (A), laser power (P), and cutting speed (v) are considered as variables in Eq. (3.231). As an example for the curve analysis, the properties of steel are given in Table 3.2.

The above analysis for the kerf width estimation can be simulated to resemble the laser oxygen-assisted cutting of mild steel in accordance with Table 3.2. The affecting parameters are selected based on the previous study [10]. The parameters include laser output power, laser scanning speed (cutting speed), and the energy-coupling factor at the workpiece surface. The energy coupling at the workpiece surface plays an important role in the laser cutting process. When the high-powered laser beam is absorbed by the workpiece material, it results in solid state heating, melting, and evaporation of the workpiece material. The melted and the evaporated material, with the existence of oxygen, due to impinging gas, initiate a high-temperature exothermic reaction. This ionizes the evaporation front slightly and generates partially-ionized surface plasma, which scatters the laser beam and modifies the focusing conditions of the incident laser beam. In addition, partially-ionized surface plasma acts as a heat source, enlarging the size of the melt in the kerf. Moreover, when the incident beam is partially blocked by the surface plasma, less incident laser energy reaches the workpiece. This reduces the mass removal rate and the size of the surface plasma reduces, which results in more laser energy reaching the workpiece surface. This later accelerates the mass removal rate [11]. Consequently, the high-temperature oxidation reaction, in most cases, results in a transiently varying energy coupling mechanism at the workpiece surface. Therefore, the energy coupling needs to be considered as one of the affecting parameters in the analysis.

In the analysis, the difference of kerf width size (Δw_k) due to each affecting parameter is also considered. The difference of kerf width size is determined from:

$$\Delta w_k = (w_k)_{P_n} - (w_k)_{P_{(n+1)}} \tag{3.228}$$

where P_n is any affecting parameter at nth level. The variation of Δw_k due to a percentage change of affecting parameters is also considered. In this case, any of the affecting parameters is increased gradually and the corresponding kerf width change is determined. Then, the amount of increment is normalized by the magnitude of the affecting parameter, i.e., $\Delta P_n/P_n$, where ΔP_n is the increment in the parameter at nth level, and P_n is the parameter at nth level. This arrangement identifies the change in Δw_k due to a certain percent of change in the parameter. If the laser processing parameters (affecting parameters) vary slightly during the cutting process, the striation is formed on kerf surface. This is because the change of Δw_k during the cutting process, i.e., the variation of affecting parameters results in varying kerf width. Moreover, variation of kerf size due to variation of any affecting parameter identifies the significance of the parameter in the striation formation.

Figs. 3.10 and 3.11 show the variation of kerf width with anenergy coupling factor (A) and the kerf width difference (Δw_k) with a percentage of energy coupling factor (dA_n/A_n). The kerf width increases with an increasing energy coupling factor. This is because the energy transfers to the workpiece, which increases as the energy coupling factor increases. The energy-coupling factor defines the rate of the energy transfer to the workpiece at the surface due to laser beam energy and the energy released during the exothermic reaction. Consequently, increasing the energy-coupling factor

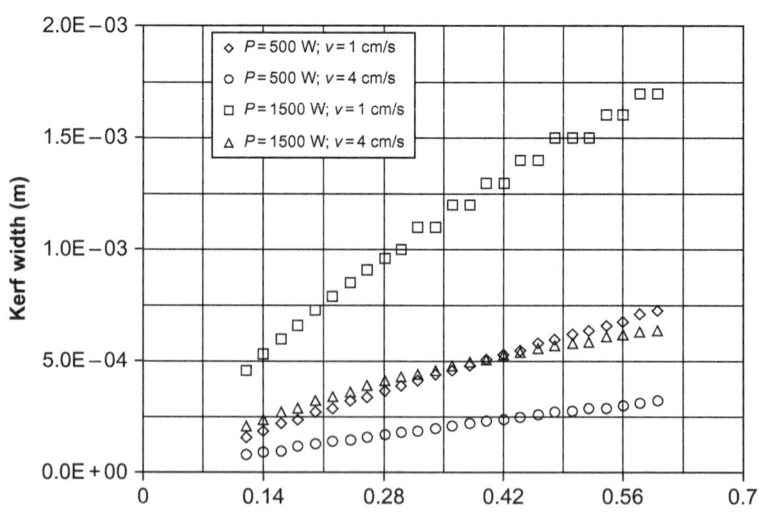

Fig. 3.10 Kerf width with an energy coupling factor (A) for different laser powers and cutting speeds [10].

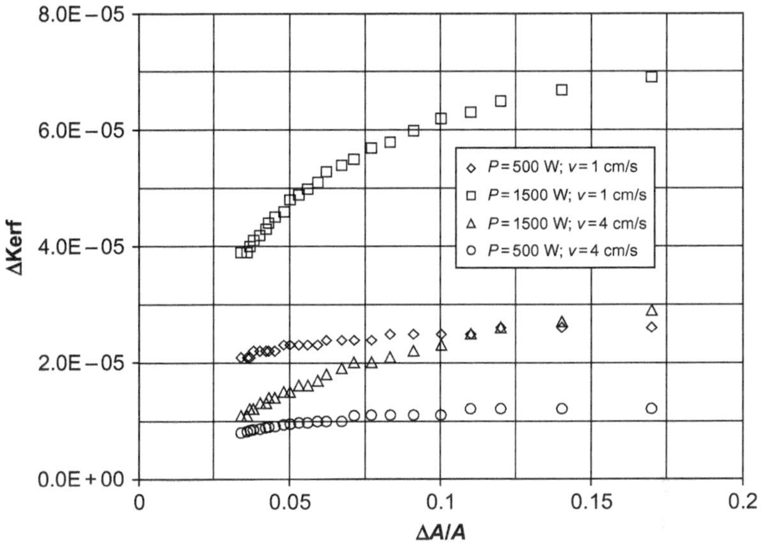

Fig. 3.11 Variation in the kerf width with an energy coupling factor variation ($\Delta A/A$) [10].

increases the energy available at the workpiece surface during the cutting process. In this case, the size of the kerf width, as a consequence of a large melt pool, increases. Moreover, the kerf size increases considerably as the laser power input increases while the cutting speed reduces, i.e., reducing the laser cutting speed increases the duration at which the high temperature oxidation reaction takes place at the workpiece surface. In this case, the combination of high laser power input and low cutting speed enhances the energy coupling at the workpiece surface and increases the kerf width considerably. As the cutting speed increases the kerf width reduces. This is true for high laser power input (1500 W) and a large energy coupling factor case. In the case of Δw_k (Fig. 3.11), increase in the percentage of energy coupling (dA_n/A_n) increases Δw_k at all levels of affecting parameters. This indicates that an increase in dA_n/A_n results in increased energy available at the workpiece surface which, in turn, increases the size of the melt zone in the kerf. The large increase in Δw_k is observed for the condition corresponding to a low cutting speed and a high power intensity. This suggests that at high power and low speed cutting conditions, when dA_n/A_n varies 5%, the size of the kerf width changes almost 1%–10% depending on the energy-coupling factor. Consequently, as the energy coupling at the workpiece surface oscillates due to an uncontrolled high temperature oxidation reaction, the kerf width also oscillates accordingly. This initiates the striation formation and increases the size of the striation. Therefore, care should be taken during the high-power laser cutting process such that the assisting gas pressure should be regulated and controlled properly in order to minimize the oscillation in energy coupling at the workpiece surface due to high temperature oxidation reaction.

Figs. 3.12 and 3.13 show the kerf width with the cutting speed, and Δw_k with the percentage of cutting speed variation (dv_n/v_n). In general, kerf width reduces with

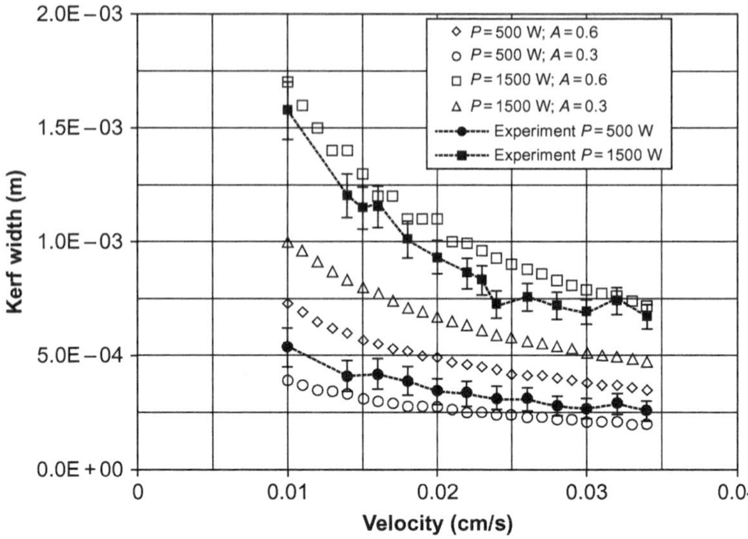

Fig. 3.12 Kerf width with cutting speed for different energy-coupling factors and laser powers [10].

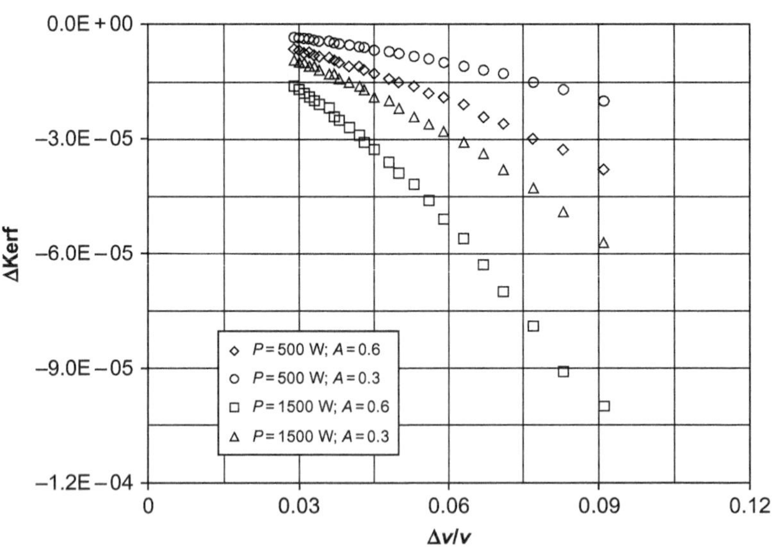

Fig. 3.13 Variation in kerf width with cutting speed variation for different energy-coupling factors and laser powers [10].

increased cutting speed. The high laser power results in a small kerf width size provided that the decrease in the kerf width is non-linear with the cutting speed. When comparing the experimental results with these predictions, they are in reasonably good agreement. In the case of Δw_k (Fig. 3.13), high laser power with a high energy-coupling factor results in large Δw_k. Moreover, a 5% variation in dv_n/v_n results in

2%–4% variations in the kerf width for the condition of 1500 W laser power and a 0.6 energy-coupling factor. As dv_n/v_n increases Δw_k increases considerably. During the laser cutting process, if the cutting speed varies without regulating the laser output power, the size of the striation amplifies. This is especially true when laser cutting odd-shaped curves or sharp corners.

In laser oxygen-assisted cutting, oxygen in the kerf width undergoes a reaction with the molten metal in the kerf. This alters the flow behavior of the molten material and influences the kerf surface morphology. The probable oxide formation in the vicinity of the cut edge, during the oxygen-assisted laser cutting process, may be expressed as:

$$Fe + \frac{1}{2}O_2 \rightarrow FeO \tag{3.229}$$

The process occurred as high temperature exothermic oxidation reactions, in which case, the solubility of oxygen in iron is very low. The cut edge may be distinguished as oxide layers having different oxygen content; consequently, two oxide layers can be distinguished. These include: (i) FeO/Fe inner layer (close to base metal), and (ii) an oxygen-rich FeO layer which is an outer layer next to an inner layer. It is probable that the oxygen behavior in the outer layer may be described using the diffusion equation [12], which is:

$$\frac{\partial x}{\partial t} = \frac{\partial}{\partial z}\left(D\frac{\partial x}{\partial z}\right) \tag{3.230}$$

where x is the oxygen concentration, D is the diffusion coefficient, and z is the axial distance. The solution of Eq. (3.230) gives an exponential decay in oxygen concentration as the distance from the outer layer increases toward the inner layer. The typical oxygen diffusion in the molten layer at the kerf surface is shown in Fig. 3.14.

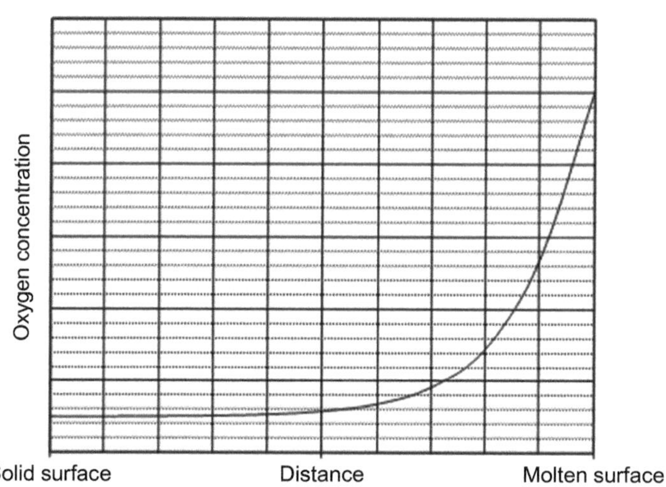

Fig. 3.14 Oxygen concentrations in the molten layer.

The formation of FeO occurs in the outer layer because free oxygen is available in this layer due to assisting gas. Moreover, once FeO is formed, the existing oxygen molecules supplied by the assisting gas initiates further reaction of $FeO+O_2$, in this case, an oxygen rich layer (Fe_2O_3) may develop at the very edge of the outer layer. It should be noted that the high-temperature oxidation reaction occurs according to the local pressure of oxygen molecules. The growing mechanism for this reaction may be expressed in terms of Gibbs free energy [12], i.e.,

$$\Delta G = RT \log P_{O_2} \tag{3.231}$$

where P_{O_2} is oxygen partial pressure and R is the gas constants of oxygen. When the temperature exceeds melting temperature of the substrate (1800°K), the critical pressure for ceasing the oxygen reaction is approximately 10^{-7} bar. Therefore, it is expected that the oxygen reaction ceases as the distance from the kerf edge toward the base material increases.

Viscosity of the molten metal and oxide layer becomes important when mixing these two components is considered. Viscosity of the molten metal is given by Brandes [13]. In this case, considerably low viscosity oxide molten metal may occur at high temperatures. The material removal by the assisting gas should be substantial and it is expected that the mixing of oxide species in the molten layer is possible due to the viscosity gradient in the inner and outer layers of the melt.

On the other hand, exothermic oxidation reactions taking place in the vicinity of the cutting kerf provides extra available energy in this region. Exothermic reactions due to oxygen-assisting gas in the laser cutting process results in massive oxidation in the melt layer [14]. Moreover, the critical pressure for ceasing the oxygen reaction may indicate that as the distance from the kerf edge toward the solid base material increases, the probability of oxygen reaction decreases. However, the mixing of the highly oxidized species with molten metal is possible due to the viscosity gradient and substantial removal of the molten metal by the assisting gas. Consequently, the mass removal rate, as well as the mixing of the molten metal that has different oxygen content may improve oxygen diffusion toward the solid base material.

3.10 Analytical assessment of assisting gas

In laser cutting many factors influence the cutting process. One of the attracting phenomena taking place in laser cutting is the formation of striations, which strongly affects the quality of laser cutting. However, the forming mechanism of stria is not well understood. Several different explanations for a stria occurrence are available in the literature, but no general consensus exists on this dynamic effect. One explanation for this formation is that at cutting speeds less than the speed of the moving molten layer, sideways burning occurs due to oxidation, which results in the formation of stria. When modeling the laser gas-assisting cutting process, difficulty arises due to the complex nature of the phenomena. However, useful assumptions simplifying the physical processes enable the formulation of the heating problem. Nevertheless, the laser gas-assisting cutting process allowing the momentum effects and chemical contribution of the assisting gas needs to be introduced in the analysis to capture the physical processes involved in the laser cutting process. Therefore, in this chapter, the gas-assisted cutting process is modeled after considering a turbulent boundary layer within the gas-liquid interface. Fluid equations governing the flow field in one dimension is considered. The effects of chemical reactions on the skin friction coefficient and heat-transfer coefficient for the reacting turbulent gas boundary layer is derived and calculated. Later, mass diffusion in the interface zone is allowed to determine the effect of the mass diffusion process on the heat-transfer coefficient. Consequently, the heating problem associated with the laser cutting process and findings are presented in line with the previous studies [15,16].

The main interest for developing the model is to treat the complete problem of interaction of the melting surface material with the boundary layer at which chemical reactions take place due to impinging gas, and seek to describe the behavior of the melting layer under the influence of the turbulent boundary layer developing within the gas-liquid interface. To model the flow of a liquid layer and the effect of assisting gas on the melt flow, including the chemical reactions, two considerations can be made. In the first, the flow situation at the molten layer can be considered to be laminar, and the physical process can be formulated accordingly, which is named a "laminar boundary layer approach" [15]. The second includes the effects of turbulence on the molten flow and chemical reactions, which is named "turbulence boundary layer approach" [16]. In this section, both the laminar boundary and turbulent layer approaches are introduced in line with the previous studies [15,16].

3.10.1 Laminar boundary layer consideration

If a jet of gas, which produces an exothermic reaction, is so directed that it flows on to the region of the workpiece at which a laser beam is focused, not only is there an increase in the rate of cutting, but the accuracy and the fineness of the cut are partly affected by the addition of the gas stream for certain gas stream velocities. This is partly due to the fact that the portion of the workpiece directly affected by the laser beam is an externally narrow strip, and also that the material in this stream is raised very rapidly to a temperature at which it melts or vaporizes, resulting in a very steep temperature gradient in the workpiece. There is, therefore, some minor spreading of the zone, which is at a temperature high enough to take part in the exothermic reaction. For metal cutting, a laser beam is concentrated on a workpiece so that it is necessary for it to be moved relative to the beam. A jet of gas capable of taking part in an exothermic reaction thereby increases the heating of the workpiece and the movement of the jet sweeps away the products of combustion.

In an attempt to investigate the forces exerted by an inert gas jet on the thin molten layer, the equations of motion of the gas flow were solved. In the analysis, the gas flow was assumed to be laminar and the chemical reaction contribution was disregarded. Therefore, an extension of the previous models became necessary. In the light of the previous studies [15–18], this chapter introduces the chemical reaction effects on the heat transfer mechanism occurring in the laminar boundary layer during the laser cutting process. Consequently, this work consists of two parts. These include:

1. The effects of the chemical reaction on heat transfer in the compressible laminar boundary layer are studied theoretically during the laser cutting process. This section is mainly concerned with the effect of a laterally introduced gas upon the laminar boundary layer, disregarding the method by which the gas arrived at the gas-liquid interface. The equations developed account for any number of simultaneously occurring gas phase reactions. Consequently, the theoretical study combines the effect of mass transfer and chemical reactions upon the heat transfer to or from the liquid surface. The study is extended up to sonic speeds of discharging gases, because the shock effect is excluded in this study.
2. The interaction mechanism during the gas assisting CO_2 laser cutting process was experimentally investigated. To achieve this, the gas-assisting cutting action was monitored using a fast shutter speed video camera. A fiber-optic probe was used to record, simultaneously, light emitted from the cutting zone corresponding to entrance and exit surfaces of the workpiece. This provided synchronization of the time base for the video recording. To investigate the effect of the assisting gas pressure on the cutting process, oxygen gas pressure was varied from 75 to 180 kPa. This also made it possible to investigate the chemical reactions occurring during the cutting process. Mild steel and stainless steel sheets at two thicknesses were used as workpieces.

3.10.1.1 Frozen boundary layer approximations

The heat transfer from a boundary layer composed of reacting gas is independent of the location of the reaction zone within the boundary layer to a first order of approximation. This approximation includes the assumption that the transport properties are independent of the boundary layer gas mixture, or at least their variation with the composition is of secondary importance compared to their variation with temperature. Because it is assumed that the chemical reactions are taking place at the liquid-gas interface and the rest of the gas phase is frozen, the composition of the gas mixture throughout the boundary layer will be determined by convection and diffusion of the products and reactants through the boundary layer, the gas species present in the external stream, and the gas species at the interface. Defining a quantity I_f, the sum of the partial enthalpies, as:

$$I_f = \sum_i C_i \left(\int_0^T C_{Pi} dT \right) + \frac{U^2}{2} \tag{3.232}$$

then the total enthalpy is given by:

$$I = I_f + \sum_i C_i h_i^0 \tag{3.233}$$

Differentiating Eq. (3.232) with respect to y (distance normal to the free surface of the target) gives:

$$\frac{\partial I_f}{\partial y} = C_{Pf}\frac{\partial T}{\partial y} + U\frac{\partial U}{\partial y} + \sum_i C_i\left(\int_0^T C_{Pi}dT\right)\frac{\partial C_i}{\partial y} \tag{3.234}$$

where

$$C_{Pf} = \sum_i C_i C_{Pi} \tag{3.235}$$

For an ideal gas, it can be assumed that:

$$C_{Pi} = C_{Pf} = C_P = \text{constant} \tag{3.236}$$

Therefore,

$$\sum_i C_i\left(\int_0^T C_{Pi}dT\right)\frac{\partial C_i}{\partial y} = C_P T\sum_i \frac{\partial C_i}{\partial y} = 0 \tag{3.237}$$

Because

$$\sum_i C_i = 1 \tag{3.238}$$

Thus, to a reasonable approximation:

$$\frac{\partial I_f}{\partial y} = C_{Pf}\frac{\partial T}{\partial y} + U\frac{\partial U}{\partial y} \tag{3.239}$$

and

$$\frac{\partial I}{\partial y} = \frac{\partial I_f}{\partial y} + \sum_i h_i^0\frac{\partial C_i}{\partial y} \tag{3.240}$$

Now, because it is assumed that the chemical reactions take place at the gas-liquid interface, heat transfer to the liquid metal surface consists of two parts: conduction and diffusion, that is:

$$-\dot{q}_g = \left(k\frac{\partial T}{\partial y} + \rho D_{12}\sum_i h_i\frac{\partial C_i}{\partial y}\right)_g \tag{3.241}$$

Eqs. (3.239) and (3.240) give

$$-\dot{q}_g = \left(\frac{k}{C_{Pf}}\frac{\partial I_f}{\partial y} + \rho D_{12}\sum_i h_i\frac{\partial C_i}{\partial y}\right) \tag{3.242}$$

where the velocity of the gas at the surface of the liquid is negligible compared to the velocity at the edge of the boundary layer.

Introducing Eq. (3.240) into Eq. (3.242) gives:

$$-\dot{q}_g = \left(\frac{k}{C_{Pf}}\frac{\partial I}{\partial y} - \frac{k}{C_{Pf}}\sum_i h_i^0 \frac{\partial C_i}{\partial y} + \rho D_{12}\sum_i h_i \frac{\partial C_i}{\partial y}\right)_g \tag{3.243}$$

or

$$-\dot{q}_g = \left(\frac{k}{C_{Pf}}\frac{\partial I}{\partial y} - \frac{k}{C_{Pf}}\sum_i h_i^0 \frac{\partial C_i}{\partial y} + \frac{k}{C_{Pf}} L\sum_i h_i \frac{\partial C_i}{\partial y}\right)_g \tag{3.244}$$

and

$L = \frac{C_P \rho D_{12}}{k}$ is the Lewis number.

However,

$$h_i = \int_0^T C_{Pi} dT + h_i^0 \tag{3.245}$$

or

$$h_i^0 = h_i - \int_0^T C_{Pi} dT + h_i^0 = h_i - C_P T \tag{3.246}$$

and hence,

$$-\dot{q}_g = \left(\frac{k}{C_{Pf}}\frac{\partial I}{\partial y} - \frac{k}{C_{Pf}}\sum_i h_i \frac{\partial C_i}{\partial y} + \frac{k}{C_{Pf}}\sum_i C_{Pf} T \times \frac{\partial C_i}{\partial y} + \frac{k}{C_{Pf}} L\sum_i h_i \frac{\partial C_i}{\partial y}\right)_g \tag{3.247}$$

or

$$-\dot{q}_g = \left(\frac{k}{C_{Pf}}\frac{\partial I}{\partial y} + \frac{k}{C_{Pf}}(L-1)\sum_i h_i \frac{\partial C_i}{\partial y}\right)_g \tag{3.248}$$

It can be seen from Eq. (3.248) that when the Lewis number is unity (i.e., $Pr/S = 1$), then the heat transfer to the liquid undersurface $-\dot{q}_g$ is independent of the location of the chemical reactions if:

(i) K/C_{Pf} is independent of the composition of the mixture;

(ii) The solution of the conservation of the species equation can be found independent of the energy equation.

Fortunately, the Lewis number for most gas mixtures is close to unity and the transport properties for most gas mixtures are relatively insensitive to composition. Therefore, the parameters $C(\rho\mu/\rho_e\mu_e)$ and $f(n)$ are insensitive to composition. It may therefore be concluded that the assumption that the heat transfer is insensitive to the location of the chemical reactions in the boundary layer approximation is then tenable.

3.10.1.2 Boundary layer heat transfer

Eq. (3.10) gives the boundary layer heat transfer to the liquid interface. To calculate $\dot{q}_g$ it is essential to know the variation of I_f, partial enthalpy and C_i, the species mass-fraction, throughout the boundary layer.

The equations used to transform the complex non-linear partial differential equations to ordinary form are:

$$s=\int_0^s \rho_e\mu_e U_e ds \tag{3.249}$$

$$\eta=\frac{\rho_e U_e}{(2\bar{s})^{1/2}}\int_0^s \frac{\rho}{\rho_e}dy \tag{3.250}$$

$$U=U_e f'(\eta) \tag{3.251}$$

$$C=(C_i)_e Z_i(\eta) \tag{3.252}$$

$$I=I_e g(\eta) \tag{3.253}$$

On using Eqs. (3.249)–(3.253), the equations of conservation of species, conservation of energy, and conservation of momentum are found, as follows:

Conservation of species equation:

$$\left(\frac{C}{SZ_i'}\right)'+fZ_i'=\frac{2\bar{s}fZ_i'}{(C_i)_e}\frac{d(C_i)_e}{ds}-\frac{2\bar{s}w_i}{\left[\rho\rho_e U_e^2\mu_e(C_i)_e\right]} \tag{3.254}$$

Conservation of energy equation:

$$\begin{aligned}\left(\frac{C}{Pr}g'\right)'+fg' &= \left(\frac{2\bar{s}f'g}{I_e}\right)+\left[\left(\frac{C}{S}\frac{1}{(L-1)}\right)\left(\sum_i\frac{h_i(C_i)_e}{I_e}Z_i'\right)'\right]\\ &\quad+\frac{U_e^2}{I_e}\left[\left(\frac{1}{Pr}-1\right)Cf'f''\right]'\end{aligned} \tag{3.255}$$

Conservation of momentum equation:

$$(Cf'')'+fff''=\left(\frac{2\bar{s}}{U_e}\right)\frac{dU_e}{ds}\left[(f')^2-\frac{\rho_e}{\rho}\right] \tag{3.256}$$

where

$$C = \frac{\rho\mu}{\rho_e\mu_e} \tag{3.257}$$

The boundary conditions are:

$$\text{at } y = 0, \eta = 0 \quad \text{at } y \to \infty, \eta \to \infty \tag{3.258}$$

$$(Z_i)_0 = (Z)_g = \frac{(C_i)_g}{(C_i)_e} \quad Z_i \to 1 \tag{3.259}$$

$$f'(0) = 0, f(0) = f_g \quad f' \to 1, f \to 0 \tag{3.260}$$

$$g'(0) = g_g \quad g' \to 0 \tag{3.261}$$

$$g(0) = g_g \quad g' \to 1 \tag{3.262}$$

In order to proceed further and reduce the lengthy arguments, a quantity g_f can be introduced, being defined by:

$$g_f = \frac{I_f}{(I_f)_e} \tag{3.263}$$

Hence, Eqs. (3.233), (3.237), and (3.263) yield:

$$I = I_f + \sum_i C_i h_i^0 \tag{3.264}$$

where

$$C_i = (C_i)_e Z_i(\eta) \tag{3.265}$$

and

$$I = I_e g(\eta) \tag{3.266}$$

Therefore, using the above equations, one can write:

$$g = \left[\frac{(I_f)_e}{I_e}\right] g_f + \frac{1}{I_e} \sum_i (C_i)_e Z_i(\eta) h_i^0 \tag{3.267}$$

Using the assumptions that:

(i) frozen external stream composition, $d(C_i)_e / d\overline{s} = 0$
(ii) frozen gas phase $(\dot{w}_i = 0)$
(iii) $dI_e / d\overline{s} = 0$

They result in the following equations:

Species conservation equation

$$\left(\frac{C}{SZ_i}\right)' + fZ_i' = 0 \tag{3.268}$$

Energy conservation equation

$$\left(\frac{C}{Pr}g'\right)' + fg' = \frac{U_e^2}{I_e}\left[\left(\frac{1}{Pr} - 1\right)Cf'f''\right]' \tag{3.269}$$

Introducing Eq. (3.266) into Eq. (3.269) and substituting Eq. (3.268) in the resulting equation gives:

$$\left(\frac{C}{Pr}g_f'\right)' + fg_f' = \frac{U_e^2}{I_e}\left[\left(\frac{1}{Pr} - 1\right)Cf'f''\right]' \tag{3.270}$$

Simplifying the matter further by assuming that Pr is close to but not exactly unity, Eq. (3.270) becomes:

$$\left(\frac{C}{Pr}g_f'\right)' + fg_f' = 0 \tag{3.271}$$

For the flat plate solution $dU_e/d\overline{s} = 0$. Therefore, Eq. (3.271) becomes:

$$(Cf'')' + ff'' = 0 \tag{3.272}$$

In all the equations, primes (′) and (″) mean differentiation with respect to η, which is a variable $\eta = \eta(S, y)$. So, the equations of conservation of species, energy, and momentum all become of the form:

$$\left(\frac{C}{Z}\lambda'\right) + f\lambda' = 0 \tag{3.273}$$

where $Z = S, Pr \rightarrow 1$, for Eq. (3.273), and the boundary conditions become:

$$\text{at } \eta = 0 \quad \text{as} \quad \eta \rightarrow \infty \tag{3.274}$$

$$\lambda = \lambda(0) \quad \text{as} \quad \lambda \rightarrow 1 \tag{3.275}$$

$$\lambda'(0) = \text{constant} \tag{3.276}$$

The solution of Eq. (3.273) is given by [19]:

$$\lambda = [1 - \lambda(0)]\left[\left(\frac{G(\eta, Z)}{G(\infty, Z)}\right)\right] - \lambda(0) \tag{3.277}$$

where

$$G(\eta, Z) = \int_0^{\eta} \frac{Z}{C} \exp\left[-\int_0^{\eta'} \frac{Z}{C} f d\eta''\right] d\eta' \tag{3.278}$$

and

$$\lambda(0)' = \left(\frac{Z}{C}\right)_g \left[\frac{1-\lambda(0)}{G(\infty, Z)}\right] \tag{3.279}$$

Putting $\lambda' = Z_i' g_f'$ and f'' in Eq. (3.283) and using the boundary conditions already mentioned, it can be shown that:

$$Z_i' = \left(\frac{S}{C}\right)_g \left[\frac{1-(Z_i)_g}{G(\infty, S)}\right] \tag{3.280}$$

$$\left(g_f'\right)_g = \left(\frac{Pr}{C}\right)_g \left[\frac{1-(g_f)_g}{G(\infty, Pr)}\right] \tag{3.281}$$

$$(f'')_g = \left(\frac{1}{C}\right)_g \left[\frac{1}{G(\infty, 1)}\right] \tag{3.282}$$

where $G(\infty, Z)$ is given by Eq. (2.378) and evaluated at $\eta \to \infty$. From Eq. (3.251), it can be seen that:

$$\frac{\partial}{\partial y} = \frac{\rho U_e}{(2\bar{s})^{1/2}} \frac{\partial}{\partial \eta} \tag{3.283}$$

$\partial I_f / \partial y$ becomes (using Eq. 3.283):

$$\frac{\partial I_f}{\partial y} = \frac{\partial}{\partial y}\left[g_f (I_f)_e\right] \left[\frac{\rho U_e}{(2\bar{s})^{1/2}}\right] \left\{\frac{\partial}{\partial y}\left[g_f (I_f)_e\right]\right\} \tag{3.284}$$

or

$$\frac{\partial I_f}{\partial y} = \frac{\rho U_e}{(2\bar{s})^{1/2}} (I_f)_e g_f' \tag{3.285}$$

Similarly $\partial C_i / \partial y$ is given by:

$$\frac{\partial C_i}{\partial y} = \frac{\partial}{\partial y}\left[(C_i)_e Z_i\right] = \frac{\rho U_e}{(2\bar{s})^{1/2}} (C_i)_e Z_i' \tag{3.286}$$

Substituting Eqs. (3.279) and (3.286) in Eq. (3.248) gives:

$$-\dot{q}_g = \frac{C}{Pr}\left[\frac{\rho_e \mu_e U_e}{(2\bar{s})^{1/2}}\right]\left[(I_f)_e g_f' + L\sum_i h_i (C_i)_e Z_i'\right]_g \tag{3.287}$$

where

$$L = \frac{\rho D_{12} C_{Pf}}{K} \tag{3.288}$$

$$Pr = \frac{C_{Pf}\mu_g}{K} \tag{3.289}$$

and

$$C = \frac{\rho_g \mu_g}{\rho_e \mu_e} \tag{3.290}$$

Substituting the values of $(Z_i')_g$ and $(g_f')_g$ from Eqs. (3.280) and (3.281) gives:

$$-\dot{q}_g = \left[\frac{\rho_e \mu_e U_e}{(2\bar{s})^{1/2}}\right]\left[\frac{(I_f)_e}{G(\infty, Pr)}\right] \times \left\{\left[1 - (g_f)_g + \frac{1}{(I_f)_e}\right] \times \left[\frac{G(\infty, Pr)}{G(\infty, S)}\right] \sum_i (h_i)_g (C_i)_e \left[1 - (Z_i)_g\right]\right\} \tag{3.291}$$

Because $L = Pr/S$ and using Eqs. (3.233), (3.263), and (3.265) give:

$$I_e - I_g = (I_f)_e\left[1 - (g_f)_g\right] + \sum_i h_i^0 (C_i)_e \left[1 - (Z_i)_g\right] \tag{3.292}$$

Hence, Eqs. (3.291) and (3.292) give:

$$-\dot{q}_g = C_H \rho_e U_e (I_e - I_g) \left\{ \begin{array}{l} 1 + \dfrac{G(\infty, Pr)}{G(\infty, S) - 1} \times \\ \displaystyle\sum_i (C_i)_e h_i^0 \frac{1 - (Z_i)_g}{I_e - I_g} + \frac{G(\infty, Pr)}{G(\infty, S)} \times \sum_i (C_i)_e (h_i)_g \frac{1 - (Z_i)_g}{I_e - I_g} \end{array} \right\} \tag{3.293}$$

where

$$C_H \rho_e U_e = \frac{\rho_e \mu_e U_e}{(2\bar{s})^{1/2} G(\infty, Pr)} \tag{3.294}$$

Eq. (3.294) gives the boundary layer heat transfer rate at the surface of the liquid molten layer for a compressible laminar boundary layer of chemically reactive species. It also accounts for mass transfer effects. The effect of mass transfer enters into the equation through the parameter $G(\infty, Z)$, which is a function of the chemical reactions and mass transfer occurring at the gas-liquid interface.

3.10.1.3 Heat transfer to the liquid metal

To derive equations for heat transfer into liquid metal, which is experiencing chemical reactions at the surface (i.e., gas-liquid interface), an equation can be written for the heat balance at the interface. After considering Fig. 3.15, a heat balance at the gas-liquid interface gives:

$$\dot{q}_{LJ} - (\rho V)_g [h_E(L)]_g = \dot{q}_g - (\rho V) h_g \tag{3.295}$$

where E denotes the chemical symbol for the material undergoing a chemical reaction.

For material E:

$$[h_E(L)]_g + L_v = [h_e(g)]_g \tag{3.296}$$

where L_v = heat of vaporization of E.

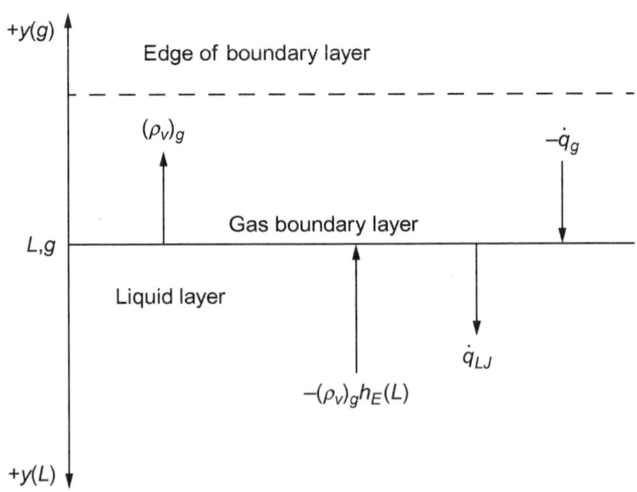

Fig. 3.15 A schematic view of liquid gas interface occurring on the workpiece during the laser cutting process.

Combining Eqs. (3.295) and (3.296) gives:

$$\dot{q}_{LJ} = \dot{q}_g - (\rho V)_g h_g + (\rho V)_g [h_E(g)]_g - (\rho V) L_v \tag{3.297}$$

Introducing Eqs. (3.295) into (3.297) results in:

$$\begin{aligned}\dot{q}_{LJ} = {} & C_H \rho_e U_e (I_e - I_g) - \sum_i h_i^0 \left[(C_i)_e - (C_i)_g \right] \\ & + \frac{G(\infty, Pr)}{G(\infty, S)} \sum_i (h_i)_g \left[(C_i)_e - (C_i)_g \right] - MT1 H_g + MT1 [h_E(g)]_g - MT1 L_v\end{aligned} \tag{3.298}$$

where $MT1$ is a mass transfer parameter defined by:

$$MT1 = \frac{(\rho V)_g}{\rho_e U_e C_H} \tag{3.299}$$

Inserting Eq. (3.295) in Eq. (3.299) gives:

$$\dot{q}_{LJ} = C_H \rho_e U_e \left\{ \begin{array}{l} (I_f)_e - (I_f)_g \\ + \dfrac{G(\infty, \mathrm{Pr})}{G(\infty, S)} \times \left[\sum_i (h_i)_g \left[(C_i)_e - (C_i)_g \right] - MT2 h_g + \dfrac{MT2}{[h_E(g)]_g} \right] - MT1 L_v \end{array} \right\} \tag{3.300}$$

where, in Eq. (3.300), $MT2$ is defined by:

$$MT2 = \frac{G(\infty, Pr)}{G(\infty, S)} MT1$$

Define another form called the chemical enthalpy potential h_c as:

$$h_C = \sum_i (h_i)_g \left[(C_i)_e - (C_i)_g \right] - MT2 \sum_i (h_i)_g (C_i)_g + MT2 [h_E(g)]_g \tag{3.301}$$

Because the enthalpy of a gas mixture is $h_C = \sum_i C_i h_i$, then,

$$h_C = \sum_{i=E} (h_i)_g \left[(C_i)_e - (1 + MT2)(C_i)_g \right] + h_E(g) \left[MT2 - (1 + MT2)(C_E)_g \right] \tag{3.302}$$

where $(C_E)_g = 0$, because all E species leaving the liquid surface are confined to the boundary layer, according to the assumption made earlier.

Combining Eqs. (3.300) and Eq. (3.302) gives:

$$\dot{q}_{LJ} = C_H \rho_e U_e \left\{ (I_f)_e - (I_f)_g + \frac{G(\infty, Pr)}{G(\infty, S)} h_c - MT1L_v \right\} \tag{3.303}$$

In Eq. (3.303), $\dot{q}_{LJ}$ is the heat transfer to the liquid metal in the presence of mass transfer occurring at the surface of the liquid. The term "he" represents the heat released or absorbed due to a chemical reaction among the gas species near the surface.

To determine the heat transfer rate, one has to find the mass fractions of various species at the surface and the edge of the boundary layer (which are used to determine h_c from Eq. 3.302). It is also necessary to develop a method to calculate $(\rho V)_g$ as a function of the surface chemistry.

To calculate $(\rho V)_g$, the following equation can be considered:

$$(\rho V)_g = -\rho_e U_e \mu_e \left[2(2\bar{s})^{1/2} f' \frac{\partial \eta}{\partial s} + \frac{f}{(2\bar{s})^{1/2}} \right] \tag{3.304}$$

At the surface of the liquid,

$$f'(0) = 0 \tag{3.305}$$

Therefore,

$$(\rho V)_g = -\rho_e U_e \mu_e \frac{f(0)}{(2\bar{s})^{1/2}} \tag{3.306}$$

where

$$f(0) = -\left(\frac{C_i}{S}\right)_g \tag{3.307}$$

because

$$C_i = \frac{\rho_i}{\rho} \text{ and } \rho_i = \rho_i \frac{k_B}{m_i} T \tag{3.308}$$

where ρ is the density of the total mixture, m_i mass of species i, and k_B is Boltzmann's constant. Once the pressure at the edge of the boundary layer and the interface temperature are known, species mass fractions C_i can be calculated. Consequently, $f(0)$ and hence $(\rho V)_g$ can be determined.

The findings from the above analysis are briskly given in line with the previous study [15]. The theoretical work proposed examines the interaction between the liquid metal and a gas jet, and develops the relationships between the various parameters influencing the cutting action. The main task of the study is to introduce chemical reactions in the laminar boundary layer developed during a gas-assisting laser cutting

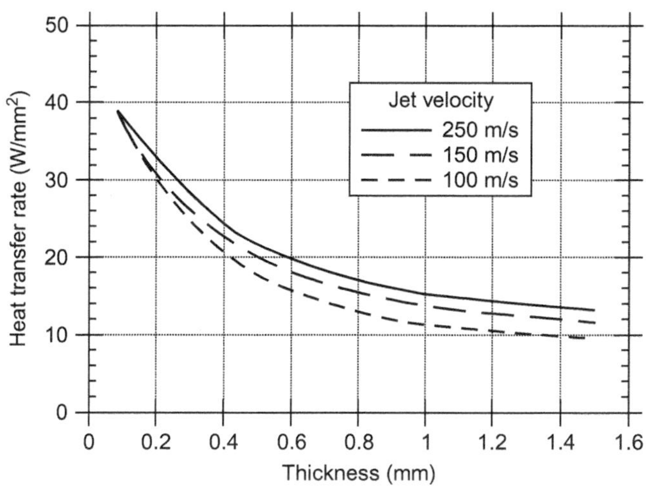

Fig. 3.16 The variation of the heat transfer rate to the liquid metal with workpiece thickness for various jet velocities [15].

process. A computer program was developed based on the analytical solution developed in the present study [15].

Fig. 3.16 shows the heat transfer rate to the liquid metal due to the overall effect (chemical reaction, cooling, and evaporation effects) with material thickness for various jet velocities and for cutting speed as a parameter. It can be seen that, for a given cutting speed, the heat transfer rate to the liquid metal decreases with increasing material thickness. This is due to the fact that, for a given cutting speed and jet velocity, the liquid layer thickness is larger for thicker materials and, consequently, the interface temperature must be maintained at high values. This results in an increased chemical reaction potential. Hence, heat loss to the outer layer is large for thick materials and this explains the downward trend of the curve. For high cutting speeds, the heat transfer rate to the liquid metal decreases for the same reason, i.e., there is a thickening of the liquid layer and the consequent interface temperature rise.

It can be observed from Fig. 3.16 that for a given cutting speed $\dot{q}_{LJ}$, the heat transfer to the liquid metal (due to the chemical reaction, evaporation, and cooling contributions) decreases with material thickness. This is due to the fact that, for a given cutting speed and jet velocity, the liquid layer thickness becomes larger for thicker materials and, consequently, the interface temperature must be maintained at higher values. This results in an increased chemical reaction potential h_c. Although the chemical reaction contribution to $\dot{q}_{LJ}$ is increased, two other factors (evaporation and cooling effects) must be considered for the downward trend of the curve for thicker materials. This is obvious when considering Eq. (3.303), which is:

$$\dot{q}_{LJ} = C_H \rho_e U_e \left\{ (I_f)_e - (I_f)_g + \frac{G(\infty, Pr)}{G(\infty, S)} h_c - MT1L_v \right\} \tag{3.309}$$

As the temperature at the interface (gas-liquid) rises for thicker materials, $(I_f)_g$ increases at a much faster rate than $h_c(I_f)_e$ remains constant because of the assumption that the edge of the boundary layer is at an atmospheric temperature. Therefore, ΔI decreases rapidly with the temperature increase. The other terms in the equation are not very sensitive to the temperature rise because L; is assumed constant, and changes in MTI are negligible. $\frac{G(\infty, Pr)}{G(\infty, S)}$ is dependent on P and S where P and S are assumed constant. Consequently, the heat loss to the outer layer is larger for thicker materials and thus explains the downward trend of the curve.

The heat transfer rate to the liquid metal (henceforth called *the chemical reaction contribution*) is integrated over each material thickness (i.e., 0.5, 1.5, 2.0, and 2.5 mm), and this area under the curve gives the total chemical reaction contribution per mm of the cut width. This quantity is plotted against thickness in Fig. 3.17 for various jet velocities. The total chemical reaction contribution over the cut width increases with increasing jet velocity. This confirms the argument presented earlier, i.e., the decreases in the liquid layer thickness with high gas jet velocity, and the consequent enhancement of conduction to the liquid and the laser cooling rate.

In order to investigate the effect of the $\dot{q}_{LJ}$ chemical reaction contribution, the total absorbed power requirements $(\dot{q}_L)$, where $\dot{q}_L = \dot{q}_S + 0.65\dot{m}(T_1 - T_m)$, is obtained from previous work [15], in which the chemical reaction contribution was disregarded. In this case, the difference $\dot{q}_B = \dot{q}_L - \dot{q}_{LJ}$ is computed. This provides the power required to be absorbed into the liquid layer from the outside source, which is the laser beam power. To calculate the relative contribution of the chemical reaction, the ratio $\dot{q}_B/\dot{q}_L$ is plotted with the material thickness. This is shown in Fig. 3.18. $\dot{q}_B/\dot{q}_{LR}$ increases with growing thickness for a given cutting speed. On the other hand, $\dot{q}_B/\dot{q}_{LR}$ also

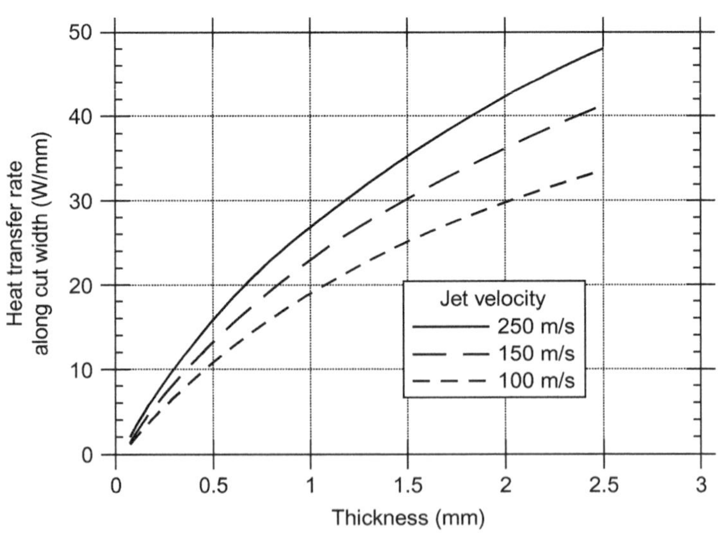

Fig. 3.17 Variation of the heat transfer rate along the cut width obtained after integration of the heat transfer rate to the liquid metal with workpiece thickness for various jet velocities [15].

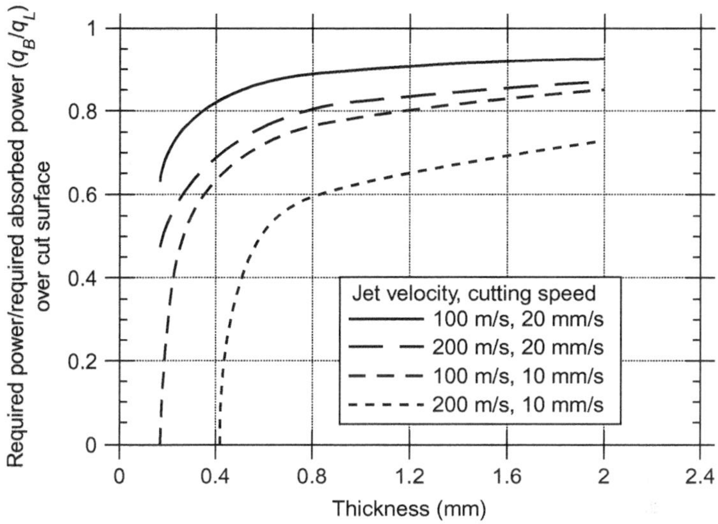

Fig. 3.18 The ratio of required power to required absorbed power $(\dot{q}_B/\dot{q}_L)$ over the cut surface with workpiece thickness for two jet velocities and cutting speeds [15].

increases with increased cutting speed for a given thickness. This indicates that the required absorbed power is large at a high cutting speed. For a given cutting speed (i.e., 10 mm/s) and thickness (e.g., 1 mm), the beam power requirement $(\dot{q}_B)$ is about 76% of the total required power for the 100 m/s case, whereas a 200 m/s jet velocity is only 60%, of the total required power (Fig. 3.18), indicating that the total chemical reaction contribution is greater for high jet velocities. The beam power intensity must be increased for low jet velocities and high cutting speeds for a given thickness. The corresponding curves showing the relationship between laser power intensity and cutting speed could not be plotted in this case because the laminar theory fails to explain the process at higher cutting speeds. The laminar theory only holds true for a speed of about 30 mm/s and for all jet velocities up to sonic.

3.10.2 Turbulence boundary layer consideration

In laser cutting process, a liquid molten layer exists on the surface metal and the assisting gas flows on the top of it, which can be seen from Fig. 3.19. On one side of this liquid metal region, that is, at the gas-liquid interface, the molten metal is dragged downstream by the impinging gas jet, and on the other side, the liquid metal is constantly being created as the solid metal melts at the liquid-solid interface. The arguments related to the turbulence boundary layer are presented in line with the previous study [16].

The solid-liquid interface, known as the melting isotherm, advances into the solid material with the melting velocity V_m. When considering the flow of gas on the liquid molten metal, it can be seen that a melting isotherm is exposed to the gas blowing

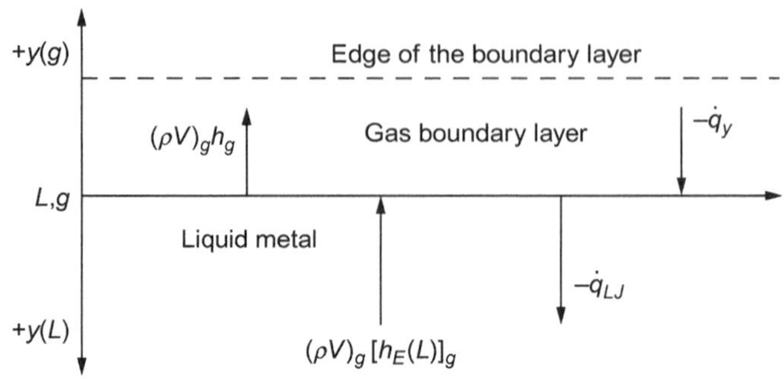

Fig. 3.19 Sketch of gas-liquid interface and boundary layer.

over it. On the top of the liquid molten layer the influence of viscous forces are evident. Because the liquid molten metal flows downstream at a velocity very much less than the gas jet velocity these viscous forces are present and become important at the gas-liquid interface. The interaction of the impinging gas jet with the liquid surface develops a complicated flow field, which results in a turbulent boundary layer on the liquid layer.

Using the model described above, the effects of chemical reactions on the skin friction coefficient and the heat-transfer coefficient can be developed for the reacting turbulent gas boundary layer flowing over the liquid molten surface. A more complex situation is introduced allowing mass transfer and diffusion of species. The effect of the exothermic reaction is treated within the framework of the method.

To develop a model describing the chemical reactions within the turbulent boundary layer created at the gas-liquid interface, the following assumptions are made: (i) all chemical reactions take place in a thin boundary layer at the interface between the gas layer and the liquid under the surface, (ii) no chemical reactions occur within the discharging gas, (iii) all chemical reactions are gas-phase reactions, and (iv) the products of chemical reactions are distributed through the gas layer by diffusion alone.

After taking the above assumptions into account as well as the boundary layer consideration, the equations of the conservation of species, momentum, and energy in the turbulent layer are as follows:

Conservation of species:

$$\rho u\frac{\partial C_i}{\partial s}+\rho v\frac{\partial C_i}{\partial y}=\frac{\partial}{\partial y}\left(\rho D_{12}\frac{\partial C_i}{\partial y}+\rho D_T\frac{\partial C_i}{\partial y}\right) \tag{3.310}$$

where $D_T(s,y)$ is the turbulent eddy diffusion coefficient defined by:

$$D_T(s,y)=\frac{\overline{(\rho v)'C_i'}}{\partial C_i/\partial y} \tag{3.311}$$

where $\overline{(\rho v)'C_i'}$ is the mass transfer due to turbulent fluctuations in the flow and,

$$C_i = \overline{C_i'} + C_i' \tag{3.312}$$

Conservation of momentum:

$$\rho u \frac{\partial C_i}{\partial s} + \rho v \frac{\partial C_i}{\partial y} = \frac{\partial}{\partial y}\left(\mu \frac{\partial C_i}{\partial y} + \varepsilon \frac{\partial C_i}{\partial y}\right) \tag{3.313}$$

where $\varepsilon(s, y)$ is the turbulent eddy viscosity defined by:

$$\varepsilon(s, y) = \frac{\overline{(\rho u)' u'}}{\partial \overline{u} / \partial y} \tag{3.314}$$

Here, $\overline{(\rho u)' u'}$ represents momentum transfer due to turbulent fluctuation in the flow, and

$$u = \overline{u} + u' \tag{3.315}$$

Conservation of energy:

Taking $\partial C_i / \partial y = 0$ outside the boundary layer, because it is assumed that the mass being introduced has the same composition as the gas layer, conservation of energy is described by:

$$\rho u \frac{\partial I}{\partial s} + \rho v \frac{\partial I}{\partial y} = \frac{\partial}{\partial y}\left[\left(\frac{\mu}{P} + \frac{\varepsilon}{P}\right)\frac{\partial I}{\partial y} + \left\{\mu\left(1 - \frac{1}{P}\right) + \varepsilon\left(1 - \frac{1}{P}\right)\right\}^{1/2}\left(\frac{\partial u}{\partial y}\right)^2\right] \tag{3.316}$$

where P_T is the turbulent Prandtl number $(C_p\varepsilon/k)$ and k is the turbulent conductivity.

To obtain the effect of chemical reactions on the heat transfer and skin friction coefficients, the following relationship can be written for the laminar sub-layer and the outer turbulent layer:

$$\frac{2C_{Hc}}{C_f} = B_1\left[(1 + B_1)\left\{1 + B_1\left(\frac{U_L}{U_e}\right)\right\}^{P-1} - 1\right]^{-1} \tag{3.317}$$

where

$$B_1 = \frac{2(\rho v)_g}{\rho_e u_e C_f} \tag{3.318}$$

and C_{Hc}, the convection heat-transfer coefficient, is defined as:

$$-\left(\dot{q}_g\right)_e = C_{Hc}\rho_e u_e\left\{\left(I_f\right)_e - \left(I_f\right)_g\right\} \tag{3.319}$$

To obtain the effect of mass transfer due to diffusion on heat transfer and skin friction coefficients, the following relationship can be written:

$$\frac{2C_{Hd}}{C_f} = B_1 \left[(1+B_1) \left\{ 1 + B_1 \left(\frac{U_L}{U_e} \right) \right\}^{S-1} - 1 \right]^{-1} \tag{3.320}$$

where $S = \mu/(kD_{12})$ (the Schmidt number) and C_{Hd} is the diffusion heat-transfer coefficient, which is defined as:

$$-\left(\dot{q}_g\right)_d = C_{Hc}\rho_e u_e \left[\sum_i h_i^o \left\{ (C_i)_e - (C_i)_g \right\} \right] \tag{3.321}$$

It should be noted that the above relationship is valid for the conditions:

(i) in the laminar sub-layer:

$0 \leq u \leq u_L; \ \varepsilon = k = 0$ and $S_i = S_L$

(ii) in the turbulent region:

$u_L \leq u \leq u_3 \ \ \mu = k = 0$ and $S_i = S_T$

3.10.2.1 Boundary layer heat transfer

In this section, determination of heat transfer from the boundary layer to the liquid surface is carried out. The heat transfer at the interface is:

$$-\left(\dot{q}_g\right) = -\left(\dot{q}_g\right)_e - \left(\dot{q}_g\right)_d \tag{3.322}$$

or

$$-\left(\dot{q}_g\right) = \frac{C_f}{2}\rho_e u_e \frac{2C_{Hc}}{C_f}\left(I_e - h_g\right) \times \left[1 + \left(\frac{C_{Hd}}{C_{Hc}} + 1 \right) \frac{\sum_i h_i^o \left\{ (C_i)_e - (C_i)_g \right\}}{I_e - h_g} \right] \tag{3.323}$$

Eq. (3.323) gives the heat transfer from the boundary layer. Here $2C_{Hc}/C_f$ is given by Eq. (3.317) and $2C_{Hd}/C_f$, is given by Eq. (3.320). Therefore, the ratio of heat-transfer coefficient due to a diffusion and chemical reaction is:

$$\frac{C_{Hd}}{C_{Hc}} = \frac{(1+B_1)\{1+B_1(U_L/U_e)\}^{P-1} - 1}{(1+B_1)\{1+B_1(U_L/U_e)\}^{S-1} - 1} \tag{3.324}$$

where S is the Schmidt number and P is the Prandtl number.

3.10.2.2 Heat transfer to the liquid undersurface

If Fig. 3.19 is considered and an energy balance is applied to the interface (gas-liquid), the net heat transfer from the gaseous boundary layer to the liquid undersurface in the absence of radiation is:

$$\dot{q}_{ij} = -\bar{\dot{q}}_g + (\rho v)_g \{h_E(L)\}_g - (\rho v)_g h_g \tag{3.325}$$

where E is the chemical symbol for the material and L and g represent the liquid and gaseous states of the material. The first term is the diffusive, convective, and conductive heat transfer from the boundary layer to the gas-liquid interface. The second is the counter-current of heat carried toward the interface because of chemical reactions and vaporization. The third term is the heat carried away from the interface into the boundary layer by mass transfer, which occurs at the interface.

Because, initially, it is assumed that the reactions occur in the gas phase, then, the interface material vaporizes and makes reactions with the flowing gas. For the vaporizing material,

$$\{h_E(g)\}_g = (h_E)_g = \{h_E(L)\}_g + L_v \tag{3.326}$$

where L_v is the heat of vaporization of the material. Combining Eqs. (3.323), (3.325), and (3.326), it yields:

$$\dot{q}_{Lj} = \rho_e u_e C_{Hc}\left((I_f)_e - (I_f)_g + \frac{C_{Hd}}{C_{Hc}} \times \left[\sum_i h_i^o \left\{ (C_i)_e - (C_i)_g \right\} - B_2 h_g + B_2 (h_E)_g \right] - B_3 L_v \right) \tag{3.327}$$

where

$$B_2 = B_3 \frac{C_{Hc}}{C_{Hd}} \tag{3.328}$$

and

$$B_3 = \frac{(\rho v)_g}{\rho_e u_e C_{Hc}} \tag{3.329}$$

Because $C_{Pi} = C_P = \text{constant}$ and $C_i = 1$,

$$\sum_i \left(\int_0^T C_{Pi} dT \right) \frac{dC_i}{dy} = 0 \tag{3.330}$$

Also

$$\sum_i h_i \frac{dC_i}{dy} - \sum_i h_i^o \frac{dC_i}{dy} = 0 \tag{3.331}$$

Therefore,

$$\dot{q}_{Lj} = \rho_e u_e C_{Hc} \Delta I \times \left\{ 1 + \left(\frac{C_{Hd}}{C_{Hc}} - 1 \right) \frac{h_c}{\Delta I} - \frac{B_3 L_v}{\Delta I} \right\} \tag{3.332}$$

where

$$\Delta I = \left(I_f\right)_e - \left(I_f\right)_g + h_c \tag{3.333}$$

and h_c is the chemical reaction enthalpy, which is defined as:

$$h_c = \sum_{1 \neq E} (h_i)_g \left\{ (C_i)_e - (1 + B_2)(C_i)_g \right\} + (h_E)_g \left\{ B_2 - (1 + B_2)(C_E)_g \right\} \tag{3.334}$$

Eq. (3.332) shows the heat transfer to the liquid metal due to chemical reactions and mass transfer occurring at the gas-liquid interface when a reacting turbulent layer flows over the liquid molten metal. This includes effects due to chemical reactions, evaporation of metal, and the cooling effect of the assisting gas jet.

3.10.2.3 Method of solution

The equation of heat transfer to the liquid metal, as shown in Eq. (3.340), is the general form for any number of simultaneous reactions. In this section a method of calculating the various quantities in Eq. (3.332) is shown for the simple case of atomic oxygen being blown over the liquid molten metal. Therefore, considering the reaction:

$$E + O \rightarrow E_0 + \Delta Q_{EO} \tag{3.335}$$

where E is the chemical symbol for the metal under investigation and ΔQ_{EO} is the heat released during the reaction. Eq. (3.332) should be utilized to find the net effect of this exothermic reaction upon the heat transfer to the bulk of the liquid metal. The steps for this calculation are as follows.

3.10.2.3.1 Chemical enthalpy potential h_c

Using the equation given in the previous work [20], it can be written that:

$$\begin{aligned} h_c = & \sum_{J=0}^{N} \Delta Q_{EJ} \left\{ \left(\overline{C}_J\right)_e - (1 + B_2)\left(\overline{C}_J - I_{J,EJ} C_{EJ2}\right)_g \right\} \\ & + \sum_{J=0}^{N} \Delta Q_{EJ2} \left\{ \left(\overline{C}_J\right)_e - (1 + B_2)\left(\overline{C}_J - I_{J,EJ2} C_{EJ2}\right)_g \right\} \\ & + \sum_{J=0}^{N} \tfrac{1}{2} \Delta Q_{J2} \left\{ \left(\overline{C}_{J2}\right)_e - (1 + B_2)\left(\overline{C}_{J2}\right)_g \right\} \end{aligned} \tag{3.336}$$

where $B_2 = B_3 C_{Hc} / C_{Hd}$.

For the reaction considered above $J=0$ and J_2 is absent. Therefore, Eq. (3.336) for the present case becomes:

$$h_c = \Delta Q_{EO}\left\{\left(\overline{C}_o\right)_e - (1+B_6)\left(\overline{C}_o - I_{E,EO}C_{EO}\right)_g\right\} \quad (3.337)$$

To calculate h_c, the mass-transfer parameter B_6 and the species mass concentration C must be calculated. The species of mass concentrations (or fractions) C_{EO} and C_O are related to the element mass fraction C_O by the following equations:

$$\left(\overline{C}_O\right)_E = (C_O)_E + I_{O,EO}(C_{EO})_E \quad (3.338)$$

and

$$\left(\overline{C}_O\right)_g = (C_O)_g + I_{O,EO}(C_{EO})_g \quad (3.339)$$

3.10.2.3.2 Species mass fractions

Assuming that the reaction goes to completion at the surface of the liquid layer and that the equilibrium concentrations determine the parameter B_2, the species mass fractions can be found as follows. The equilibrium constant for the reaction is:

$$\frac{P_E P_O}{P_{EO}} = k_1(T) \quad (3.340)$$

for the vaporization of E. Thus:

$$P_E = k_2(T) \quad (3.341)$$

where P indicates the partial pressure.

Combining the above equations, and using Dalton's law of partial pressure, yields:

$$P_E = k_2 \quad (3.342)$$

$$P_{EO} = k_2 P_e \quad (3.343)$$

$$P_O = \frac{k_i P_e - k_1 k_2}{k_1 + k_2} \quad (3.344)$$

The equilibrium constants k can be computed from the relationship:

$$\Delta G^O = -RT \ln k \quad (3.345)$$

where ΔG^O is the change in the standard free energy for the reaction. Once P_E, the pressure at the edge of the boundary layer and the interface temperature are known species, mass fractions C_i can be calculated from the above equations.

3.10.2.3.3 Mass-transfer parameter B_2

B_2 in terms of concentrations at g, that is the interface, can be written as:

$$B_2 = \frac{\left(\overline{C}_E\right)_g}{1-\left(\overline{C}_E\right)_g} = \frac{\left(C_E + I_{E,EO}C_{EO}\right)_g}{1-\left(C_E + I_{E,EO}C_{EO}\right)_g} \tag{3.346}$$

Therefore, if the species concentrations are known B_2 can be calculated. The derivation of B_2 is presented below.

From the equation for the conservation of mass at the interface (gas-liquid):

$$(\rho v)_g + D_{12}\left(\frac{\partial \overline{C}_E}{\partial y}\right)_g = (\rho v)_g\left(\overline{C}_E\right)_g \tag{3.347}$$

The conservation of species yields:

$$\frac{d}{dy}\left\{(\rho v)_g C_i - \rho(D_{12}+D_T)\frac{dC_i}{dy}\right\} = 0 \tag{3.348}$$

Integrating and evaluating the constant of integration at $y=0$ gives:

$$(\rho v)_g C_i - \rho(D_{12}+D_T)\frac{dC_i}{dy} = (\rho v)_g (C_i)_g - \left(\rho D_{12}\frac{dC_i}{dy}\right)_g \tag{3.349}$$

However,

$$\overline{C}_k = \sum I_{i,k} C_i \tag{3.350}$$

Combining Eqs. (3.349) and (3.350), it can been seen that:

$$(\rho v)_g \overline{C}_k - \rho(D_{12}+D_T)\frac{d\overline{C}_k}{dy} = (\rho v)_g \left(\overline{C}_k\right)_g - \left(\rho D_{12}\frac{d\overline{C}_k}{dy}\right)_g \tag{3.351}$$

However, from the equation of motion:

$$(\rho v)_g - \rho(D_{12}+D_T)\frac{dz}{dy} = (\rho v)_g - \left(\rho D_{12}\frac{dz}{dy}\right)_g \tag{3.352}$$

Comparing Eqs. (3.351) and (3.352) and integrating these equations (between e and g) and arrangements yield:

$$\overline{C}_k - \frac{\left(\overline{C}_k\right)_e - \left(\overline{C}_k\right)_g}{z_e - z_g} + \frac{\left(\overline{C}_k\right)_g z_e - \left(\overline{C}_k\right)_e z_g}{z_e - z_g} \tag{3.353}$$

Differentiating Eq. (3.353) and evaluating at $y=0$ for $k=E$ and multiplying by ρD_{12} gives:

$$\left(\rho D_{12}\frac{d\overline{C}_E}{dy}\right)_g = \frac{\left(\overline{C}_E\right)_e - \left(\overline{C}_E\right)_g}{z_e - z_g}\left(\rho D_{12}\frac{dz}{dy}\right)_g \tag{3.354}$$

Heat transfer by diffusion is:

$$\left(-\dot{q}_g\right)_d = \left(\sum h_i^0 C_i\right)_e\left(\rho D_{12}\frac{dz}{dy}\right)_g \tag{3.355}$$

and

$$\left(\dot{q}_g\right)_d = C_{Hd}\rho_e u_e \sum_i h_i^0\left\{(C_i)_e - (C_i)_g\right\} \tag{3.356}$$

Introducing Eqs. (3.355) and (3.356) in Eq. (3.354) results in:

$$\left(\rho D_{12}\frac{d\overline{C}_E}{dy}\right)_g = \left\{\left(\overline{C}_E\right)_e - \left(\overline{C}_E\right)_g\right\}C_{Hd}\rho_e u_e \tag{3.357}$$

Now, combining Eqs. (3.347) and (3.357) gives B_2 in terms of concentrations at g, which is the interface:

$$B_2 = \frac{\left(\overline{C}_E\right)_g}{1-\left(\overline{C}_E\right)_g} = \frac{\left(C_E + r_{E,EO}C_{EO}\right)_g}{1-\left(C_E + r_{E,EO}C_{EO}\right)_g} \tag{3.358}$$

3.10.2.3.4 Determination of value of C_{Hd}/C_{Hc}

B_3 can be written as:

$$B_3 = \frac{C_{Hd}}{C_{Hc}}B_2 \tag{3.359}$$

and the relation between B_3 and B_1 is:

$$B_1 = \frac{C_{Hc}}{C_f/2}B_3 = \frac{C_{Hd}}{C_f/2}B_2 \tag{3.360}$$

Eqs. (3.324), (3.359), and (3.360) can be combined to give:

$$B_2 = \left(1 + B_3\frac{C_{Hc}}{C_f/2}\right)\left(1 + B_3\frac{C_{Hc}}{C_f/2}\frac{u_L}{u_e}\right)^{S-1} - 1 \tag{3.361}$$

$$B_3 = \left(1 + B_3 \frac{C_{Hc}}{C_f/2}\right)\left(1 + B_3 \frac{C_{Hc}}{C_f/2} \frac{u_L}{u_e}\right)^{P-1} - 1 \tag{3.362}$$

and u_L/u_e is given by:

$$\frac{u_L}{u_e} = 12.4\left(\frac{\rho_e}{\rho_g}\frac{C_L}{2}\right)^{1/2} \tag{3.363}$$

From Eqs. (3.346) and (3.359), C_{Hd}/C_{Hc} can be found from the following formula. B_2 is calculated from Eq. (3.346) and Eqs. (3.361) and (3.363) are used to solve for $\left(2B_3C_{Hc}/C_f = B_1\right)$. This is substituted in Eq. (3.366) to obtain B_3. Then C_{Hd}/C_{Hc} is calculated from Eq. (3.359). Other quantities in Eq. (3.332) are, therefore, all known.

3.10.2.4 Findings of turbulent boundary treatment

When examining the contribution of chemical reactions, evaporation, and cooling effects in laser gas-assisted cutting, it is assumed that the reactions between vaporized metal molecules and the assisting gas takes place in a very thin gas boundary layer at the interface between the gas and the liquid metal. It is further assumed that all of the available oxygen reacts with metallic vapor. Therefore, the quantities of heat $\dot{q}_{LJ}$, due to a chemical reaction, evaporation, and cooling are the maximum quantities that are transferred to the liquid metal. The findings of the above analysis are presented below in line with the previous studies:

The variation of $\dot{q}_{LJ}$, the heat transfer to the liquid metal, with material thickness is shown in Fig. 3.20 for different cutting speeds; $\dot{q}_{LJ}$ decreases with increasing material thickness. This is due to the fact that, for a given cutting speed and jet velocity, the liquid layer thickness δ_L is larger for thicker materials and, consequently, the interface temperature (T_L) must be maintained at higher values. This results in an increased chemical reaction potential h_c. However, as the chemical reaction contribution to $\dot{q}_{LJ}$ is increased, two other factors must be considered for the downward trend of the curve for thicker materials. This can be seen more clearly by examining the actual equation for $\dot{q}_{LJ}$, which is:

$$\dot{q}_{Lj} = \rho_e u_e C_{Hc} \Delta I \times \left\{1 + \left(\frac{C_{Hd}}{C_{Hc}} - 1\right)\frac{h_c}{\Delta I} - \frac{B_3 L_v}{\Delta I}\right\} \tag{3.364}$$

where

$$\Delta I = \left(I_f\right)_e - \left(I_f\right)_g + h_c \tag{3.365}$$

As the temperature (T_L) at the interface (gas-liquid) rises for thicker materials, $(I_f)_g$ increases at a much faster rate than h_c. $(I_f)_E$ remains constant because of the assumption that the edge of the boundary layer is at atmospheric temperature. Therefore ΔI

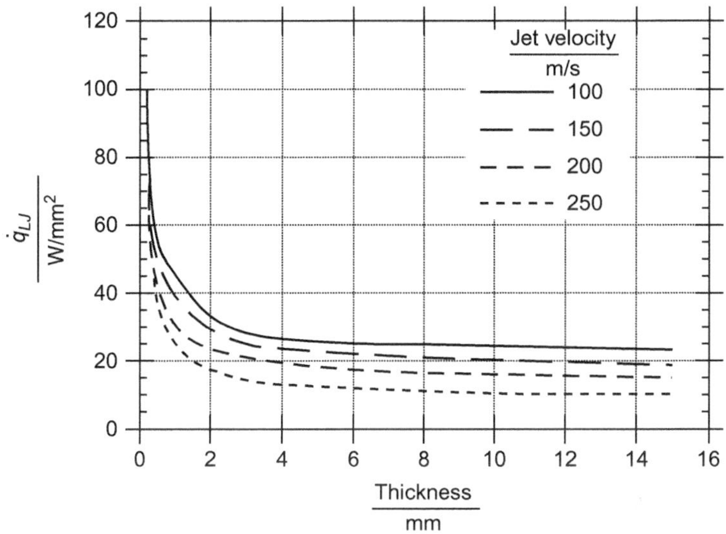

Fig. 3.20 Variation of $\dot{q}_{LJ}$ with the thickness for different jet velocities. Mild steel, oxygen-assisting gas, turbulent flow, $u_L/u_e = 0.6$ [16].

decreases rapidly with temperature increases. The other terms in the equation are not very sensitive to the temperature rise because L_v is assumed constant and changes in B_3 are negligible. The term C_{Hd}/C_{Hc} is dependent on P, S, and B, where P and S are assumed constant. Therefore, the heat loss to the outer layer is larger for thicker materials and thus explains the downward trend of the curve. The quantity $\dot{q}_{LJ}$ decreases for higher cutting speeds for the same reasons; that is, thickening of the liquid layer and consequently an interface temperature (T_L) rise; $\dot{q}_{LJ}$ increases with increasing jet velocities. This confirms the previous discussion of a decrease in liquid layer thickness with a higher gas-jet velocity and the consequent enhancement of conduction to the liquid and lesser cooling rate.

The quantity $\dot{q}_{LJ}$ (henceforth called the chemical reaction contribution) is integrated over each material thickness and this area under the curve gives the total chemical reaction contribution $\dot{q}_{LJT}$ per mm of the cut width. This quantity $\dot{q}_{LJT}$ is again plotted with thickness in Fig. 3.21, and similarly, the area under the required absorbed power intensity ($\dot{q}_L$), the material thickness curve is calculated. This gives the total absorbed power requirement $\dot{q}_{LR}$ also per mm of cut width. In both cases, integrations were performed for several jet velocities (i.e., 100, 150, 200, and 250 m/s). The difference $(\dot{q}_{LR} - \dot{q}_{LTJ})$ is then found. This difference $(\dot{q}_B = \dot{q}_{LR} - \dot{q}_{LTJ})$ is the power required to be absorbed into the liquid layer from the outside source that is the laser beam.

To understand the relative contribution of the chemical reaction and the laser beam, the ratio $(\dot{q}_B/\dot{q}_{LR})$ is plotted with material thickness in Figs. 3.22 and 3.23 for two jet velocities. It can be seen that the beam power requirement ($\dot{q}_B$) is about 72.5% of the total required power for the 100 m/s case, whereas for 250 m/s jet velocity it is only 46% of the total required power. This indicates that the total chemical reaction

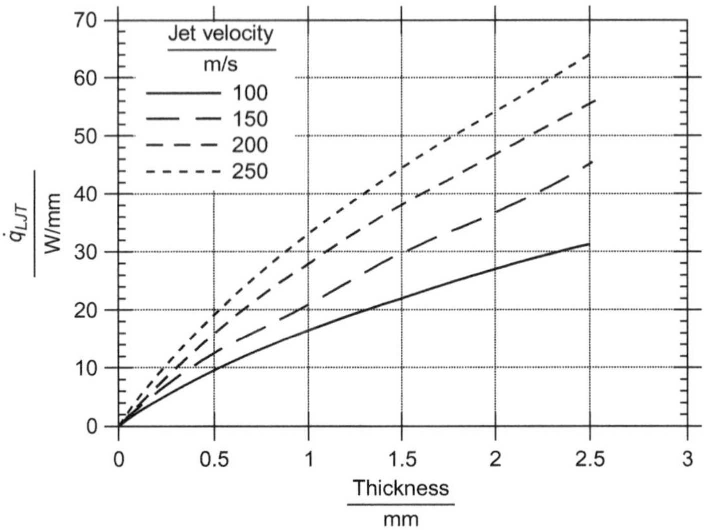

Fig. 3.21 Variation of $\dot{q}_{LJT}$ with the thickness for different jet velocities. Mild steel, oxygen-assisting gas, turbulent flow, $u_L/u_e = 0.6$ [16].

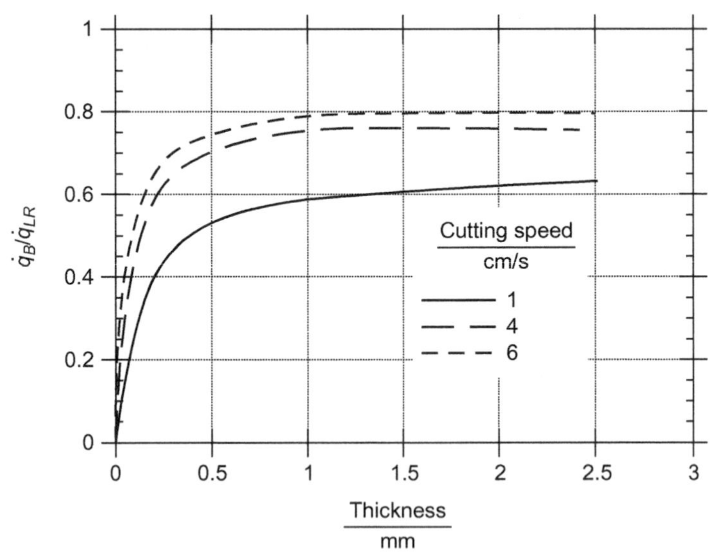

Fig. 3.22 Variation of $\dot{q}_B/\dot{q}_{LR}$ with the thickness for different cutting speeds, $u_e = 100\text{m/s}$, $u_L/u_e = 0.6$ [16].

contribution is substantial for high jet velocities. At higher cutting speeds this ratio ($\dot{q}_B/\dot{q}_{LR}$) is high because the required absorbed power is larger at high cutting speeds, where q_{LJT} is almost independent of the cutting speed. Therefore, for a 6 cm/s cutting speed and for a constant material thickness (i.e., 1 mm), the corresponding values of $\dot{q}_B$ for jet velocities of 100 and 250 m/s are 96% and 91% of the total required

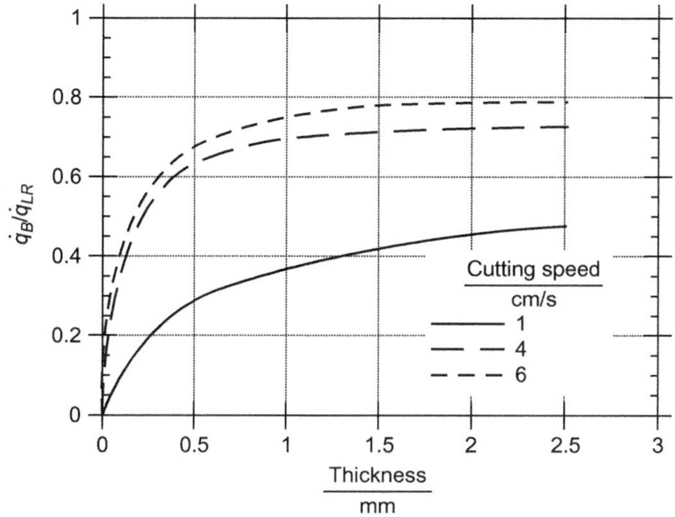

Fig. 3.23 Variation of $\dot{q}_B/\dot{q}_{LR}$ with the thickness for different cutting speeds, $u_e = 250\text{m/s}$, $u_L/u_e = 0.6$ [16].

absorbed power ($\dot{q}_{LR}$), respectively. Similar arguments hold for other material thicknesses and cutting speeds, as can be seen from Figs. 3.21 and 3.22.

Fig. 3.24 shows the variation of required laser power intensity distribution over the cut surface ($\dot{q}_{BC}$), which is found by multiplying the required absorbed power intensity ($\dot{q}_L$) by the ratio ($\dot{q}_{BC}/\dot{q}_L$), with a cutting speed for a material thickness of 1.5 mm and for a particular gas-jet velocity (i.e., 100 m/s). An increment in cutting speed requires

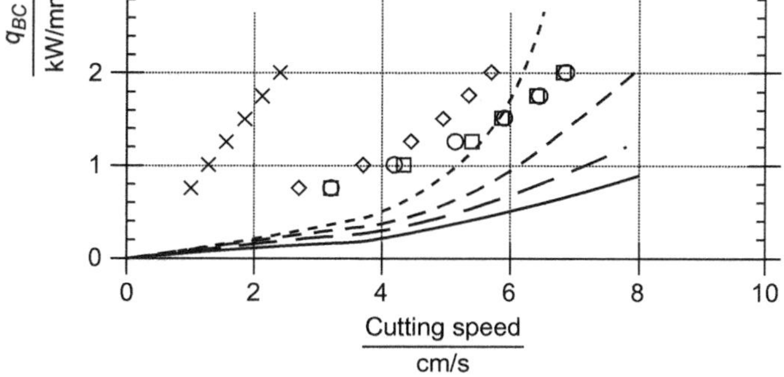

Fig. 3.24 Variation of $\dot{q}_{BC}$ with cutting speed for different jet velocities. Mild steel, oxygen-assisting gas, turbulent flow, thickness = 1.5mm [16].

an increased laser power intensity ($\dot{q}_{BC}$). An increase in power intensity ($\dot{q}_{BC}$) gives rise to a substantial increment in cutting speed initially (up to 6 m/s). Subsequently, the rise continues at a slower rate until a point is reached where the cutting speed saturates for the given jet velocity. This trend is in agreement with experimental results and is attributed to the fact that the mass flow of oxygen is insufficient to oxidize the molten metal at higher speeds, or to remove the molten metal from the cutting zone. As the gas-jet velocity is increased, the laser power requirement is smaller for a given cutting speed and material thickness than in the case of lower jet velocity. This indicates a larger chemical reaction contribution at higher jet velocities. It is evident from experimental observations that some scattering in cutting speed with power intensity occurs. When comparing the experimental results with the theoretical predictions it is evident that experimental results corresponding to an assisting gas-jet velocity of 100 m/s result in relatively low cutting speeds. This may be due to the fact that at low assisting gas-jet velocity the rate of oxidation in the kerf is lower than that of the high assisting gas-jet velocities. In the theoretical model this probabilistic reaction rate is not considered and it is also assumed that all the melting metal in the kerf reacts with the oxygen during the cutting process, which results in a high rate of exothermic heat release, enhancing the cutting process. However, in reality, only some portion of this melting metal reacts with oxygen. Consequently, it results in less exothermic heat being available in the kerf to enhance the cutting process. On the other hand, when the assisting gas-jet velocity increases, this reaction rate increases; therefore, experimental and theoretical results converge. Nevertheless, both curves resulting from theory and experiment generally show similar trends with increasing $\dot{q}_{BC}$.

Fig. 3.25 shows the relationship between $\dot{q}_{BC}$ and cutting speeds for several gas-jet velocities. A similar discussion may be made here as it was for Fig. 3.23. The only

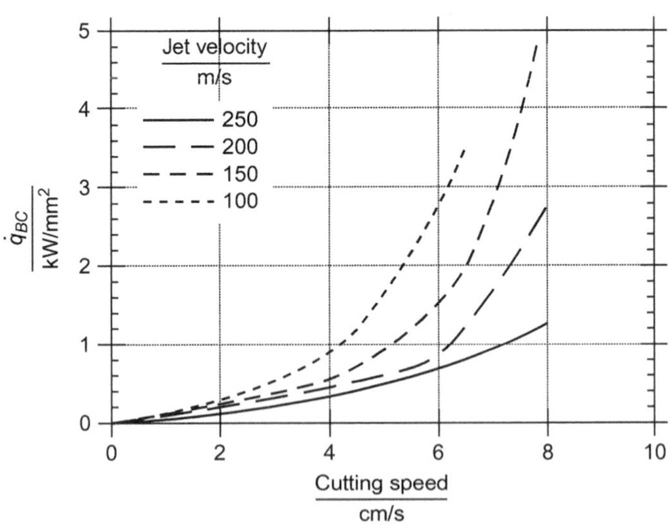

Fig. 3.25 Variation of $\dot{q}_{BC}$ with cutting speeds for different jet velocities. Mild steel, oxygen-assisting gas, turbulent flow, thickness = 2mm [16].

exception is that the laser power ($\dot{q}_{BC}$) requirements are larger than in the 1.5 mm case for a corresponding cutting speed and gas-jet velocity.

3.11 Analysis of dross formation during cutting

In the laser cutting process, an assisting gas is used to either protect the cutting section from the high temperature oxidation reactions or enhance the oxidation reactions to release extra energy in the cutting section. In combustible material processing, such as plastics and wood, inert gas is usually selected. This prevents the exothermic reactions and minimizes the combustion in the cutting sections, in which case, sideways burning is minimized or controlled. In the case of metal processing, oxygen is used as an assisting gas to elevate the energy release, via high temperature exothermic reactions, enhancing the cutting process. In both cases, the assisting gas impinges into a section processed and drags away the molten metal in the cutting section. Depending on the cutting parameters, the molten layer thickness varies. Moreover, the dross of jet consisting of small metal droplets is formed at the onset of molten leaving the cutting kerf. The size of the droplet depends on the liquid metal Reynolds number as well as liquid and gas Weber numbers. Because the mass removal rate from the kerf is associated with the molten layer and dross ejection, investigation into the liquid layer and dross formations during laser cutting becomes essential. The phase change process involving melting and evaporation, as well as the convective and chemical contribution of the assisting gas, can be solved numerically through using the differential form of conservation equations. This requires the solution of transport equations for mass, momentum, and energy, including the species transport. However, such modeling requires comprehensive examination of the problem and, due to unknown properties of species for metallic substrates (vapor properties), this approach is difficult to comprehend. Moreover, a simplified approach based on the equilibrium equations can be introduced when modeling the melting and assisting gas effect. This approach forms the basis for the lump parameter analysis.

In this section, the melt depth and dross size are formulated using a Lump parameter analysis. The findings of the laser cutting experiments are also included to compare the dross size with the predictions in line with the previous study [21].

3.11.1 Formulation of liquid layer thickness and velocity

The influence of assisting gas, including the cooling and exothermic reaction contribution on the cutting process, needs to be modeled prior to dross formulation. This is because the liquid layer film and the liquid film velocity at the liquid film-assisting gas interface should be known prior to droplet diameter predictions. In the case of oxygen as assisting gas, the exothermic reaction contributes to the energy transport process at the interface. Moreover, during the melting process, the assisting gas forms a boundary layer flow over the liquid film surface (a molten metal surface). Heat transfer from the liquid surface to the boundary layer flow occurs at the assisting gas-liquid interface because the assisting gas is at room temperature, and heat transfer from the liquid surface to the solid substrate occurs at the liquid-solid interface. A high -temperature oxidation reaction (the exothermic reaction during which metal oxides are formed) at the melt surface provides excess energy to the laser irradiated region during the machining operation. It should be noted that, in general, nitrogen is used as an assisting gas in the laser machining operation; however, existence of oxygen in the air results in high temperature exothermic reactions taking place in the cutting section.

The species formed after the chemical reactions (the product of exothermic reactions) contribute to heat transfer taking place at the gas-liquid interface. In this case, the heat transfer at the gas-liquid interface due to the high temperature oxidation reaction can be described through the ratio of dimensionless heat transfer coefficients due to diffusion (C_{H_d}) and chemical reaction (C_{H_c}), which was presented as [16]:

$$\frac{C_{Hd}}{C_{Hc}} = \frac{\left[1 + B_1\left(\frac{U_L}{U_e}\right)\right]^{\mathrm{Pr}-1} - 1}{\left[1 + B_1\left(\frac{U_L}{U_e}\right)\right]^{Sc-1} - 1} \tag{3.366}$$

where

$B_1 = \frac{2(\rho U)_g}{\rho_e U_e C_f}$ and Pr and Sc ($Sc = \frac{\upsilon}{D}$, where υ is the thermal diffusivity and D is the diffusion coefficient) are turbulent Prandtl and Schmidt numbers, respectively. U_e is the gas jet velocity at the outer edge of the assisting gas boundary layer (free stream velocity). U_L is the liquid velocity at interface, and C_f is the skin friction coefficient. The rate of heat transfer ($\dot{q}_C$) per unit area of the molten metal excluding the cooling effect while including the chemical reaction contribution of the assisting gas can be written as [16]:

$$\dot{q}_c = \rho_e U_e C_{H_c}\left[\left(\frac{C_{H_d}}{C_{H_c}} - 1\right)h_c\right] \tag{3.367}$$

where U_e is the gas velocity and ρ_e is the gas density at the edge of the boundary layer, h_c is the chemical reaction enthalpy. The ratio $\frac{C_{H_d}}{C_{H_c}}$ is dependent on P , Sc, and B_1 where Pr and Sc are constant [16].

The energy balance associated with the melting process can be simplified through investigating the melting process by lumped parameter technique. Therefore, it is considered that the melt layer generated flows steadily in the direction of the assisting gas due to the drag force developed at the assisting gas-liquid interface.

Consider the melt layer generated at the solid surface during the steady laser heating process (Fig. 3.26). Assume a small fraction (β) of molten metal evaporates from the surface at the assisting gas-melt interface (Fig. 3.27) during the laser heating process. The rate of energy required ($\dot{E}_{req}$) to generate a melt flow rate ($\dot{m}_L$) at the surface of the solid substrate can be written as:

$$\dot{E}_{req} = \dot{m}_L[Cp_s(T_m - T_i) + L_m + \beta L_{ev} + 1.65Cp_m(T_{ev} - T_m)] \tag{3.368}$$

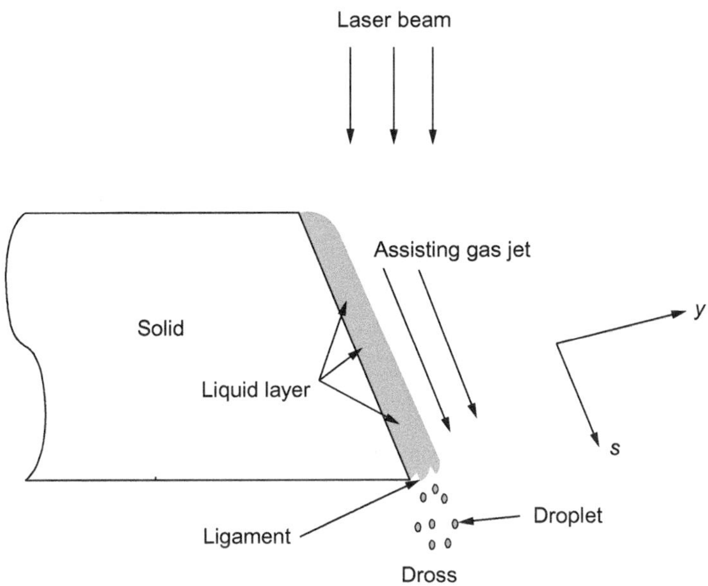

Fig. 3.26 A schematic view of laser cut section and dross ejection [21].

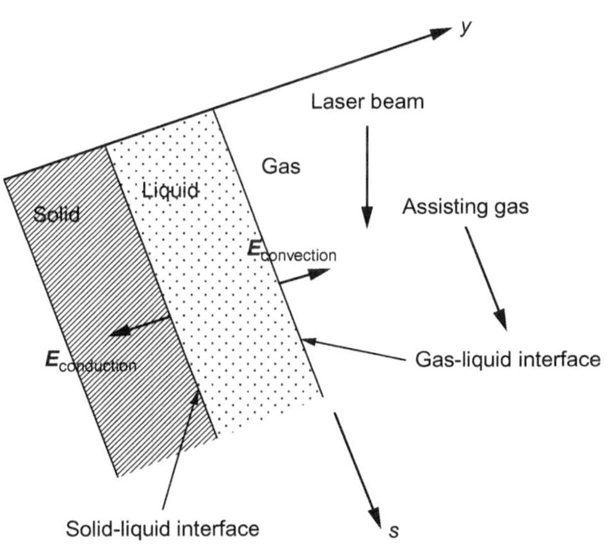

Fig. 3.27 A schematic view of melt cross-section at the kerf surface [21].

where T_i and T_m are initial and the melting temperatures of the solid substrate. L_m and L_{ev} are the latent heating of melting and evaporation of the substrate material. Cp_m and Cp_s are the specific heat capacity of the liquid and solid phases of the substrate. It should be noted that because the liquid layer flows along the surface due to a shear force across the gas-liquid interface, it stores an extra rate of energy. Therefore, the

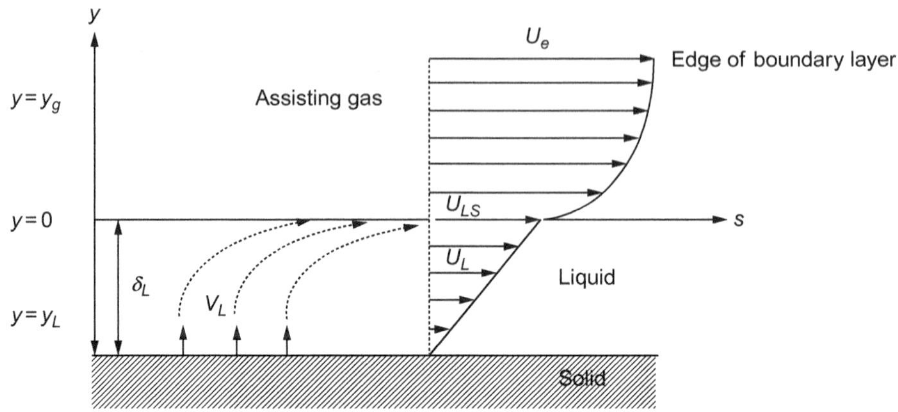

Fig. 3.28 A schematic view of liquid and gas side velocities in the coordinate system.

term $0.65Cp_m(T_{ev}-T_m)$ represents the extra rate of energy due to melt layer flow at the surface, as approximated in the previous study [21].

Moreover, in the laser melting process, the rate of mass generated from solid into liquid at the solid surface can be written as:

$$\lim_{\Delta t\to 0}\frac{\Delta m}{\Delta t}=\frac{dm}{dt}=\dot{m}_L=\frac{d}{dt}(\rho\forall)=\rho_L V_L A \quad (3.369)$$

where $\forall$ is the volume, A is the cross sectional area, ρ_L is the density of melt, and V_L is the liquid velocity along the y-axis (V_L is the velocity of melt generated from solid into liquid as depicted in Fig. 3.28).

The rate of energy convection ($\dot{E}_{Conv}$) from the surface due to the assisting gas is:

$$\dot{E}_{Convective}=\dot{m}_g C_H\left(h_o-h_g\right) \quad (3.370)$$

where $\dot{m}_g$ is the assisting gas mass flow rate, h_o is the total enthalpy ($h_o=\int_{T_{ref}}^{Toe} CpdT+1/2U_e^2$) at the edge of the boundary layer and h_g is the total enthalpy ($h_g=\int_{T_{ref}}^{Tev} CpdT+1/2U_{LS}^2$) at the assisting gas-liquid interface (where U_{LS} is the assisting gas-liquid interface velocity). C_H is the heat transfer factor, which can be obtained from the Reynold's analogy, i.e., [22]:

$$C_H=\frac{1}{2}\frac{C_f}{Pr^{2/3}} \quad (3.371)$$

where C_f is the skin friction coefficient and Pr is the Prandtl number. The evaluation of the skin friction coefficient is based on the type of assisting gas flow. In the case of turbulent boundary layer, C_f can be evaluated as [16]:

$$C_f = 0.0576Re^{-1/5} \tag{3.372}$$

where Re is the Reynolds number of the assisting gas flowing over the molten metal.

Therefore, the rate of energy input ($\dot{E}_{in}$) for melting at the assisting gas-liquid interface can be written as:

$$\dot{E}_{in} = P_o + \int \dot{q}_c dA - \dot{E}_{Convection} \tag{3.373}$$

where P_o is the laser power available at the assisting gas-liquid interface. It should be noted that the reflectivity of the surface is included in the P_o, i.e., P_o represents the laser power available after the reflection. Therefore,

$$\dot{E}_{in} = P_o + \int \dot{q}_c dA - m_g C_H (h_o - h_g) \tag{3.374}$$

However, the rates of heat transfer from liquid surface to assisting gas ($\dot{m}_g C_H (h_o - h_g)$) can be written as:

$$\dot{E}_{gas} = \dot{m}_g C_H (h_o - h_g) = \rho_g U_e A C_H \left[Cp_g (T_{ev} - T_{oe}) + \frac{1}{2}(U_e^2 - U_{LS}^2) \right] \tag{3.375}$$

where A is the area, ρ_g is the assisting gas density, Cp_g is the specific heat capacity of the assisting gas at the edge of the boundary layer, C_H is the heat transfer parameter, and T_{oe} is the gas temperature at the edge of the boundary layer.

Consider the rate of heat transfer from liquid to solid substrate across the liquid-solid interface (Fig. 3.26); in this case, continuity of heat flux at the interface can be employed, i.e.:

$$\dot{E}_{Conduction} = k_m \left[\frac{dT}{dx} \right]_m = Ah(T_m - T_{oi}) \tag{3.376}$$

where k_m and dT/dx is the thermal conductivity of the melt layer and the temperature gradient at the melt-solid interface. A is the area, h is the heat transfer coefficient at the interface, and T_{oi} is the solid temperature at infinitely large depth from the interface. It should be noted that the conjugate heat transfer equation employing Neumann boundary condition (continuity of heat flux and temperature) can be used across the liquid and gas surfaces (interface). However, the Neumann and radiation boundary conditions are omitted, because conjugate solution of the heating situation is simplified using a lump parameter analysis. Therefore, a simplified approach is introduced employing a Dirichlet boundary at the interface (a boundary condition of first of kind).

The rate of energy required for melting is ($\dot{E}_{melt}$):

$$\dot{E}_{melt} = P_o + \int \dot{q}_c dA - \dot{m}_g C_H (h_o - h_g) - Ah(T_m - T_{oi}) \tag{3.377}$$

Setting the rate of energy balance across the melt per unit area yields:

$$\frac{\dot{E}_{req}}{A}=\frac{\dot{E}_{melt}}{A} \tag{3.378}$$

Therefore:

$$V_L=\frac{1}{\rho_L}\frac{\frac{P_o}{A}+\dot{q}_C-h(T_m-T_{oi})-\rho_g U_e A C_H\left[Cp_g(T_{ev}-T_{oe})+\frac{1}{2}\left(U_e^2-U_{LS}^2\right)\right]}{\left[Cp_s(T_m-T_i)+L_m+\beta L_{ev}+1.65Cp_m(T_{ev}-T_m)\right]} \tag{3.379}$$

Consider the flow system as shown in Fig. 3.27. For the steady production of liquid substrate while neglecting the evaporation from the liquid surface during laser heating, continuity equation for incompressible flow can be written as:

$$\frac{\partial U_L}{\partial s}+\frac{\partial V_L}{\partial s}=0 \tag{3.380}$$

where U_L is the velocity of the liquid layer in the s direction. It should be noted that only a small fraction (β) of melt evaporates.

At the assisting gas-liquid interface, shear stresses should be the same. In this case, $\tau_g=\tau_L$ where τ_g the shear stress is exerted by an assisting gas on the liquid layer and τ_L is the shear stress exerted by the liquid layer on the gas layer. If the liquid layer flow is assumed to be treated as extremely small liquid layer thickness, then the shear stress (τ_L) is:

$$\tau_L=\mu_L\frac{\Delta U_L}{\delta_L} \tag{3.381}$$

where δ_L is liquid layer thickness. For zero velocity at the solid surface yields: $\Delta U_L=U_{LS}$, where U_{LS} is the liquid layer velocity at the assisting gas-liquid interface. Eq. (3.389) becomes:

$$\tau_L\simeq\mu_L\frac{U_{LS}}{\delta_L} \tag{3.382}$$

In the case of flow due to inclined surface with pressure variation, the velocity distribution in the liquid layer can be obtained from the Navier-Stokes equation. After assuming the liquid layer depth is considerably smaller than its width, then the Navier-Stokes equation for two-dimensional flow reduces to:

$$\frac{\partial^2 U_L}{\partial y^2}=\frac{1}{\mu}\frac{d}{ds}(P+\gamma z) \tag{3.383}$$

where the inclination angle of the surface is:

$$\sin\theta = -\frac{dz}{ds} \tag{3.384}$$

assuming a linear variation of pressure in the liquid layer with constant slope along the s direction due to considerably thin layer of liquid. It results:

$$\frac{d}{ds}(P+\gamma z) = \text{constant} \tag{3.385}$$

Solving Eq. (3.383) with appropriate boundary condition yields the velocity distribution in the liquid layer, i.e.:

$$U_L = \frac{1}{2\mu}\frac{d}{ds}(P+\gamma z)\left(y^2 - y\delta\right) + U_{LS}\left(1-\frac{y}{\delta}\right) \tag{3.386}$$

Because $\frac{d}{ds}(P+\gamma z) = \text{constant}$ along the S direction, therefore:

$$\frac{\partial}{\partial s}\left(\frac{U_L}{U_{LS}}\right) = \left(1-\frac{y}{\delta_L}\right) \tag{3.387}$$

The continuity equation (Eq. 3.380) can also be written as:

$$\int_0^y \frac{\partial V_L}{\partial s} dy = -\int_0^y \frac{\partial U_L}{\partial s} dy \tag{3.388}$$

or

$$\int_{V_L(0)}^{V_L(y)} dV_L = -\frac{d(\delta_L U_{LS})}{ds}\int_0^y \frac{U_L}{U_{LS}}\frac{dy}{\delta_L} \tag{3.389}$$

Inserting Eq. (3.387) into Eq. (3.389) and after the mathematical arrangements, Eq. (3.389) yields:

$$V_L(y) - V_L(0) = -\frac{d}{ds}\left\{\delta_L U_{LS}\left[\frac{y}{\delta_L} - \frac{1}{2}\left(\frac{y}{\delta_L}\right)^2\right]\right\} \tag{3.390}$$

Because at $y=0$ (at the assisting gas-liquid interface) $\Rightarrow V_L(0)=0$ and $y=-\delta_L$ (at the liquid-solid interface) $\Rightarrow V_L(0)=V_L$, Eq. (3.390) reduces to:

$$V_L = -\frac{d}{ds}\left(\frac{1}{2}\delta_L U_{LS}\right) \tag{3.391}$$

Introducing V_L from Eq. (3.379) into Eq. (3.390), it yields:

$$-\frac{d}{ds}\left(\frac{1}{2}\delta_L U_{LS}\right) = \frac{1}{\rho_L}\frac{\frac{P_o}{A}+\dot{q}_C - h(T_m - T_{oi}) - \rho_g U_e C_H\left[Cp_g(T_{ev}-T_{oe})+\frac{1}{2}(U_e^2 - U_{LS}^2)\right]}{[Cp_s(T_m - T_i)+L_m+\beta L_{ev}+1.65Cp_m(T_{ev}-T_m)]} \tag{3.392}$$

or:

$$-\frac{d}{ds}\left(\frac{1}{2}\delta_L U_{LS}\right) = C_1 - \frac{\rho_g U_e C_H\left[Cp_g(T_{ev}-T_{oe})+\frac{1}{2}(U_e^2 - U_{LS}^2)\right]}{C_2} \tag{3.393}$$

where

$$C_1 = \frac{1}{\rho_L}\frac{\frac{P_o}{A}+\dot{q}_C - h(T_m - T_{oi})}{[Cp_s(T_m - T_i)+L_m+\beta L_{ev}+1.65Cp_m(T_{ev}-T_m)]} \tag{3.394}$$

and

$$C_2 = \rho_L[Cp_s(T_m - T_i)+L_m+\beta L_{ev}+1.65Cp_m(T_{ev}-T_m)] \tag{3.395}$$

Due to the assisting gas flow over the melt surface the heat transfer characteristics are modified accordingly. In this case, the rate of heat transfer from the liquid surface to assisting gas is modified as [22]:

$$\dot{E}_{gas} = Ns^{-1/5} \tag{3.396}$$

where N is a function varying with $\dot{E}_{gas}$ and considered to be constant along S [23]. Therefore Eq. (3.393) becomes:

$$-\frac{d}{ds}\left(\frac{1}{2}\delta_L U_{LS}\right) = C_1 - \frac{Ns^{-1/5}}{C_2} \tag{3.397}$$

Integration of Eq. (3.397) yields:

$$-\left(\frac{1}{2}\delta_L U_{LS}\right) = \int\left[C_1 - \frac{Ns^{-1/5}}{C_2}\right] \tag{3.398}$$

or:

$$-\left(\frac{1}{2}\delta_L U_{LS}\right) = C_1 - \frac{5}{4}\frac{Ns^{4/5}}{C_2} + C_3 \tag{3.399}$$

where C_3 is a constant to be determined from the boundary conditions.

Consider at the tip of the cutting edge, the velocity of the liquid surface is approximately zero, because the liquid layer thickness is negligibly small in this region. Therefore, at $s=0: U_{LS}=0$; therefore, $C_3=0$.

Replacing $\dot{E}_{gas}=Ns^{-1/5}$ in Eq. (3.399) yields:

$$-\left(\frac{1}{2}\delta_L U_{LS}\right)=C_1 s-\frac{5}{4}\frac{\rho_g U_e C_H\left[Cp_g(T_{ev}-T_{oe})+\frac{1}{2}\left(U_e^2-U_{LS}^2\right)\right]s}{C_2} \tag{3.400}$$

Let's define $C_4=\rho_g U_e C_H\left[Cp_g(T_{ev}-T_{oe}\right)$

Eq. (3.400) yields:

$$-\left(\frac{1}{2}\delta_L U_{LS}\right)=C_1 s-\frac{5}{4}\frac{C_4 s}{C_2}+\frac{1}{2}\frac{\left(U_e^2-U_{LS}^2\right)s}{C_2} \tag{3.401}$$

Re-arrangement of Eq. (3.401) yields:

$$\frac{s}{2C_2}U_{LS}^2-\frac{1}{2}\delta_L U_{LS}=C_1 s-\frac{\left(5C_4+2U_e^2\right)s}{4} \tag{3.402}$$

Moreover, to obtain the liquid layer thickness, consider the shear stresses at the assisting gas-liquid interface, which is:

$$\tau_g=\tau_L \tag{3.403}$$

From Eq. (3.382): $\tau_L=\mu_L\frac{U_{LS}}{\delta L}$; and τ_g can be written as:

$$\tau_g=\frac{C_f}{2}\rho_g U_e^2 \tag{3.404}$$

Inserting the shear stresses into Eq. (3.402), the liquid layer thickness becomes:

$$\delta_L=\mu_L\frac{U_{LS}}{U_e}\frac{2}{C_f\rho_g U_e} \tag{3.405}$$

Combining Eqs. (3.402) and (3.405), it yields:

$$U_{LS}=\sqrt{\frac{C_1}{C_5}s-\frac{5}{4}\frac{\left(C_4+2U_e^2\right)s}{C_5}} \tag{3.406}$$

where

$$C_5=\frac{s}{2C_2}-\frac{1}{2}\mu_L\frac{1}{C_f\rho_g}\frac{2}{U_e^2}$$

Substituting Eq. (3.406) into Eq. (3.405) gives the liquid film thickness in the cutting section, i.e.:

$$\delta_L = \mu_L \frac{\sqrt{\frac{C_1}{C_5}s - \frac{5}{4}\frac{(C_4 + 2U_e^2)s}{C_5}}}{\frac{1}{2}C_f \rho_g U_e^2} \tag{3.407}$$

Eqs. (3.406) and (3.407) are used to compute the liquid layer velocity and liquid layer thickness in the cutting section.

3.11.2 Dross formation

The dross formation depends on the liquid properties, such as viscosity, density and surface tension as well as laser and cutting properties such as assisting gas velocity, kerf size, and liquid layer thickness. Moreover, a laminar flow of liquid film breaks up into droplets into an orderly and periodic manner for the liquid Reynolds number less than 1000 [24]. In practice, small satellite droplets can also be formed in between the main droplets due to high aerodynamic forces and instabilities associated with the breakup process. When the disintegration of liquid film occurs, due to the stretching actions, a network of ligaments is formed moving in the direction of the flow. Moreover the existence of a velocity difference between the ligaments and the assisting gas, the droplet formed are comparable with the size of ligaments. The relative importance of viscous and surface tensions, process resisting breakup, and droplet formation is associated with the Ohnesarge number (Z), $Z = \frac{\sqrt{We_L}}{Re_L}$, where We_L and Re_L are liquid Weber and Reynolds numbers, respectively. We_L and Re_L are calculated based on liquid film thickness. Moreover, once the ligaments of liquid are formed, surface tension is the only force that acts to form spherical drops. Consequently, the breakup process is slowed down by high viscosity, and the formation of non-spherical particles becomes unavoidable [23]. The secondary breakup of the initially formed droplet is possible as the assisting gas velocity increases. This process depends on the gas Weber number, which is defined as: $We_g = \frac{\rho_g \Delta U^2 D}{\sigma}$, where Δu is the gas drop velocity, D is the droplet diameter, ρ_g is the gas density, and σ is the surface tension.

The liquid sheet formed at the cut surface is dragged away by the assisting gas. This results in the disintegration of the liquid sheet to form ligaments, as shown in Fig. 3.29. In such a situation, ligaments are produced from a liquid sheet breakdown through dilatation waves [24]. Moreover, the liquid film disintegration and ligament formation were formulated by Dombrowski et al. [25] after assuming the waves grew until they had an amplitude equal to the thickness of the ligament, then a drop was produced. The formulation of drop formation based on the ligament disintegration can be appropriate to a laser cutting process. Droplet diameter (d_D) formulated earlier is adopted herein [24]:

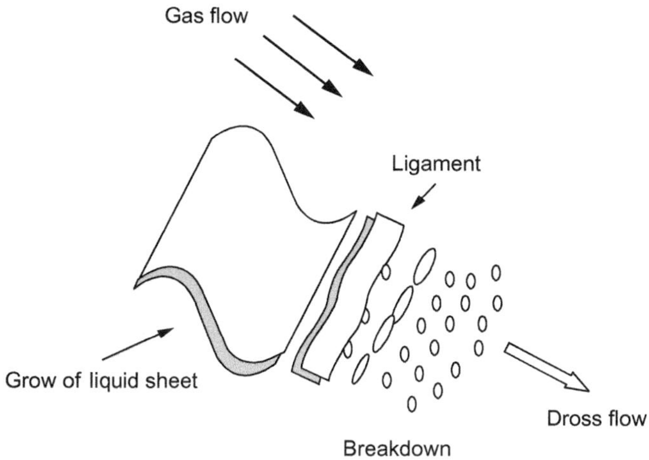

Fig. 3.29 Growth of a liquid sheet and fragmentation, formation of ligament, and breakdown of ligaments into drops [21].

$$d_D = \left(\frac{3\pi}{\sqrt{2}}\right)\delta_L\left(1+\frac{3\mu_L}{\sqrt{\rho_L\delta_L}}\right)^{1/6} \tag{3.408}$$

If a droplet is formed at an early stage in the kerf, it may break up into smaller droplets due to the resistance in a high velocity environment. This process is involved with a secondary atomization. Moreover, it was reported that for small values of Z, ligaments are sheared from the outer part of the droplet to give an octopus-like appearance; in which case, the initial stage involves flattening the droplet into a disc shape [24]. This process results in small size droplets moving downstream of large droplets. The secondary atomization ceases and droplets remain stable below a critical Ohnesorge number, which is $Z^2 < 0.1$, particularly for metallic liquids [24]. The diameter of the droplets (d_s) formed due to the secondary atomization is [24]:

$$d_S = d_D 5.17\frac{Z^{\frac{1}{3}}}{We^{-\frac{2}{3}}} \tag{3.409}$$

where

$$Z = \frac{\sqrt{We_L}}{Re_L} = \frac{\mu_L}{\sqrt{\rho_L\sigma\delta_L}} \tag{3.410}$$

Eqs. (3.408) and (3.409) are used to compute the droplet diameters.

Table 3.3 **Workpiece and assisting gas properties used in the simulations**

Boiling temperature	3133 K
Melting temperature	1810 K
Density of assisting gas	1.97 kg/m^3 at 150 kPa
Density of workpiece	7880 kg/m^3
Fraction of evaporation contribution (β)	0.1
Specific heat capacity of solid	330 J/kg K
Specific heat capacity of melt	330 J/kg K
Specific heat capacity of gas	918 J/kg K
Surface tension	1.5 N/m
Thermal conductivity of molten metal	27 W/mK
Thermal conductivity of solid	52 W/mK
Latent heat of melting	2.72×10^5 J/kg
Latent heat of boiling	6.10×10^6 J/kg
Viscosity	0.9×10^{-2} N s/m^2

3.11.3 Findings of dross ejection

The liquid film formed at the solid surface due to a laser irradiation pulse is modeled and dross formation is analyzed using a Lump parameter analysis. Table 3.3 gives the material properties used. The findings are discussed in line with the previous study [21].

Fig. 3.30 shows the liquid layer thickness predicted with laser power intensity for different assisting-gas velocities. In general, increasing the power intensity enhances the melt thickness, showing that energy absorbed is high and the rate of melting of the solid substrate increases. Moreover, increasing the assisting gas velocity lowers the

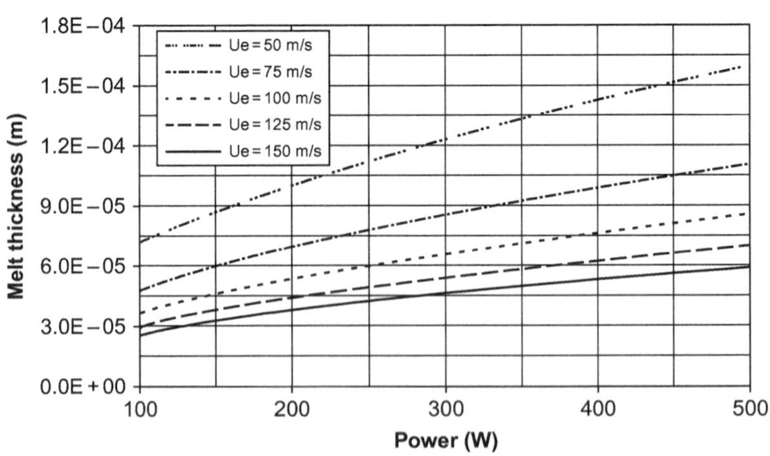

Fig. 3.30 Melt thickness with laser output power for different assisting-gas velocities (Ue) [21].

melt thickness, which is more pronounced at high assisting-gas velocities. It should be noted that the assisting gas has two effects: (i) due to the shearing force between the assisting gas and at the melt surface the melt velocity in the direction of the assisting gas velocity increases, and (ii) it reduces the temperature at the melt surface due to convection heat transfer. Therefore, increasing the melt velocity at the melt surface lowers the melt thickness. In this case, recession velocity of the solid surface remains almost steady due to melt production for constant absorbed irradiation energy by the solid. Consequently, the constant melt production rate, as well as the increasing melt removal rate, due to the increasing assisting gas velocity, lowers the melt thickness. The liquid laser thickness becomes considerably small for the laser power intensity at around 100 W. In this case, the laser power close-to-threshold value for cutting the melt production is considerably small and the liquid layer thickness becomes less sensitive to the laser power within the range of threshold.

Fig. 3.31 shows melt surface velocity with the laser power intensity for different assisting gas velocities. In general, increasing the power intensity improves the rate of melt production, which accelerates the melt flow in both axial and radial directions. However, the transverse speed of the workpiece is associated with the laser beam energy available in the cutting section; in which case, a high cutting speed results in less energy available at the cutting section at an instant of time when the laser beam irradiates the surface in the cutting section. Moreover, the influence of assisting-gas velocity on the melt velocity at the melt surface is more significant than the influence of the laser power intensity. In this case, the melt velocity at the melt surface enhances significantly with increasing assisting-gas velocity. This is because of the drag force generated by the assisting gas at the melt surface, i.e., the momentum transfer to the liquid film by the assisting gas at the interface results in the acceleration of the melt surface lowering the melt thickness as evident from Fig. 3.30.

Fig. 3.32 shows the dross diameter predicted from Eq. (3.408) with laser output power for different assisting gas velocities. Experimentally, obtained dross (droplet)

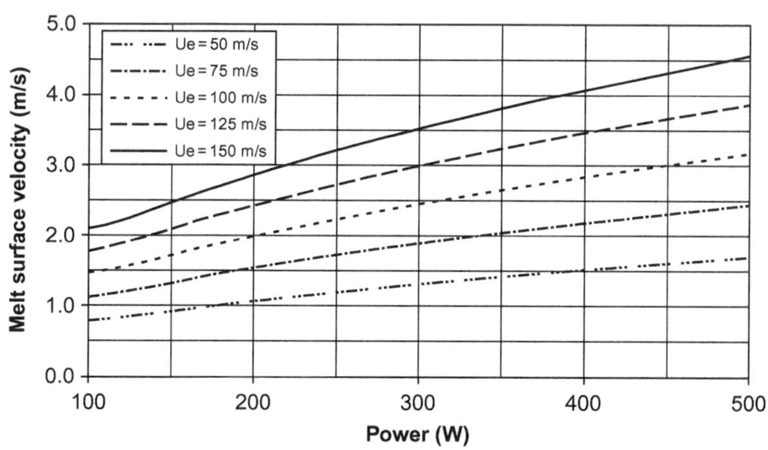

Fig. 3.31 Melt surface velocity with laser output power for different assisting-gas velocities [21].

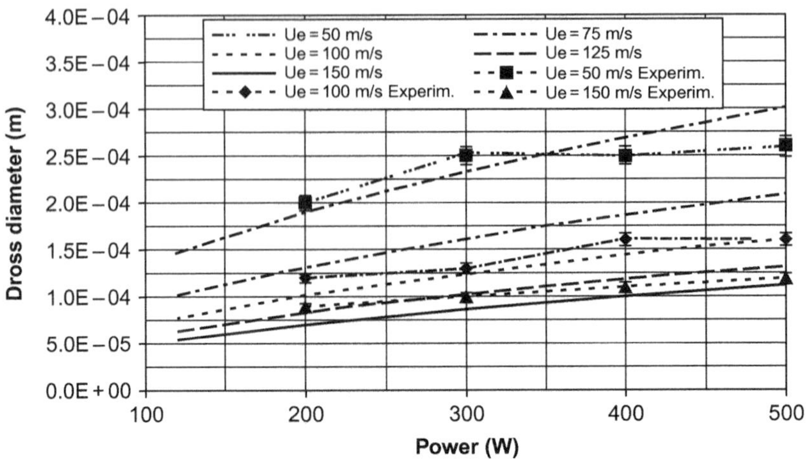

Fig. 3.32 Dross diameter variation with laser output power for different assisting gas velocities due to predictions and experiment [21].

diameters are given for comparison. It should be noted that the droplet diameter presented in the figure is obtained by taking the average diameter of the droplets. The droplet diameter is mainly affected by the droplet Weber and Reynolds numbers, i.e., $Z=\frac{\sqrt{We_L}}{Re_L}$. In this case, Z is on the order of 0.005. Consequently, a ligament is formed prior to the droplet formation. Moreover, increasing the power intensity increases the liquid film thickness, which enhances the droplet diameter. This is particularly true for a low assisting-gas velocity, i.e., increasing the assisting-gas velocity lowers the liquid film thickness. When comparing the average droplet diameter measured and the predictions of droplet diameters, both results agree well. However, at low assisting-gas velocities, some small discrepancies occur. This may be associated with the instabilities (the Kelvin-Helmholtz instability) associated with the droplet formation where $Z=0.0025$ and the assumptions made in the analysis for droplet formation. In the actual situation some droplets may form prior to ligament formation while in the analysis, this situation is overruled and liquid formation prior to droplet formation is assumed.

References

[1] M. Kalyon, B.S. Yilbas, Analytical solution for laser evaporative heating process: time exponentially decaying pulse case, J. Phys. D. Appl. Phys. 34 (2001) 3303–3311.
[2] B.S. Yilbas, M. Kalyon, Analytical approach for entropy generation during a laser pulse heating process, AICHE J. 52 (2006) 1941–1950.
[3] M. Kalyon, B.S. Yilbas, Repetitive laser pulse heating analysis: pulse parameter variation effects on closed form solution, Appl. Surf. Sci. 252 (2006) 2242–2250.
[4] B.S. Yilbas, A closed form solution for temperature rise inside solid substrate due to time exponentially varying pulse, Int. J. Heat Mass Transf. 45 (2002) 1993–2000.

[5] M. Abramowitz, I.A. Stegun, Handbook of Mathematical Functions, Dover, New York, 1965.
[6] B.S. Yilbas, N. Ageeli, M. Kalyon, Laser induced thermal stresses in solids: exponentially time decaying pulse case, Lasers Eng. 14 (1) (2004) 81–101.
[7] B.S. Yilbas, N. Ageeli, Thermal stresses due to exponentially decaying laser pulse and a convection boundary at the surface, Lasers Eng. 16 (2006) 235–265.
[8] B.S. Yilbas, The analysis of CO_2 laser cutting, Proc. Inst. Mech. Eng. B 211 (1996) 223–232.
[9] A. Kar, J.A. Rothenflue, W.P. Latham, Scaling laws for thick-section cutting with a chemical oxygen-iodine laser, J. Laser Appl. 9 (6) (1997) 279–286.
[10] B.S. Yilbas, Experimental investigation into CO_2 laser cutting parameters, J. Mater. Process. Technol. 58 (1996) 323–330.
[11] B.S. Yilbas, Z. Yilbas, Effects of plasma in CO_2 laser cutting quality, Opt. Lasers Eng. 9 (1988) 1–12.
[12] A. Ivarson, J. Powell, J. Kamalu, C. Maginsson, The oxidation dynamics of laser cutting of mild steel and the generation of striations on the cut edge, J. Mater. Process. Technol. 40 (1994) 359–374.
[13] E.A. Brandes, Smithells Metals Reference Book, sixth ed., Butterworths, London, 1983.
[14] B.S. Yilbas, J. Nickel, A. Coban, Effect of oxygen in laser cutting process, Mater. Manuf. Process. 12 (6) (1997) 1163–1175.
[15] B.S. Yilbas, A.Z. Sahin, Oxygen assisted laser cutting mechanism—a laminar boundary layer approach including the combustion process, Opt. Lasers Technol. 27 (3) (1995) 175–184.
[16] B.S. Yilbas, Z. Sahin, Turbulent boundary layer approach allowing chemical reactions for CO_2 laser oxygen-assisted cutting process, Proc. Inst. Mech. Eng. C 208 (1994) 275–284.
[17] M. Vicanek, G. Simon, Momentum and heat transfer of an inert gas jet to the melt in laser cutting, J. Phys. D Appl. Phys. 20 (1987) 1191–1196.
[18] B.S. Yilbas, R. Davies, Investigation into development of liquid layer and formation of surface plasma during CO_2 laser cutting process, Proc. Inst. Mech. Eng. B J. Eng. Manuf. 206 (1992) 287–298.
[19] D. Rosenthal, The theory of moving sources of heat and its applications to metal treatments, Trans. ASME 68 (1946) 849–866.
[20] C.S. Lee, Parametric studies of pulsed-laser cutting of thin metal plates, J. Appl. Phys., Trans. Am. Inst. Phys. 58 (3) (1985) 1339–1343.
[21] B.S. Yilbas, B.J. Abdul Aleem, Dross formation during laser cutting process, J. Phys. D Appl. Phys. 39 (2006) 1451–1461.
[22] W.H. Dorrance, Viscous Hypersonic Flow, McGraw-Hill Company, New York, 1992.
[23] A.J. Yule, J.J. Dunkly, Atomization of Metals for Powder Production and Spray Deposition, Oxford University Press, Oxford, 1994. pp. 15–87.
[24] H. Lubanska, Correlation of spray ring data for gas atomization of liquid metals, JOM 22 (1970) 45–48.
[25] N. Dombrowski, W.R. Johns, The aerodynamic instability and disintegration of viscous liquid sheets, Chem. Eng. Sci. 18 (1963) 203–214.

Laser cutting quality assessment and numerical methods for modeling of cutting

4

4.1 Introduction

In practical applications of laser cutting, the quality of the cut surfaces is the critical factor, and recently, methods of studying the influence on quality of the main process have been developed, which seek to improve quality rather than explain the cutting mechanism. One of the important aspects of the laser cutting section is the formation of striations during the cutting process. However, the formation of striations has received considerable attention, because it is closely related to the laser cutting quality. Striations occur due to the nonsteady nature of the laser cutting process. One of the explanations of this phenomenon may include sideways burning due to low cutting speeds. This effect was observed through high-speed photography of the process. However, some researchers suggested that this was not the case because the melt developed in the kerf was blown out by the assisting gas jet. Moreover, striations can vary in size and frequency along the cut edge; therefore, striation formation cannot be so simply accounted for. It is also proposed that the pulsation in the molten layer prior to it being blown down from the kerf may cause periodic striations to form; in this case, the fluctuations in absorbed laser power could cause both the temperature and liquid layer thickness to oscillate. It was shown in the previous study that absorbed beam-power fluctuations could be the result of the formation of surface plasma. Consequently, pulsating the laser beam power at a similar frequency of striations may minimize the formation of striation patterns. Therefore, it is expected that the cut quality improves when pulsing the laser output power with some order of striation frequency during the cutting process. On the other hand, some deviations in cutting quality can be attributed to slow-process drifts and disturbances. There appear to be many reasons for such drifts including workpiece property variations, assisting gas pressure fluctuations, laser power spatial distribution fluctuations, and optical integrity perturbations. When laser cutting is applied, it is the end product (i.e., is the cut edges) that matters. However, modeling of cutting gives insight into the process, but qualitative assessment of the end product is left obscure. Consequently, quality measurement of the cutting process becomes important when the laser is used as a machine tool; when it's involved with the assessment of kerf width and striation patterns, as well as determining the effecting parameters on measurable variables in terms of cutting geometry in relation to laser cut quality. The cutting parameters include cutting speed, assisting gas pressure, workpiece thickness, and laser-pulsing frequency, while measurable variables are considered as waviness, flatness, and metallurgical changes at cut surface. An international standard for thermal cutting is used when assessing the resulting cut quality.

The Laser Cutting Process. http://dx.doi.org/10.1016/B978-0-12-812982-1.00004-7

4.2 Striation pattern assessments

Although the edge quality of laser cut steel is generally superior to that achieved by mechanical sawing, the periodic striations place a limit on the applicability of laser cutting to the production of finished component edges. Therefore, investigation into striations becomes essential. In this study, an attempt is made to measure the striation frequency at different CO_2 laser cutting parameters. To achieve this, an optical method is introduced to detect the dross ejection characteristics during cutting process. To validate the results of the optical method, surface roughness of the cut edges is measured. The study is extended to include the classification of striation patterns, formed at different laser settings, through the use of a neural network. Consequently, a neural network is developed and its training data is generated mathematically relevant to striation patterns developed. In addition, the microphotography of the resulting cut widths and cut surfaces are accomplished.

Striation patterns can be classified using the neural networks [1]. Consequently, a neural network using a backpropagation algorithm is used to classify the striation patterns in line with the previous study [1].

The topology of the network includes input, hidden and output layers, that is, three layers network. The number of processing elements in the input layer can be selected as 50 while 27 processing elements are employed in the hidden layer [1]. The selection of the processing elements in the hidden layer is based on trial and error, in which case, the network performance is the highest for the training data. According to the possible striation patterns, which may result in different cutting settings, four processing elements are set in the output layer. A typical schematic view of neural network is shown in Fig. 4.1.

The training data is generated using the mathematical formulas, providing that the data generated covers all the striation patterns, which are possibly produced after the cutting process. Therefore, four possible patterns are considered. These include normal, increasing trends, decreasing trends, and cyclic patterns. It is possible to produce a series of numbers resembling the specific patterns. Consequently, 4×240 pattern

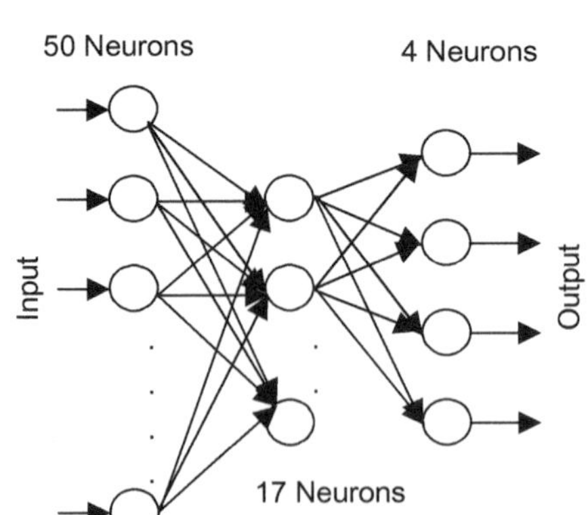

Fig. 4.1 Schematic view of neural network used to classify the striation patterns [1].

strings and their desired outputs are generated. Pattern strings 4×120 are used to train the network and remaining strings kept for testing the network performance. In training and testing processes, both data structures were selected independently. The mathematics involved for generating the patterns are given below:

(1) Normal pattern:

$$R(t) = \xi + r(t)^{*}\varphi \tag{4.1}$$

(2) Increasing trend:

$$R(t) = \xi + r(t)^{*}\varphi + gt \tag{4.2}$$

(3) Decreasing trend:

$$R(t) = \xi + r(t)^{*}\varphi - gt \tag{4.3}$$

(4) Cyclic pattern:

$$y(t) = \xi + r(t)^{*}\varphi + A^{*}\sin(2^{*}\pi^{*}t/T) \tag{4.4}$$

where ξ: mean value of the process variable being monitored (taken as 75); φ: standard deviation of the process (taken as 5); A: amplitude of cyclic variation (taken as 49 or less); g: magnitude of the gradient of the trend (taken as being in the range 0.2–0.5); r: normally distributed random number (between −3 and 3); t: discrete time at which the pattern is sampled (taken as being within the range 0–49); T: period of a cycle (taken as being in the range 3–11 sampling intervals); $R(t)$: sample value at time t.

Patterns in the training set are scaled between 0 and 1 accordingly, with maximum and minimum of input values. Scaling procedure may be given as:

$$\mathrm{S} = (\mathrm{HIGH} - \mathrm{LOW})/(\mathrm{MAX} - \mathrm{MIN})$$

$$\mathrm{O} = (\mathrm{MAX}^{*}\mathrm{LOW}) - (\mathrm{MIN}^{*}\mathrm{HIGH})/(\mathrm{MAX} - \mathrm{MIN})$$

$$\mathrm{SCLINP}(i) = \mathrm{S}^{*}\mathrm{INP}(i) + \mathrm{O}$$

where S: scaling factor; O: offset of the scaling; INP(i): input value for processing element i; SCLINP(i): scaled input value for processing element i; HIGH: maximum of the scaled range (equals 1); LOW: minimum of the scaled range (equals 0); MAX: maximum allowed value for inputs (equals 105); MIN: minimum allowed value for inputs (equals 45). Typical scaled form of patterns generated is shown in Fig. 4.2.

To improve the laser cut quality, evaluation of characteristics governing the laser cutting process is necessary. These characteristics include kerf width and waviness, dross appearance, elemental changes in the melted zone, and microstructural variation in the heat-affected zone. Depending on the ultimate use of the part produced by a laser beam cutting process, evaluation of one or all of these characteristics may apply. The formation of striation is strongly related to the quality of a laser cutting process. However, the mechanisms affecting the striation are not well established. Striations

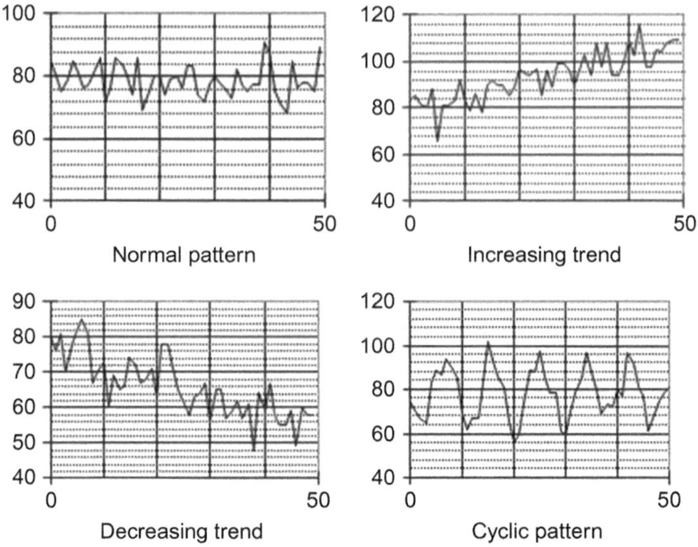

Fig. 4.2 Control chart patterns used in the classification of striation patterns [1].

may result from nonsteady nature of laser cutting. Consequently, the following discussions mainly cover the formation of striation patterns in line with the previous study [1].

Fig. 4.3 shows striation patterns measured along the kerf surfaces for different laser pulsing frequencies. The periodic variation of patterns is evident for all frequencies. The possible explanation of this variation should include the oscillation of the surface liquid layer and surface plasma effects; in this case, the surface plasma may absorb incident laser, which results in less beam power reaching the workpiece surface. Therefore, the dross production rate reduces. However, surface plasma heated via absorption of incident beam may expand rapidly resulting in sufficient beam power reaching the surface for proceeding the cutting process. Consequently, this plasma absorption may be the possible explanation for the periodic oscillation of striation. The frequency of this oscillation is compared with the averaged frequency measured by the periodic structure that appeared on the cut edges. The result of this comparison shows that the striation frequency obtained in both measurements are in good agreement; that is, the optical method gives 250 Hz, while the measurement of periodic structure from the photographs gives 230 Hz for the cutting speed of 5 cm/s. It should be noted here that the measurement of the average wavelength of the striation is achieved by knowing the cutting speed at which the cuts are performed. On the other hand, when the cutting speed reduces to 3 cm/s, the striation frequency also reduces. This may be due to the excessive side burnings which increase the kerf width and the size of the striation.

The network is tested with the mathematically generated test data. The performance of the network is shown in Fig. 4.4. It should be noted that the performance of the network is defined as:

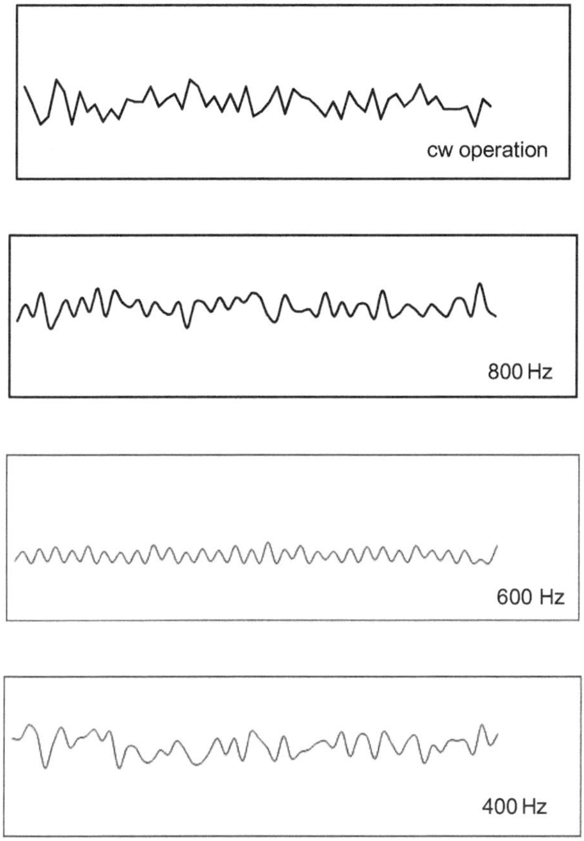

Fig. 4.3 Typical striation traces obtained at different laser pulsing frequencies [1].

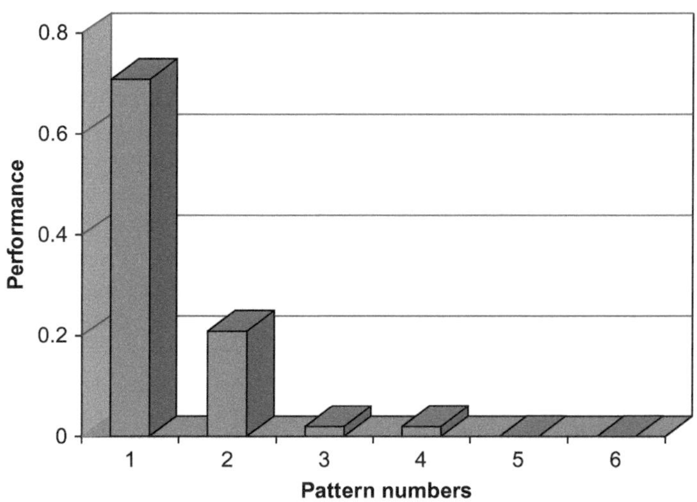

Fig. 4.4 Neural network performances with pattern numbers. Pattern numbers 1, 2, 3 and 4 are normal, cyclic, and increasing and decreasing trends, respectively [1].

$$\text{Performance} = \frac{\text{Number of patterns classified correctly}}{\text{Total number of patterns}} \tag{4.5}$$

The performance of the network is considerably high ($\asymp$98%), indicating that the network learns all the patterns generated, as shown in Fig. 4.4.

After obtaining considerable numbers of striation patterns corresponding to different cutting conditions, they are tested through the network training with the data generated mathematically. In general, at low oxygen pressure and cutting speeds, the striation patterns appear to fit either the increasing or decreasing trends. In this case, a deviation from the normal striation pattern is evident. On the other hand, when cutting speed is high while oxygen pressure is low, the striation patterns, in general, show the indication of a cyclic pattern. In addition, the network confuses the cyclic pattern with an increasing trend. This may be due to the overlapping of the striation patterns occurring at this cutting condition. However, the normal pattern behavior occurs at high cutting speeds and oxygen pressures. In this case, the cutting quality may improve resulting in less sideways burning and the formation of parallel kerf edges.

4.3 Factorial analysis for laser cutting quality assessments

To determine the significance levels of the cutting parameters, a statistical analysis was carried out. To analyze the effect of laser cutting parameters on the resulting cut quality (measurable variables of waviness, flatness, metallurgical changes at the cut surface, and overall quality), it was necessary to design an experiment in such a way that the influence of each parameter could be examined individually, and an interaction. This required a complete factorial design in which selected factors are varied at appropriately selected levels. The mathematical analysis of the factorial design is given in [2]. It should be noted that in the qualitative analysis, the main effects are zero-order effects; that is, independent of each other. However, first order interactions demonstrate that the effects of the interacted factors on observation are dependent on each other [2].

The factors affecting the cut quality may include cutting speed, assisting gas pressure, workpiece thickness, and laser pulsing frequency. An F test was conducted to identify the significant effects of the parameters and their interactions on the various observations describing the cut quality. The terms "most significant," "very significant," and "significant" refer to the minimum correlation levels of 0.99, 0.99 to 0.95 and 0.95 to 0.90 of the factors, respectively.

On the other hand, the qualitative and quantitative analyses can be carried out to identify the affecting laser cutting parameters on the resulting cutting geometry. The mathematical arrangements of these analyses are given below.

4.3.1 *Qualitative analysis*

Consider the case where four factors at four levels of laser cutting parameters are identified. The mathematical expression representing the factorial design can be expressed as:

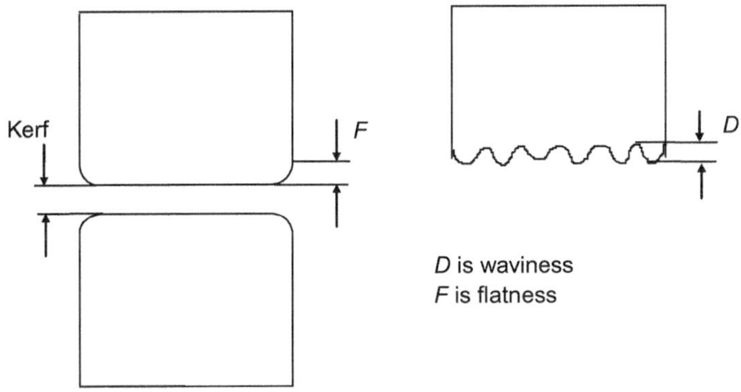

Fig. 4.5 Geometric features of the laser cut section.

$$X_{i,j,k,l} = P_i - F_j - E_k - T_l - P_iF_j - P_iE_k - P_iT_l - F_jE_k - F_jT_l - E_kT_l \\ - P_iE_kF_j - P_iE_kT_l - F_jE_kT_l - \varepsilon_{ijkl} \quad (4.6)$$

where X_{ijkl} is the response (value given to the cut geometric features Fig. 4.5), P_i, F_j, E_k, T_k are the main effects of parameters selected, P_iF_j, P_jE_k, P_iT_l, F_jE_k, F_jT_l, E_kT_l, $P_iE_kF_j$, $P_iE_kT_l$, $F_jE_kT_l$ are the first and second order interactions of selected factors, and ε_{ijkl} is the random error in the experiment provided that i, j, k, l represent the levels of pressure, focus settings, laser output energy, and thickness, respectively. It should be noted that:

$$P_i \times F_j = F_j \times P_i \\ P_i \times F_j \times E_k = F_j \times P_i \times E_k \quad (4.7)$$

The above argument is valid for all the other interactions.

To calculate the main effects of each factor and the effects of interaction, it is necessary to determine the mean corrected sum of squares of response at all levels of factors. However, the mathematical arrangements of qualitative analysis are not given here, but it is referred to [2].

The significance of the effect of factors and interactions is tested by making the initial hypothesis that the mean squares of the effects or interactions are from the same populations as the error mean square, which is that all the mean squares are on a comparable basis with respect to error variance. It should be noted that the mean squares of effects and interactions are linear functions of observation. Because there are no repeats, the remaining mean squares are representing the higher order interactions being used as an estimate of the error variance to test the significance of the effects calculated by means of the F-test [2]. The mean square of the effects is determined by dividing the corrected sum of squares by an appropriate number of degrees of freedom. In this case, the variance ratio is:

$$F_{r,s} = \frac{\text{Mean sum of squares}}{\text{Error mean sum of squares}} \quad (4.8)$$

where r and s are the degrees of freedom of the mean sum of squares and the error mean sum of squares, respectively.

If the calculated variance ratio of any effect exists in the value in the F-test table at the corresponding r and s degrees of freedom, the effect is significant at that level [2].

4.3.2 Quantitative analysis

A quantitative factor is a factor for which the levels may be represented in terms of continuous points. The effect of such a factor is a mathematical function of the levels of the factor. This can be written in a polynomial form [2] as:

$$X = a_o y + a_1 y^2 + a_2 y^3 + \cdots + a_n y^{n+1} \tag{4.9}$$

where X is the response, y is the level of factor, and n is the number of levels selected. The coefficients in the polynomial equation are estimated by the least square technique. The actual values of the constants in the orthogonal form of the polynomial are derived from $\Sigma\beta'X$, $\Sigma\gamma'X$, $\Sigma\delta'X$ multiplying the corresponding quantity of η/γ [2]. However, during the calculations of the variances, it is convenient to drop the multiplier γ/η, which is needed only to be introduced when the actual numerical values of the constants are required.

The details of the mathematical arrangements of the quantitative analysis are given in [2]. The variance ratio is determined similar to that obtained for the quantitative analysis. The variance ratios due to qualitative and quantitative analyses can be determined using the computer codes.

4.3.3 Evaluation of cut quality

The cut quality may successfully be evaluated in terms of waviness, flatness (Fig. 4.5) and overall quality provided that the assessment for the cut quality is based on the international standards for thermal cuts (DIN2310). Height of waviness and flatness can be measured with a reading microscope.

To determine the overall quality of the cut, scores are assigned to waviness, flatness, and metallurgical changes. The scoring scheme includes:

(i) Waviness (D):

$D \leq 30\mu m$ score of 3; $30 \leq D \leq 60\mu m$ score of 2;
$D \geq 60\mu m$ score of 1 are assigned.

(ii) Flatness (F):

$F \leq 20\mu m$ score of 3; $20 \leq F \leq 40\mu m$ score of 2;
$F \geq 40\mu m$ score of 1 are assigned.

(iii) Metallurgical changes (Mm): When no degradation and/or degradation of less than 1% results in any element at the cut surface, a score of 1 is assigned or/else a score of 0 is assigned.

The overall quality is then defined as:

$$Q_{overall} = D + F + Mm \tag{4.10}$$

The main effects of all the parameters selected appear to be significant, moreover, the effects of cutting speed and assisting gas pressure are the most significant. The first order interaction of the parameters such as cutting speed, assisting gas pressure, workpiece thickness, laser pulsing frequency, and thickness have a waviness, flatness, and overall quality are found to be significant. When considering the second order interaction, only cutting speed, assisting gas pressure, and laser pulsing frequency have a significant effect on waviness and overall quality. Consequently, cutting speed, assisting gas pressure, and laser pulsing appear to be the most important parameters in the laser cutting process. In addition, the workpiece thickness has an influence on the cut quality, because its first order interactions are found to be significant.

In the quantitative analysis, the main effects are also zero-order effects where polynomial expressions are introduced for the effect of the factors. In the case of waviness, flatness and overall quality, the linear effects of thickness, assisting gas pressure, the quadratic effects of cutting speed, and laser pulsing frequency are found to be most significant; that is, the relationship between observation and thickness as well as assisting gas pressure are linear, but a quadratic relationship is obtained for cutting speed and laser pulsing frequency. Interactions indicate that the main effects of factors vary at different levels of interacting parameters. For example, in the case of overall quality, the interaction of the linear effect of the thickness and quadratic effect of cutting speed is found to be significant. However, similar arguments can apply for flatness and metallurgical changes. It should be noted that in the case of metallurgical changes, the linear effects of cutting speed and assisting gas pressure is found to be significant, moreover, the first order interactions indicate that only the linear effects of cutting speeds and laser pulsing frequency is found to be significant. The overall quality increases at certain parameter level combinations. However, this combination of parameter levels changes as the workpiece thickness varies. Consequently, the optimum values of the parameters exist for improved overall quality and it depends upon the levels of the parameters used in the cutting process.

A typical example for the F-test results is given in Table 4.1 for waviness (D), out-of-flatness (F), and overall quality for laser cut geometry of Inconel 800 HT alloy [3]. The main effects and first order interactions of parameters (X) are given, and second and higher order interactions are combined with the common error in the analysis. U, P, T, and F represent the cutting speed, laser output power, workpiece thickness, and focus setting of focusing lens, respectively.

Cutting speed: The main effect of cutting speed on the out-of-flatness and overall quality is found to be the most significant while its effect on the waviness is very significant. This indicates that increasing or decreasing cutting speed mostly influences the out-of-flatness as well as overall quality. Moreover, first order interaction of cutting speed and laser power output has a very significant effect on the waviness and the overall quality. The other first order interaction of cutting speed has only a significant effect on the waviness, out-of-flatness, and the overall quality. It should be noted that

Table 4.1 Quantitative analysis results for laser cut geometry of Inconel 800 HT alloy

	Waviness (*D*)	Out-of-flatness (*F*)	Overall quality
U	0.95	0.99	0.99
P	0.99	0.95	0.95
T	0.95	0.99	0.95
F	0.95	0.95	0.95
UXP	0.95	0.9	0.95
UXT	0.9	0.9	0.9
UXF	0.9	0.95	0.9
PXT	0.9	0.9	0.9
PXF	0.95	0.95	0.95
TXF	0.9	0.9	0.9

the first order interactions represent the influence of both interacting parameters on the measurable response or assessing property. Consequently, cutting speed and laser power together influence the waviness considerably.

Laser power intensity: Laser power intensity has a significant effect on the waviness and a very significant effect on the out-of-flatness and overall quality. This may indicate that the influence of laser output power on the waviness is more pronounced as compared to its effect on the out-of-flatness and overall quality. The first order interaction of the laser power intensity and cutting speed as well as pulsing frequency have a very significant effect on the waviness, out-of-flatness, and overall quality. This indicates that both laser power intensity and pulsing frequency have a coupled effect on the waviness, out-of-flatness, and overall quality.

Thickness: Thickness of the workpiece influences mostly the out-of-flatness; its effect on the waviness and overall quality is very significant. However, the first order interactions of thickness with other parameters have a significant effect on the waviness, out-of-flatness, and overall quality.

Pulsing frequency: Pulsing frequency has a significant effect on the waviness, but a very significant effect on the out-of-flatness and overall quality. The first order interactions of pulsing frequency and power intensity have a very significant effect on the waviness, out-of-flatness, and overall quality. Because the pulsing frequency and laser power intensity are related to each other (intensity pulsing with the frequency), it is expected that the coupling of laser power intensity and pulsing frequency has influence on the waviness, out-of-flatness, and overall quality. It can also be noted from the optical measurement that the striation frequency varies once the laser power intensity and pulsing frequency are varied.

4.4 Numerical investigation of the laser cutting process

Lasers can be considered to be an effective tool for cutting metallic materials. Depending on the laser pulse energy and duration, laser cutting can be classified as a nonconduction limited heating processes. However, in the solid phase heating, prior to key-hole formation, the heating process can be described as conduction-limited heating. In the case of conduction-limited heating, the substrate material surface is heated below the melting temperature of the substrate material and a high rate of cooling at the surface modifies the microhardness and stress levels in the heated surface vicinity. However, phase change takes place during the laser nonconduction limited heating of the substrate material. On the other hand, the thermal stress distribution and its levels can be controlled via introducing multibeam heating at the surface. In this case, the location of the laser spots and their intensities at the workpiece surface become important for the stress distribution and its levels in the irradiated workpiece. Experimental investigation of laser heating and thermal stress formation is difficult and expensive due to its small irradiated area and requirement of state-of-art equipment. However, model studies provide insight into the physical process during the laser treatment process and require less effort and low cost. Consequently, modeling of laser multibeam heating of metal surfaces becomes fruitful.

Formulation and numerical analysis related to laser heating in relation to laser cutting is presented below. The findings are discussed in line with the previous study [4].

4.4.1 Heating analysis

The laser heating is formulated using the Fourier heating law for constant speed workpiece, The enthalpy equation governing the laser heating process can be written as:

$$\nabla \cdot (\rho U H) = \nabla \cdot (k \nabla T) + S_{o1} + S_{o2} + S_{o3} \tag{4.11}$$

where H is the enthalpy, ρ is the density, U is the laser beam scanning velocity, k is the thermal conductivity, and S_{o1} is is the volumetric source term, resembling the absorption of the laser beam due to the first laser spot, S_{o2} is the source term due to the second spot, and S_{o3} is the source term due to the third spot, respectively.

The laser beam axis overlays along the z-axis (Fig. 4.6) and the laser beam scans at the surface along the x-axis. The volumetric source resembling the absorption of the incident beam is:

$$S_{o1} = I_{o1}\delta(1 - r_f)\exp(-\delta z)\exp\left(-\left(\frac{\sqrt{(x-x_1)^2 + (y-y_1)^2}}{a}\right)^2\right)$$

and

$$S_{o2} = I_{o2}\delta(1 - r_f)\exp(-\delta z)\exp\left(-\left(\frac{\sqrt{(x-x_2)^2 + (y-y_2)^2}}{a}\right)^2\right)$$

and

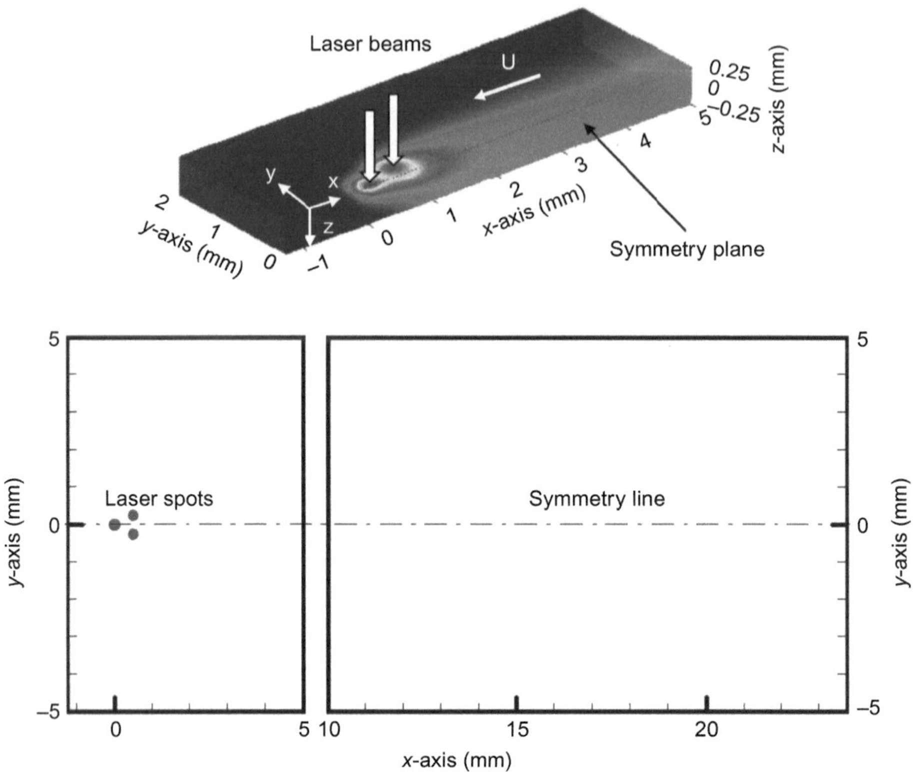

Fig. 4.6 Schematic view of three-beam laser heating of a moving sheet and the coordinate system [4].

$$S_{o3} = I_{o2}\delta\left(1 - r_f\right)\exp\left(-\delta z\right)\exp\left(-\left(\frac{\sqrt{(x-x_3)^2 + (y-y_3)^2}}{a}\right)^2\right) \tag{4.12}$$

where I_{o1} is the peak intensity at the first, and I_{o2} are peak intensities at the second and third spots, δ is the absorption coefficient, a is the Gaussian parameter, r_f is the surface reflectivity, x and y are the axes (Fig. 4.6), x_1 is the distance between the origin of the coordinate system and the first beam center, and x_2 is the distance between the origin of the coordinate system and the second beam center along the x-axis (Fig. 4.6). Moreover, y_1 is the distance between the origin of the coordinate system and the second beam center along the y-axis, and y_2 is the distance between the origin of the coordinate system and the third beam center along the y-axis. It should be noted that $|y_1| = |y_2|$. The laser beam axis is the z-axis (Fig. 4.6). The laser intensity is kept the same for the second and the third spots in the simulations.

In order to solve Eq. (4.1), two boundary conditions for each principal axis should be specified. Convection heat loss is the same as the conduction flux at the surface; that is, the convection heat flux across the irradiated surface is:

$$k\frac{\partial T(x,y,0,t)}{\partial z}=h_t(T_s-T_\infty) \tag{4.13}$$

where h_t is the heat transfer coefficient at the free surface. The heat transfer coefficient predicted earlier [5] is used in the present simulations across the heated spots (h_t = 3000 W/m^2 K) due to the jet impingement and the other regions at the surface, the natural convection is considered (h_t = 10 W/m^2 K) to account for the cooling. It should be noted that the high value of the heat transfer coefficient at the free surface of the workpiece is due to the assisting gas jet impingement at the surface.

The other boundary conditions, therefore, are:

$$z \text{ at bottom surface}: \; z=z_{th}: \frac{\partial T(x,y,z_{th},t)}{\partial z} = \frac{h_t}{k}(T_{s,b}-T_\infty) \text{ (specified)} \tag{4.14}$$

where z_{th} is the workpiece thickness, h_t is the heat transfer coefficient due to natural convection (h_t = 10 W/m^2 K), and $T_{s,b}$ is the surface temperature at the workpiece bottom surface.

$$x \text{ at infinity } (\pm\infty): \; x=\pm\infty: T(\pm\infty,y,z,t)=T_o \text{ (specified)} \tag{4.15}$$

$$y \text{ at outer edge}: \; y=y_{width}: \frac{\partial T(x,y_{width},z,t)}{\partial y}=\frac{h}{k}(T_{s,w}-T_\infty) \tag{4.16}$$

where y_{width} is the half width of the workpiece and $T_{s,w}$ is temperature at the outer edge of the workpiece.

$$y \text{ at the symmetry plane}: \; y=0: \frac{\partial T(x,0,z,t)}{\partial y}=0 \tag{4.17}$$

4.4.2 Thermal stress analysis

The elastic strain increment vector, $\{\Delta\varepsilon^{el}\}$ is related to the stress increment vector, $\{\Delta\sigma\}$ by Hooke's law:

$$\{\Delta\sigma\}=[D]\{\Delta\varepsilon^{el}\} \tag{4.18}$$

where $[D]$ contains the elastic constants related to temperature-dependent elastic modulus, E and Poisson's ratio, ν. The total strain vector, $\{\Delta\varepsilon\}$, may be expressed as follows:

$$\{\Delta\varepsilon\} = \{\Delta\varepsilon^{el}\} + \{\Delta\varepsilon^{th}\} + \{\Delta\varepsilon^{pl}\} \tag{4.19}$$

where $\{\Delta\varepsilon^{el}\}$ is the elastic strain increment vector, $\{\Delta\varepsilon^{th}\}$ is the thermal strain increment vector, $\{\Delta\varepsilon^{pl}\}$ is the plastic strain increment vector. The incremental thermal strain vector, $\{\Delta\varepsilon^{th}\}$ arises from the volume changes that accompany the temperature increment, ΔT, which is calculated by the thermal analysis. It is normally accounted for in stress analyses through a temperature-dependent differential thermal expansion coefficient, $\alpha(T)$. The differential and total thermal expansion (α_T) coefficients are related to each other through:

$$\alpha_T(T) = \frac{1}{T - T^o}\int_{T^o}^{T} \alpha(T)dT \tag{4.20}$$

where T^o is a reference temperature designating the point at which the material exhibits no dilatational strain.

The coupled thermal stress analysis required to identify the displacement-strain relations, which are expressed in the dimensionless form, are as follows [6]:

$$\overline{\varepsilon}_{xx} = \frac{\partial\overline{u}}{\partial\overline{x}},\quad \overline{\varepsilon}_{yy} = \frac{\partial\overline{v}}{\partial\overline{y}},\quad \overline{\varepsilon}_{zz} = \frac{\partial\overline{w}}{\partial\overline{z}} \tag{4.21}$$

$$\overline{\varepsilon}_{xy} = \frac{1}{2}\left(\frac{\partial\overline{u}}{\partial\overline{y}} + \frac{\partial\overline{v}}{\partial\overline{x}}\right),\quad \overline{\varepsilon}_{yz} = \frac{1}{2}\left(\frac{\partial\overline{v}}{\partial\overline{z}} + \frac{\partial\overline{w}}{\partial\overline{y}}\right),\quad \overline{\varepsilon}_{zx} = \frac{1}{2}\left(\frac{\partial\overline{u}}{\partial\overline{z}} + \frac{\partial\overline{w}}{\partial\overline{x}}\right) \tag{4.22}$$

An exact implementation of Newton's method involves a nonsymmetrical Jacobian matrix, which is a stress-strain relation in dimensionless form, illustrated in the following matrix representing the coupled equations [6]:

$$\begin{Bmatrix} \overline{\sigma}_{xx} \\ \overline{\sigma}_{yy} \\ \overline{\sigma}_{zz} \\ \overline{\sigma}_{yz} \\ \overline{\sigma}_{zx} \\ \overline{\sigma}_{xy} \end{Bmatrix} = \frac{\overline{E}}{(1+v)(1-2v)} \begin{bmatrix} 1-v & v & v & 0 & 0 & 0 \\ v & 1-v & v & 0 & 0 & 0 \\ v & v & 1-v & 0 & 0 & 0 \\ 0 & 0 & 0 & 1-2v & 0 & 0 \\ 0 & 0 & 0 & 0 & 1-2v & 0 \\ 0 & 0 & 0 & 0 & 0 & 1-2v \end{bmatrix} \begin{Bmatrix} \overline{\varepsilon}_{xx} \\ \overline{\varepsilon}_{yy} \\ \overline{\varepsilon}_{zz} \\ \overline{\varepsilon}_{yz} \\ \overline{\varepsilon}_{zx} \\ \overline{\varepsilon}_{xy} \end{Bmatrix} - \begin{Bmatrix} 1 \\ 1 \\ 1 \\ 0 \\ 0 \\ 0 \end{Bmatrix} \frac{\overline{\alpha}\overline{E}\overline{T}}{1-2v} \tag{4.23}$$

Solving this system of equations requires the use of the unsymmetrical matrix storage and solution scheme. The von Mises (equivalent stress) stress can be written as:

$$\sigma_{eqv} = \frac{1}{\sqrt{2}}\left[\left(\sigma_x - \sigma_y\right)^2 + \left(\sigma_y - \sigma_z\right)^2 + \left(\sigma_z - \sigma_x\right)^2 + 6\left(\tau_{xy}^2 + \tau_{yz}^2 + \tau_{zx}^2\right)\right]^{1/2} \quad (4.24)$$

Furthermore, the mechanical and thermal equations are solved simultaneously.

4.4.3 Numerical solution

A finite element model is used to solve governing equations of heat transfer and thermal stress. COMSOL Finite Element Code [7] is used in the simulations. A nonuniform rectangular grid is used with $100 \times 150 \times 100$ cells. The grids are dense near the two heat sources in order to accurately resolve for temperature distribution. The solver was run until the converged results were obtained. In this case, the residual error for the energy equation was less than the limit set in the simulations. It should be noted that the error related to the predictions is minimized through setting the residual error in the energy equation in COMSOL [7].

In order to demonstrate temperature and stress fields in the heated region, a case study is introduced; in which case, a laser delivering three beams with varying intensity is used and the intensity at the first, the second, and third spots are given in Table 4.2. The steel sheet is used in the simulations, and Table 4.3 gives the thermal properties of the steel used in the simulations.

Fig. 4.7A shows the temperature distribution along the x-axis, and the y-axis location is at $y=0.25$ mm, where at the center of the second and third spots, (Fig. 4.6), different laser intensities appear at the irradiated spots. The temperature increases

Table 4.2 Laser parameters used in the simulations

First spot coordinate	Second spot coordinate	Third spot coordinate	Feed rate (m/s)	Intensity (W/m^2)	Gaussian parameter "a" (m)
$x_1=0$ $y_1=0$ $z_1=0$	$x_2=0.5$ mm $y_2=0.25$ mm $z_2=0$	$x_3=0.5$ mm $y_3=-0.25$ mm $z_3=0$	0.40	0.1×10^{12}– 0.2×10^{12}	0.0002

Table 4.3 Thermal properties of the workpiece used in the simulations [8]

Specific heat capacity (J/kg K)	$=109.2+2.57 \times T-0.00653 \times T^2$
Thermal conductivity (W/mK)	$=6.742+0.0286 \times T$
Density (kg/m^3)	$=7945.3-0.2 \times T$
Absorption coefficient ($1/m$) $\times 10^7$	6.17
Thermal expansion coefficient (1/K)	$=1.356 \times 10^{-5}+6.95 \times 10^{-9} \times T$
Elastic modulus (Pa)	$=2.24x \times 10^{11}-8.93 \times 10^7 \times T$
Poisson's ratio	$=0.265+8.22 \times 10^5 \times T$

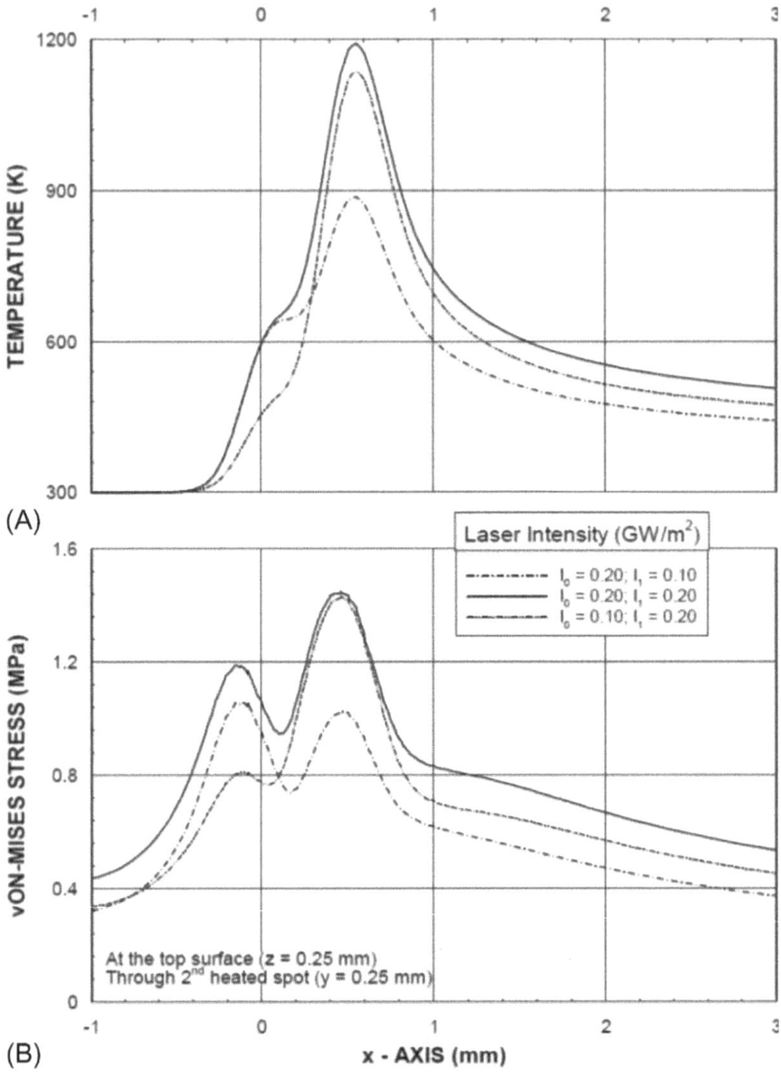

Fig. 4.7 Temperature and the von Mises stress variation along the *z*-axis [4].

to reach its maximum at the second spot center. The presence of the first spot modifies a slight temperature rise around the second spot. In this case, the rise of temperature between the first and second spots is influenced by the first spot. Consequently, the temperature gradient varies in this region. As the intensity at the first spot reduces, this influence diminishes gradually. The location of the peak temperature along the x-axis is modified by the power intensity at the irradiated surface. Therefore, reducing the power intensity at the second spot results in the shift of the temperature peak along the x-axis. In addition, decay of temperature from its peak value is sharper along the

x-axis. This is because of the high cooling rates behind the second spot as compared to that of the first spot.

Fig. 4.7B shows the von Mises stress distribution along the x-axis at the y-axis location; $y=0.25$ mm for different beam intensities at the irradiated spots. The von Mises stress distribution has two peaks. The first peak corresponds to the high temperature gradient due to first and second spots, and the second peak is because of the temperature gradient due to the second peak alone. Because the y-axis location is at $y=0.25$ mm, the temperature gradient is high around the second spot, which triggers the formation of high levels of von Mises stresses in this region. The location of peak von Mises stress is not same as the location of the peak temperature. As the laser intensity reduces at the second peak, the von Mises stress reduces due to attainment of the low-temperature gradients. Moreover, location of the peak stress due to the first spot moves away from the first spot center as the intensity reduces at the first spot.

Fig. 4.8A shows the temperature distribution along the z-axis at the location $x=0$, and $y=0$ for different power intensities at the first and second spots. Temperature decay is sharp in the surface region and temperature decay becomes gradual as the depth below the surface increases toward the solid bulk. The sharp decay of temperature in the surface region is attributed to convection cooling at the surface. Moreover, the high temperature gradient at the surface enhances the conduction heat transfer from the surface region toward the solid bulk, which contributes to sharp decay of temperature in the surface region. As the laser intensity at the first spot reduces, the value of the maximum temperature reduces, giving rise to relatively small temperature gradients than that at high intensities. In addition, a low temperature gradient suppresses conduction heat transfer from the surface region toward the solid bulk. Consequently, temperature decay becomes more gradual along the x-axis than that of high intensities.

Fig. 4.8B shows the von Mises stress distribution inside the substrate along the z-axis at x and y-axes locations at $x=0$ and $y=0$ for different laser intensities. The von Mises stress reduces sharply from its maximum value as the depth below the surface increases. In this case, the von Mises stress distribution follows almost the same as temperature distribution in the surface region. However, as the depth below the surface increases, decay of the von Mises stress becomes gradual. This is attributed to the thermal strain developed in this region. Consequently, in this region, the von Mises stress does not follow temperature distribution. The influence of laser intensity on the von Mises stress is significant; in which case, increasing intensity enhances the temperature gradient while increasing the von Mises stress levels, particularly in the surface region.

Fig. 4.9A shows the temperature distribution along the z-axis at x, and y-axis locations at $x=0$ and $y=0.25$ m. The y-axis location corresponds to the center of the second irradiated spot. The maximum temperature of $z=0$ remains high, which is attributed to the location of the second spot, as it is on the scanning direction of the first spot. Therefore, the region of the second spot is heated by the first spot during the laser scanning at the workpiece surface. Increasing intensity at the second spot enhances the temperature increase in the surface region; however, increasing intensity at the first spot modifies the temperature distribution inside the substrate material slightly.

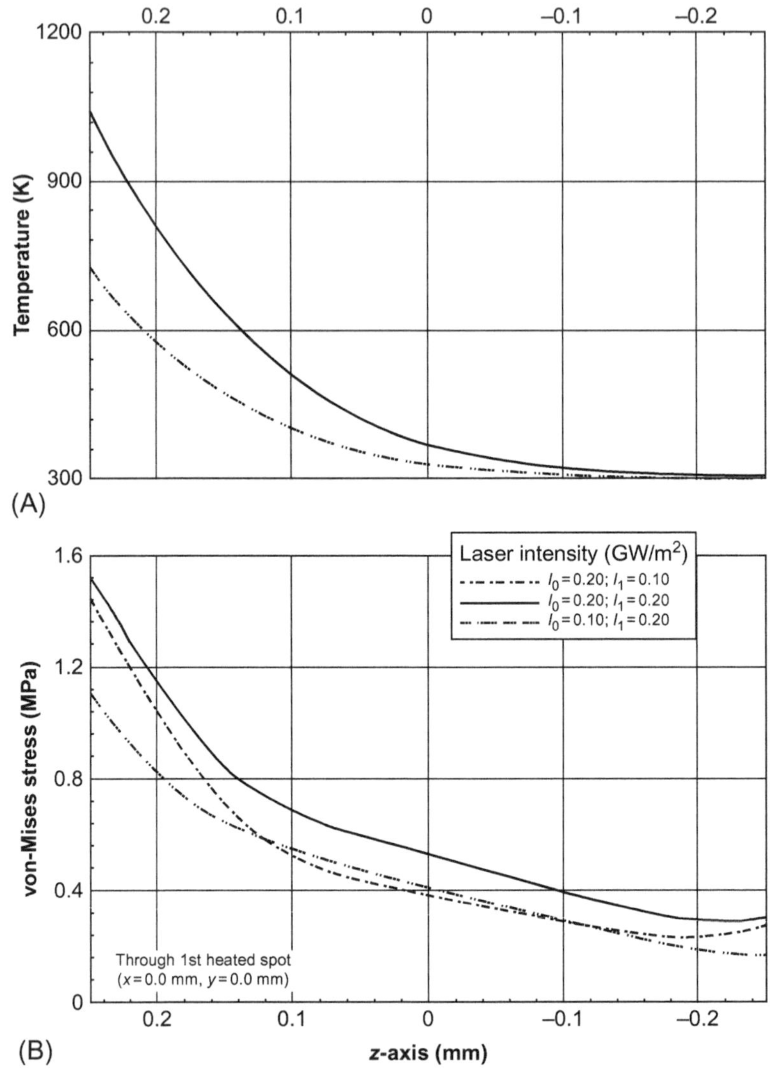

Fig. 4.8 Temperature and the von Mises stress variation along the *z*-axis [4].

Fig. 4.9B shows the von Mises stress distribution along the *x*-axis at *x* and *y*-axis location at $x = 0$ and $y = 0.25$ mm as similar to those shown in Fig. 8A. The von Mises stress reduces sharply in the surface region, which is more pronounced at high laser intensities at the second spot. The von Mises stress remains almost the same value at the same depth below the surface for all intensities used in the simulations. This is attributed to thermal strain developed in this region. It should be noted that tensile stress is formed at the surface and stress becomes compressive in the region below the surface. Because the von Mises stress incorporates normal stress compounds and shear stresses, the von Mises stress becomes almost constant in this region.

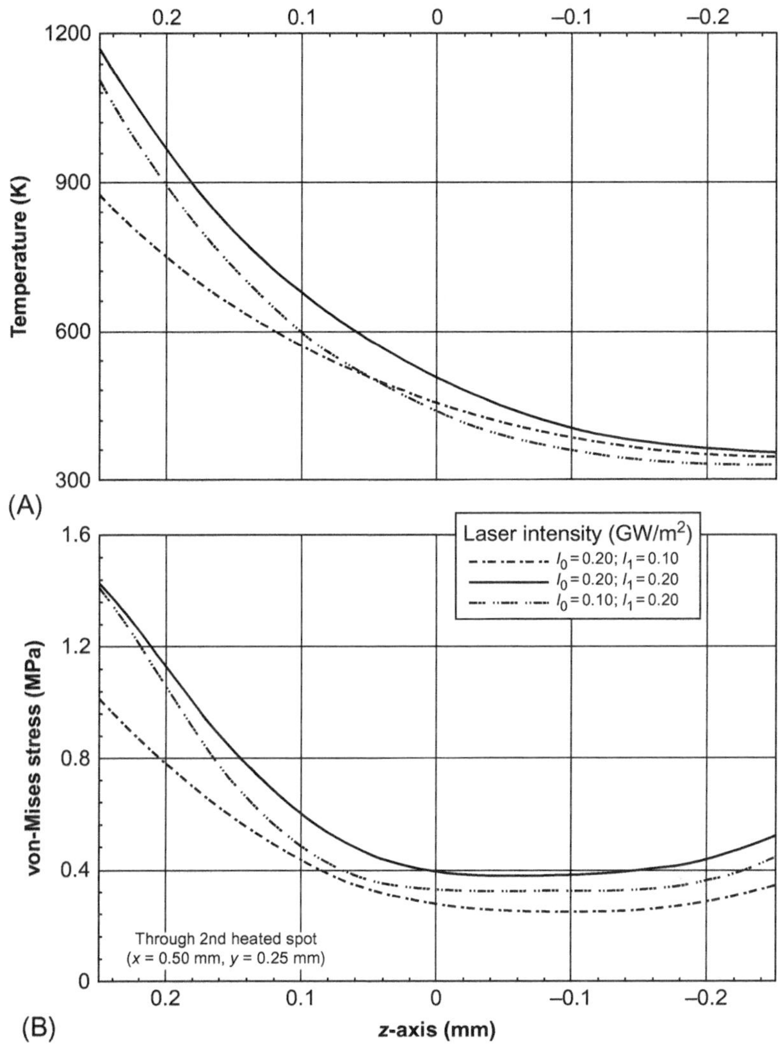

Fig. 4.9 Temperature and the von Mises stress variation along the *z*-axis [4].

4.5 Three-dimensional heating considerations for laser cutting

Laser cutting process requires three-dimensional considerations in the heating analysis. This is because of the nonsymmetric situation created due to the laser cutting operation along one direction. However, the cutting situation becomes complicated and expensive in terms of computation efforts and time when the mass removal from the kerf region is incorporated in the program. This is mainly because of the excessively large mesh size requirement for the simulation to capture the physical process involved. In addition, the process involves multiphysics including phase change to high temperature oxidation reactions. Therefore, the heating situation can be simplified after assuming three-dimensional kerf is formed and the assisting gas impinging onto the kerf while the cutting is progressing. In this case, the initial key-hole formation is avoided in the simulations. The formulation of the problem and findings will be presented with the previous findings [9–12].

4.5.1 Mathematical formulation of the heating situation

The jet impingement onto a surface can be considered as an axisymmetric and the problem may reduce to a two-dimensional case such as laser drilling of circular holes. However, the jet impingement onto a kerf, in relation to cutting, cannot be reduced to a two-dimensional case because the slot has a rectangular geometry rather than a circular geometry. Therefore, consideration of the axisymmetric heating is not possible in relation to the laser cutting process and the problem remains a three-dimensional heating problem. The surface temperature of the kerf wall is kept at 1500°K to resemble the laser-produced kerf. The flow equations are solved using the control volume approach and the RNG k-ε model is incorporated to account for the turbulence.

The Favre-Averaged transport equations in three-dimensional space are used to model the impinging gas jet on the kerf. Because air is used as an assisting gas (working fluid), the compressibility is accounted for; that is:

Continuity equation:

$$div\left(\overline{\rho}\underline{\widetilde{U}}\right) = 0 \tag{4.25}$$

Favre-Averaged transport equations:

$$\begin{aligned}
div\left(\overline{\rho}\widetilde{U}\underline{\widetilde{U}}\right) &= -\frac{\partial \overline{P}}{\partial x} + div\left(\mu grad\widetilde{U}\right) + \left[-\frac{\partial\left(\overline{\overline{\rho}u'^2}\right)}{\partial x} - \frac{\partial\left(\overline{\overline{\rho}u'v'}\right)}{\partial y} - \frac{\partial\left(\overline{\overline{\rho}u'w'}\right)}{\partial z}\right] + S_{Mx} \\
div\left(\overline{\rho}\widetilde{V}\underline{\widetilde{U}}\right) &= -\frac{\partial \overline{P}}{\partial y} + div\left(\mu grad\widetilde{V}\right) + \left[-\frac{\partial\left(\overline{\overline{\rho}u'v'}\right)}{\partial x} - \frac{\partial\left(\overline{\overline{\rho}v'^2}\right)}{\partial y} - \frac{\partial\left(\overline{\overline{\rho}v'w'}\right)}{\partial z}\right] + S_{My} \\
div\left(\overline{\rho}\widetilde{W}\underline{\widetilde{U}}\right) &= -\frac{\partial \overline{P}}{\partial z} + div\left(\mu grad\widetilde{W}\right) + \left[-\frac{\partial\left(\overline{\overline{\rho}u'w'}\right)}{\partial x} - \frac{\partial\left(\overline{\overline{\rho}v'w'}\right)}{\partial y} - \frac{\partial\left(\overline{\overline{\rho}w'^2}\right)}{\partial z}\right] + S_{Mz}
\end{aligned} \tag{4.26}$$

Scalar transport equation (ϕ represents any scalar quantity such as energy)

$$div\left(\overline{\rho}\widetilde{\Phi}\widetilde{\underline{U}}\right) = div\left(\Gamma_{\Phi} grad\,\widetilde{\Phi}\right) + \left[-\frac{\partial\left(\overline{\rho u'\phi'}\right)}{\partial x} - \frac{\partial\left(\overline{\rho v'\phi'}\right)}{\partial y} - \frac{\partial\left(\overline{\rho w'\phi'}\right)}{\partial z}\right] + S_{\Phi} \quad (4.27)$$

where the over-bar indicates a time-averaged variable, the under-bar indicates a vector quantity, and the tilde indicates a Favre-Averaged (density-weighted) variable. Γ_{Φ} is a relevant diffusion coefficient [13].

The energy equation is:

$$div\left(\overline{\rho}\widetilde{\underline{U}}\widetilde{c_p}\widetilde{T}\right) = div\left(k_{eff}\, grad\,\widetilde{T}\right) + \left[-\frac{\partial\left(\overline{\rho c_p u'\theta'}\right)}{\partial x} - \frac{\partial\left(\overline{\rho c_p v'\theta'}\right)}{\partial y} - \frac{\partial\left(\overline{\rho c_p w'\theta'}\right)}{\partial z}\right] + S_T \quad (4.28)$$

where k_{eff} is the effective thermal conductivity.

4.5.1.1 Turbulence modeling

To evaluate the turbulence kinetic energy generation and dissipation, the RNG k-ε model is used. In this case, the turbulence kinetic energy, k, and its rate of dissipation, ε are [13,14]:

$$div(\rho k\underline{U}) = div\left(\alpha_k \mu_{eff}\, grad\, k\right) + G_k - \rho\varepsilon - Y_M \quad (4.29)$$

$$div(\rho\varepsilon\underline{U}) = div\left(\alpha_\varepsilon \mu_{eff}\, grad\, \varepsilon\right) + C_{1\varepsilon}\frac{\varepsilon}{k}(G_k) - C^*_{2\varepsilon}\rho\frac{\varepsilon^2}{k} \quad (4.30)$$

where α_k is the inverse effective Prandtl number for k, α_ε is the inverse effective Prandtl number for ε, $\mu_{eff}=\mu+\mu_t$ is the effective viscosity, G_k the generation of turbulence kinetic energy due to the mean velocity gradients, Y_M is the dilatation dissipation term, $C_{1\varepsilon}=1.42$ is a constant, and $C^*_{2\varepsilon}$ is a modified constant. Y_M represents the contribution of the fluctuating dilatation in compressible turbulence to the overall dissipation rate ($Y_M = 2\rho\varepsilon M_t^2$, M_t is the turbulent Mach number, $M_t\sqrt{k/a^2}$ and $a \equiv \sqrt{\gamma RT}$ is the speed of sound). The term on the left-hand side of equal sign of each equation (Eqs. 4.29 and 4.30) represents the transport of k or ε by convection. The first term on the right-hand side is the transport of k or ε by diffusion. The second term on the right-hand side is the rate of generation of k or ε. The third term on the right-hand side is the rate of destruction of k or ε. Because air is used as the assisting gas, contribution of buoyancy to the generation of turbulence kinetic energy is neglected.

The scale elimination procedure in RNG theory results in a differential equation for turbulent viscosity [14], which is:

$$d\left(\frac{\rho^2 k}{\sqrt{\varepsilon\mu}}\right) = 1.72\frac{\hat{v}}{\sqrt{\hat{v}^3 - 1 + C_v}}d\hat{v} \quad (4.31)$$

where $\hat{v} = \mu_{eff}/\mu$ and $C_v \asymp 100$.

It is important to note that in the high-Reynolds-number limit, Eq. (4.31) gives [15]:

$$\mu_{eff} = \mu_t = \rho C_\mu \frac{k^2}{\varepsilon} \tag{4.32}$$

with $C_\mu = 0.0845$ [14].

The inverse effective Prandtl numbers, α_k and α_ε, are computed using the following formula [15]:

$$\left|\frac{\alpha - 1.3929}{\alpha_0 - 1.3929}\right|^{0.6321} \left|\frac{\alpha + 2.3929}{\alpha_0 + 2.3929}\right|^{0.3679} = \frac{\mu}{\mu_{eff}} \tag{4.33}$$

where $\alpha_0 = 1.0$. In the high-Reynolds-number limit, $\alpha_k = \alpha\varepsilon \asymp 1.393$ [14].

$C^*_{2\varepsilon}$ is calculated from the following equation [14]:

$$C^*_{2\varepsilon} \equiv C_{2\varepsilon} + \frac{C_\mu \eta^3 (1 - (\eta/\eta_0))}{1 + \beta\eta^3} \tag{4.34}$$

where $C_{2\varepsilon} = 1.68$, $\beta = 0.012$, $\eta_0 = 4.38$, and $\eta = Sk/\varepsilon$ with $S =$ strain rate.

Moreover, G_k represents the generation of turbulence kinetic energy due to the mean velocity gradients and is calculated using the following [14]:

$$G_k = \tau_{ij} \cdot S_{ij} \tag{4.35}$$

The Reynolds stresses τ_{ij} according to Boussinesq hypothesis are proportional to the mean rates of deformation S_{ij}, that is:

$$\tau_{ij} = -\rho \overline{u'_i u'_j} = 2\mu_t S_{ij} - \frac{2}{3}\rho k \delta_{ij} \tag{4.36}$$

with

$$S_{ij} = \frac{1}{2}\mu_t \left(\frac{\partial U_i}{\partial x_j} + \frac{\partial U_j}{\partial x_i}\right) \tag{4.37}$$

Turbulent heat transport is modeled using Eq. (4.28). For the RNG k-ε model, the effective thermal conductivity is:

$$k_{eff} = \alpha c_p \mu_{eff} \tag{4.38}$$

where α is calculated from Eq. (4.33), but with $\alpha_0 = k_{th}/\mu c_p$. k_{th} and c_p are the thermal conductivity and specific heat of the gas into consideration and are temperature-dependent in the present study.

4.5.1.2 Flow boundary conditions

The following boundary conditions are considered in the simulations:

4.5.1.2.1 Solid wall

For the solid wall, the no-slip condition is assumed at the solid wall, and the boundary condition for the velocity at the solid wall, therefore, is:

$$U_i = 0. \tag{4.39}$$

However, the wall-bounded turbulent flows are affected in many ways by the presence of the wall. Toward the outer part of the near-wall region, turbulence is rapidly strengthened by the production of turbulent-kinetic energy due to large gradients in mean velocity. Consideration, therefore, needs to be given to account for the wall functions. The wall functions link the solution variables at the near-wall cells and the corresponding quantities on the wall. This eliminates the need to resolve the inner region. The enhanced wall treatment is a near-wall modeling method that solves the turbulence sublayer instead of using function to set and simulate turbulence behavior near the wall [14]. In this approach, the whole domain is subdivided into a viscosity-affected region that is completely resolved all the way to the viscous sublayer and a fully-turbulent region. The demarcation of the two regions is determined by a wall-distance-based, turbulent Reynolds number, Re_y, defined as:

$$Re_y \equiv \frac{\rho y \sqrt{k}}{\mu} \tag{4.40}$$

where y = normal distance from the wall at the cell centers.

In the fully-turbulent region ($Re_y > Re_y^*$ and $Re_y^* = 200$), the RNG k-ε model is employed [15]. The turbulent viscosity is computed from [14]:

$$\mu_{t,2layer} = \rho C_\mu l_\mu \sqrt{k} \tag{4.41}$$

The length scale (l_μ) appearing in Eq. (4.41) is computed from:

$$l_\mu = y c_l \left(1 - e^{-Re_y/A_\mu}\right) \tag{4.42}$$

The two-layer formulation for turbulent viscosity described above is used as a part of the enhanced wall treatment, in which the two-layer definition is smoothly blended with the high-Reynolds-number μ_t definition from the outer region; that is:

$$\mu_{t,enh} = \lambda_\varepsilon \mu_t + (1 - \lambda_\varepsilon)\mu_{t,2layer} \tag{4.43}$$

where μ_t is the high-Reynolds-number viscosity [14]. A blending function, λ_ε, is defined in such a way that it is equal to unity far from the walls and is zero very near to the walls. The main purpose of the blending function λ_ε is to prevent solution convergence from being impeded when the k-ε solution in the outer layer does not match with the two-layer formulation.

Similarly, temperature near the wall is obtained by blending the laminar linear sublayer and turbulence logarithmic laws [14]:

$$T^+ = e^{\Pi} T^+_{lam} + e^{1/\Pi} T^+_{turb} \tag{4.44}$$

where

$$\Pi = -\frac{a(Pry^+)^4}{1 + bPr^3 y^+} \tag{4.45}$$

and where Pr is the molecular Prandtl number, the coefficients are $a=0.01$ and $b=5$.

4.5.1.2.2 Inlet conditions

The boundary conditions for temperature and the mass flow rate at the nozzle inlet are introduced. Therefore, at the nozzle inlet: $T=$specified (300 K) and the values of the mass flow rate are: 2.34×10^{-5} and 4.68×10^{-5} kg/s. The selection of the values of the temperature and mass flow rate is due to the similar values used in the processing of the laser assisting gas. Air is used as a working fluid.

4.5.1.2.3 Outlet conditions

The flow is considered to be extended over a long domain; therefore, the boundary condition for any variable ϕ is

$$\frac{\partial \phi}{\partial x_i} = 0, \tag{4.46}$$

where x_i is the normal direction at outlet.

4.5.1.3 Solid side

4.5.1.3.1 Constant wall temperature boundary

A constant temperature boundary is set at the kerf walls to resemble the laser cut kerf surface; that is, $T_w=$constant (1500°K).

4.5.1.3.2 Solid fluid interface

The coupling of conduction within the solid, and convection within the fluid, termed "conjugation," is required for the present analysis at the solid/fluid interface. The appropriate boundary conditions are continuity of heat flux and temperature and are termed "boundary conditions of the fourth kind," that is,

$$T_{W_{solid}} = T_{W_{gas}} \quad \text{and} \quad K_{W_{solid}} \frac{\partial T_{W_{solid}}}{\partial x} = K_{W_{gas}} \frac{\partial T_{W_{gas}}}{\partial x} \tag{4.47}$$

The heat transfer coefficient is determined from:

$$h = \frac{\left(-k_{W_{solid}}\left(\frac{\partial T_{W_{solid}}}{\partial n}\right)\right)}{\left(T_{W_{gas}} - T_{ref}\right)} \tag{4.48}$$

where $(\partial T_{W_{solid}}/\partial n)$ is the temperature gradient in the solid normal to the cavity surface, while T_{ref} is the reference temperature and equals to the jet temperature at the nozzle exit. It should be noted that at the gas temperature in the vicinity of the wall $(T_{W_{gas}})$ is considered when computing the heat transfer coefficient.

The wall shear stress is determined from the rate of fluid strain at the cavity wall, which is:

$$\tau = \mu\left(\frac{\partial V}{\partial n}\right) \tag{4.49}$$

where $(\partial V/\partial n)$ is the rate of fluid strain at the cavity wall and n represents the direction normal to the cavity surface.

4.5.2 Numerical method and simulation

A control volume approach is employed to discretize the governing equations. The discretization procedure is given in the literature [16]. The problem of determining the pressure and satisfying continuity may be overcome by adjusting the pressure field so as to satisfy continuity. A staggered grid arrangement is used in which the velocities are stored at a location midway between the grid points; that is, on the control volume faces. All other variables including pressure are calculated at the grid points. Fig. 4.10 shows the grid used in the simulations. This arrangement gives a convenient way of handling the pressure linkages through the continuity equation and is known as the Semi-Implicit Method for Pressure-Linked Equations (SIMPLE) algorithm. The details of this algorithm are given in the literature [16].

The computer program used for the present simulation can handle a nonuniform grid spacing. In each direction, fine-grid spacing near the gas jet impinging point and the cavity is allocated while gradually increased the spacing for locations away from the cavity is considered. Elsewhere the grid spacing is adjusted to maintain a constant ratio of any of two adjacent spacing. The number of grid points used in the simulations for the nozzle and the kerf, including the solid part in x, y, and z directions, is $26 \times 68 \times 26$ and $41 \times 54 \times 41$, respectively. Moreover, the number of grid points used in the spacing between the nozzle exit and the kerf surface in x, y, and z directions is $41 \times 134 \times 41$. The number of grid points used in the spacing between the kerf exit and bottom outer boundary in x, y, and z directions is $41 \times 52 \times 41$. The selection of the grid sizes is based on the grid independent tests. Therefore, the grid size of $41 \times 54 \times 41$ in x, y, and z directions is selected (Fig. 4.10). It should be noted that increases in the grid size in all directions amplifies the total grid number by cubic power, which requires excessive computational time and memory.

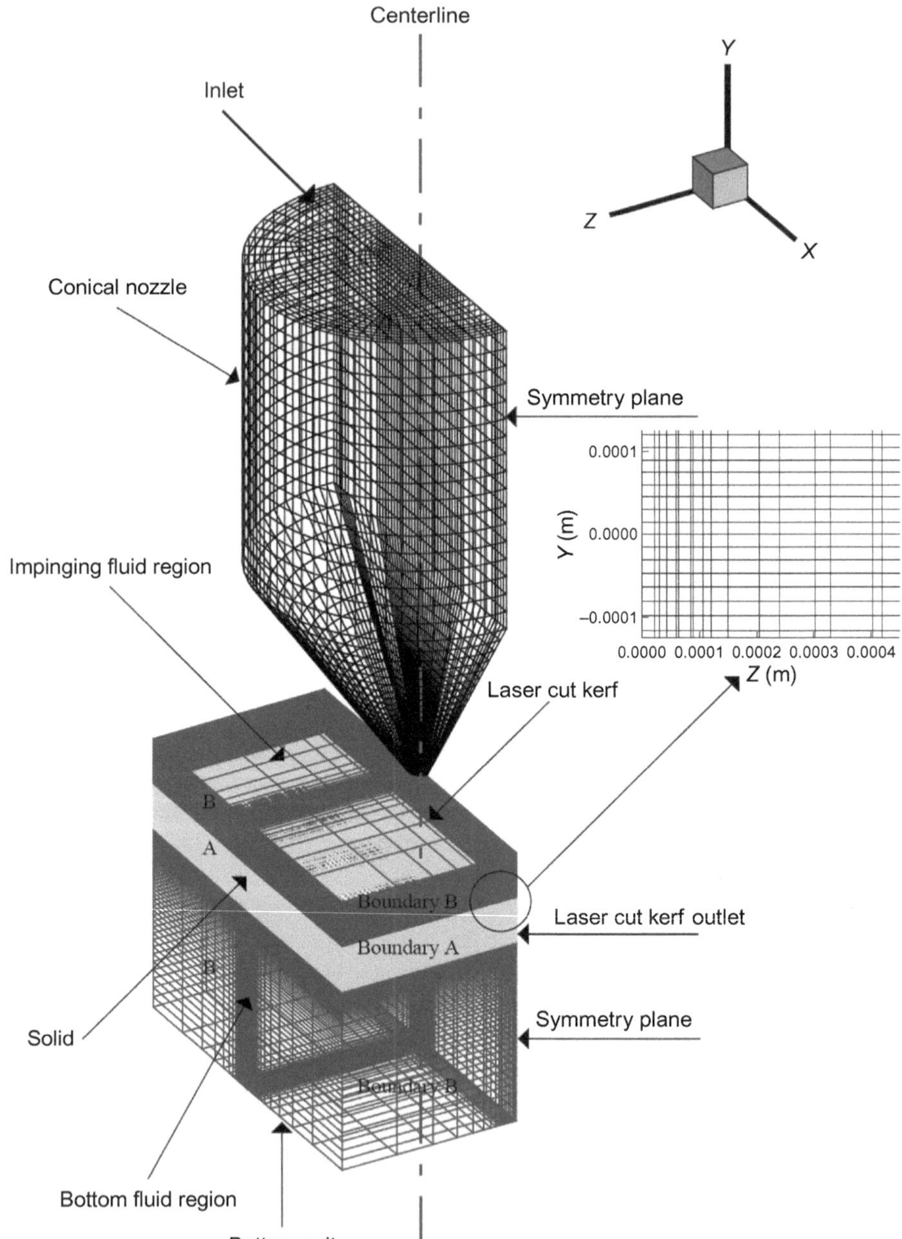

Fig. 4.10 Mesh generated in the solution domain [9].

Because the problem involves a gas jet impinging onto the workpiece surface and penetrating into narrow three-dimensional kerf, the near-wall modeling approach, combined with adequate mesh resolution in the near-wall region for success of simulations, is incorporated. The near-wall mesh is set to be fine enough to resolve the laminar sublayer, provided that the enhanced wall treatment is identical to the two-layer model described above. For capturing a laminar viscous sublayer, the non-dimensional wall coordinate y^+ is set to be equal to 1 at the wall adjacent cells [14].

A numerical code FLUENT [16] is used to simulate the jet impingement situations. The convergence criterion for the residuals is set as $|\psi^k - \psi^{k-1}| \leq 10^{-6}$ to terminate the simulations; consequently, the converged results are obtained and presented.

4.5.3 Case studies on impinging gas jet effects on kerf sites

The material properties are given for stainless steel, which is the most common material used in laser cutting. The thermal properties of stainless steel used in the simulations are given in Table 4.4.

Three cases are introduced and the findings are presented under the appropriate subheadings in line with the previous studies [9–12].

4.5.3.1 Effect of kerf wedge angle on heat transfer rates and skin friction

The effect of the kerf wedge angle on heat transfer rates and skin friction is studied for four wedge angles =0, 2, 4, and 6 degrees. The selection of the wedge angles between 0 and 6 degrees is based on the kerf wedge angle resulting from the experiments prior to the simulations. The wedge angle of 0 degree corresponds to the parallel kerf surfaces while the wedge angle of 6 degrees corresponds to the highly-inclined kerf surfaces. The kerf wall is kept at 1500°K to resemble the laser cut kerf. The averaged jet velocities at the nozzle exit accommodated in the simulations are 30 and 60 m/s, respectively, and the kerf thicknesses is kept at 0.5 mm. Air is used as an assisting gas. Moreover, to examine the variation of the Nusselt number and the skin friction along the kerf surface, locations along the kerf surface are selected as shown in Fig. 4.11.

Fig. 4.12A and B shows normalized velocity contours in the symmetry plane for kerf thickness = 0.5 mm and averaged jet velocity at the nozzle exit = 30 m/s and four kerf wedge angles. The jet emerging from the nozzle expands while forming a free

Table 4.4 Material properties used in the simulations

Temp. (K)	300	400	600	800	1000	1200	1500
Cp (J/kg K)	477	515	557	582	611	640	682
K (W/mK)	14.9	16.6	19.8	22.6	25.4	28	31.7
ρ (kg/m^3)	8018	7968	7868	7769	7668	7568	7418

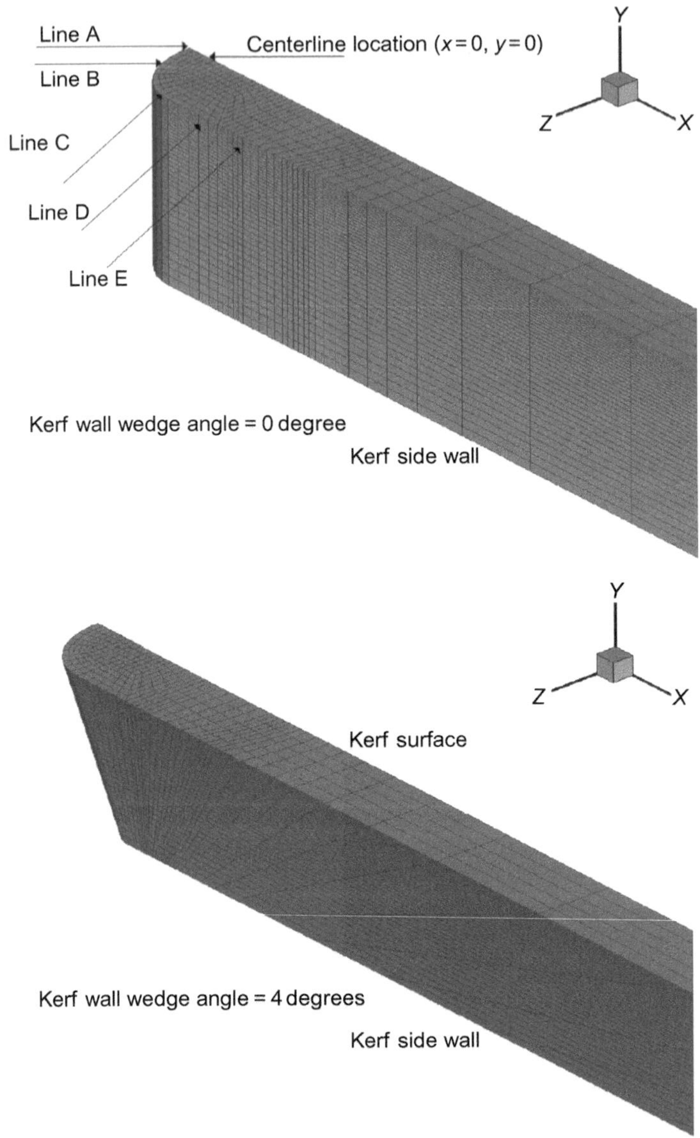

Fig. 4.11 Schematic of the three-dimensional kerf geometry for two kerf wedge angles (0 and 4 degrees). Lines A, B, C, D, and E extend from kerf surface along kerf side wall to kerf exit [9].

boundary shear layer with the surrounding ambient stagnant air as it approaches the kerf top surface. Because the jet expansion occurs above the kerf top surface (between the nozzle exit and the top surface of the kerf), no significant effect of the kerf wall wedge angle is found on the size of the expanded jet. Moreover, the shear layer in the outer region of the expanded jet forms the circulation cell in this region. The

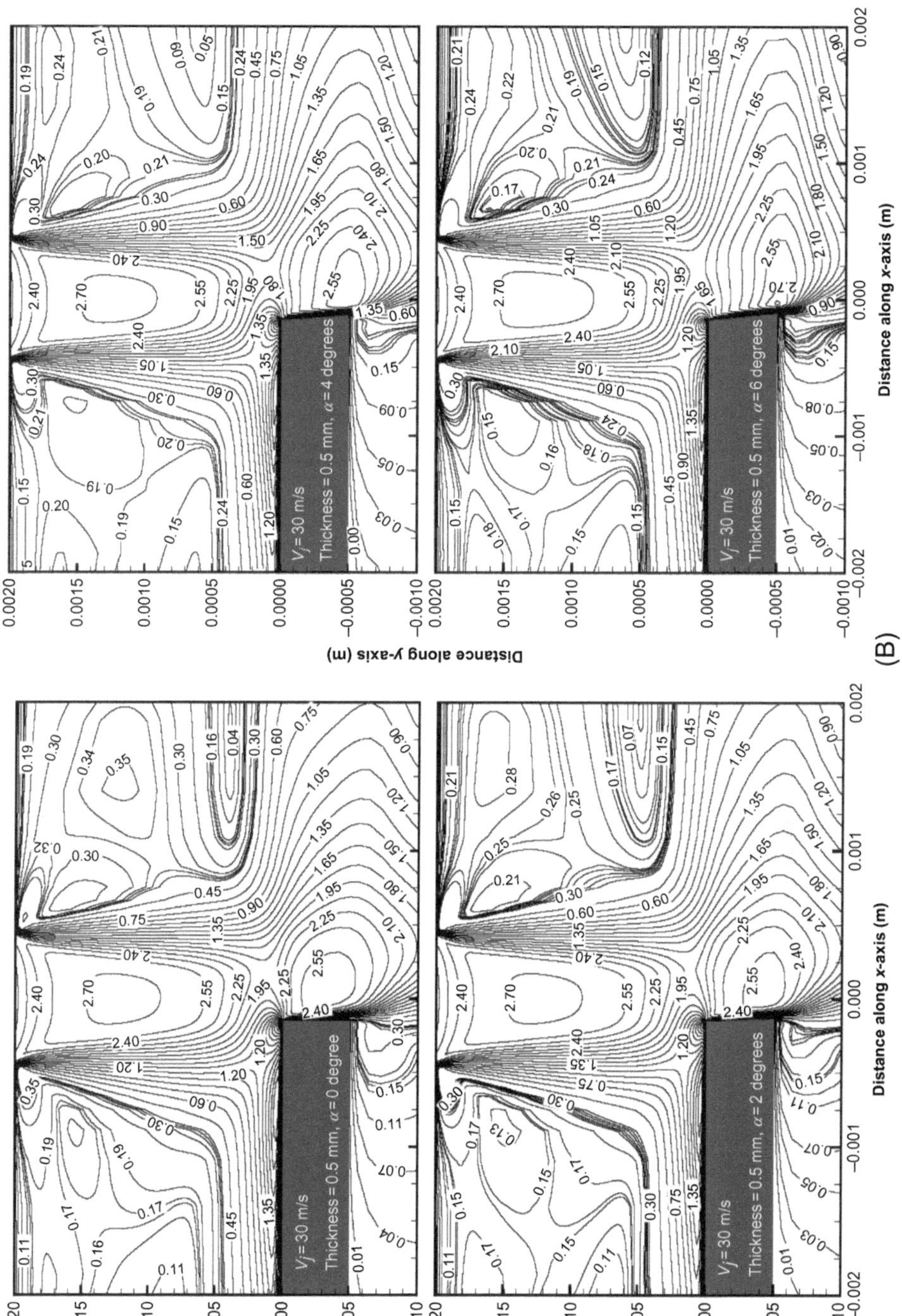

Fig. 4.12 (A) Normalized velocity magnitude (V/V_j) for jet-averaged velocity of 30 m/s, kerf thickness of 0.5 mm, and kerf wedge angles of 0 and 2 degrees [9]. (B) Normalized velocity magnitude (V/V_j) for a jet-averaged velocity of 30 m/s, kerf thickness of 0.5 mm, and kerf wedge angles of 4 and 6 degrees [9].

orientation of the circulation cell developed in the outer region of the impinging jet above the kerf top surface changes and it shifts away from the kerf surface region as the kerf wall wedge angle increases. However, upon jet penetration into the kerf, the flow undergoes a radial acceleration in the region close to the kerf inlet due to the blockage effect of the kerf wall wedge angles. This causes the radial expansion of the flow along the kerf edges. The radial expansion of the jet is more pronounced as the kerf wall wedge angle increases from 0 to 6 degrees. This is because of the blockage effect of the kerf, which increases with the increasing wedge angle. The radial acceleration of the flow is suppressed toward the exit from the kerf bottom, which results in the flow deceleration in this region. The flow velocity in the kerf increases for the large kerf wall wedge angles. However, significant increase in the velocity magnitude of the flow is observed as the flow moves toward the kerf exit in the negative y-axis direction.

Fig. 4.13 shows the velocity magnitude distribution along the nozzle symmetry axis for the kerf thickness = 0.5 mm, two averaged jet velocities, and four kerf wall wedge angles. It should be noted that the locations $y=0$, $y=0.002$, and $y=-0.0005$ represent the workpiece surface, nozzle exit, and kerf depth (workpiece thickness), respectively. The jet velocity after exiting from the nozzle increases and reaches its maximum at a distance of about one nozzle diameter, after which it reduces to a minimum just before impingement onto the kerf surface. The value of the minimum velocity is lower for the large kerf wall wedge angle. This is due to the blockage effect of the kerf wall wedge. Moreover, the flow velocity increases as the flow approaches the kerf bottom exit, and it reaches its second peak upon expanding from the kerf bottom exit. The magnitude and location of the second velocity peak are slightly affected by the kerf wall wedge angle. The location of the second peak velocity is shifted toward the kerf exit (negative direction of the x-axis) as the kerf wall wedge angle increases. The magnitude of the second velocity peak changes slightly as the kerf wall wedge angle increases, particularly for the averaged jet velocity = 60 m/s. This is due to the high axial momentum of the jet in association with the high jet velocity jet.

Fig. 4.14 shows the fluid temperature distribution along the nozzle symmetry axis for kerf thickness = 0.5, two averaged jet velocities, and four kerf wedge angles. Temperature increases with the onset of the fluid entering the kerf region. In addition, the fluid temperature increases as the kerf wall wedge angle increases in this region. The value and location of the maximum fluid temperature depend on the kerf wall wedge angle, provided this dependence is minimal for the small kerf wall wedge angle; that is, increasing the kerf wall wedge angle from 0 to 2 degrees, it has a negligible effect on the maximum temperature. As the wedge angle increases from 2 to 6 degrees, the value of the maximum temperature changes and its location moves closer to the kerf surface region.

Fig. 4.15 shows the Nusselt number distribution along the normalized kerf thickness for four different kerf wedge angles. The kerf thickness = 0.5 mm and the averaged jet velocity = 30 m/s. The kerf thickness is normalized with the maximum thickness; in which case, the normalized kerf thickness = 0 corresponds to the kerf inlet, and normalized kerf thickness = −1 corresponds to the kerf bottom exit.

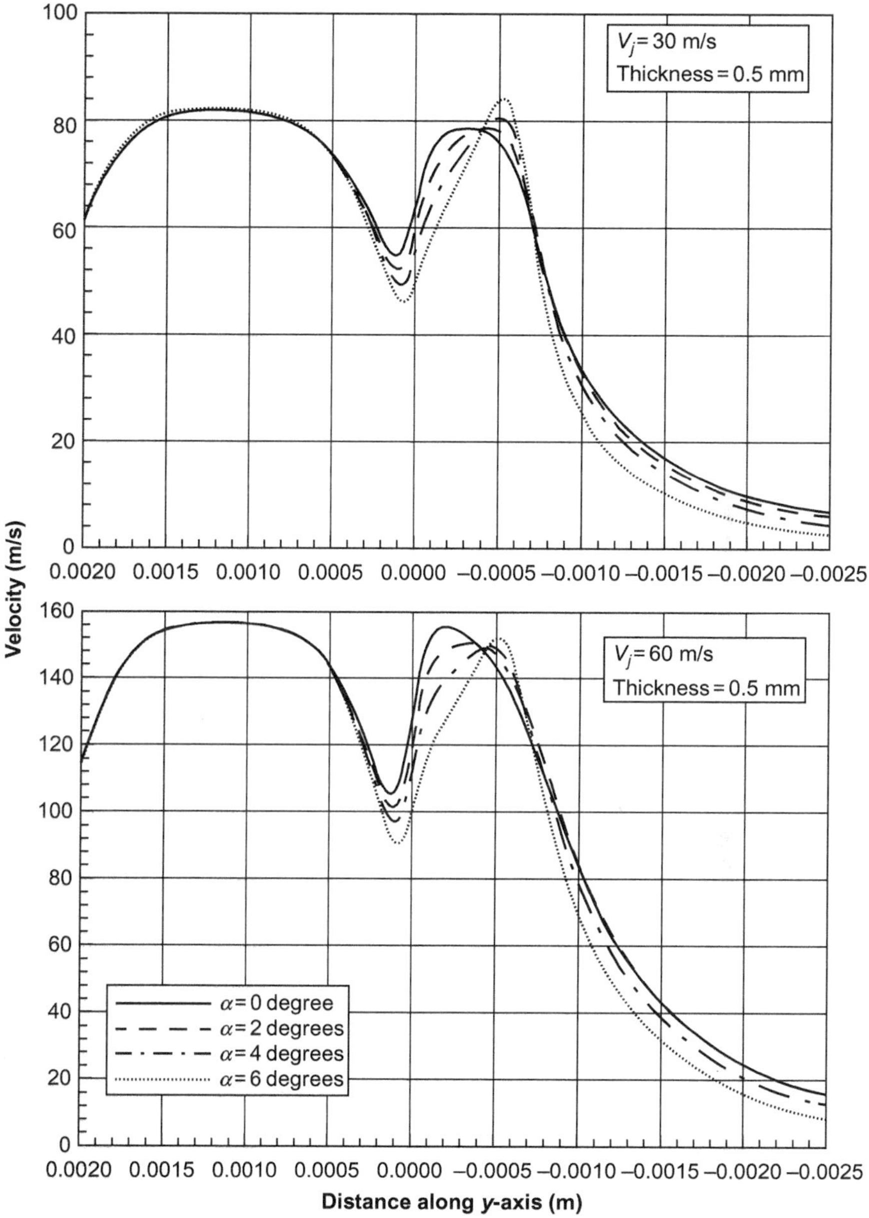

Fig. 4.13 Velocity magnitude along laser beam axis for two different averaged jet velocities, kerf thickness of 0.5 mm, and four kerf wedge angles [9].

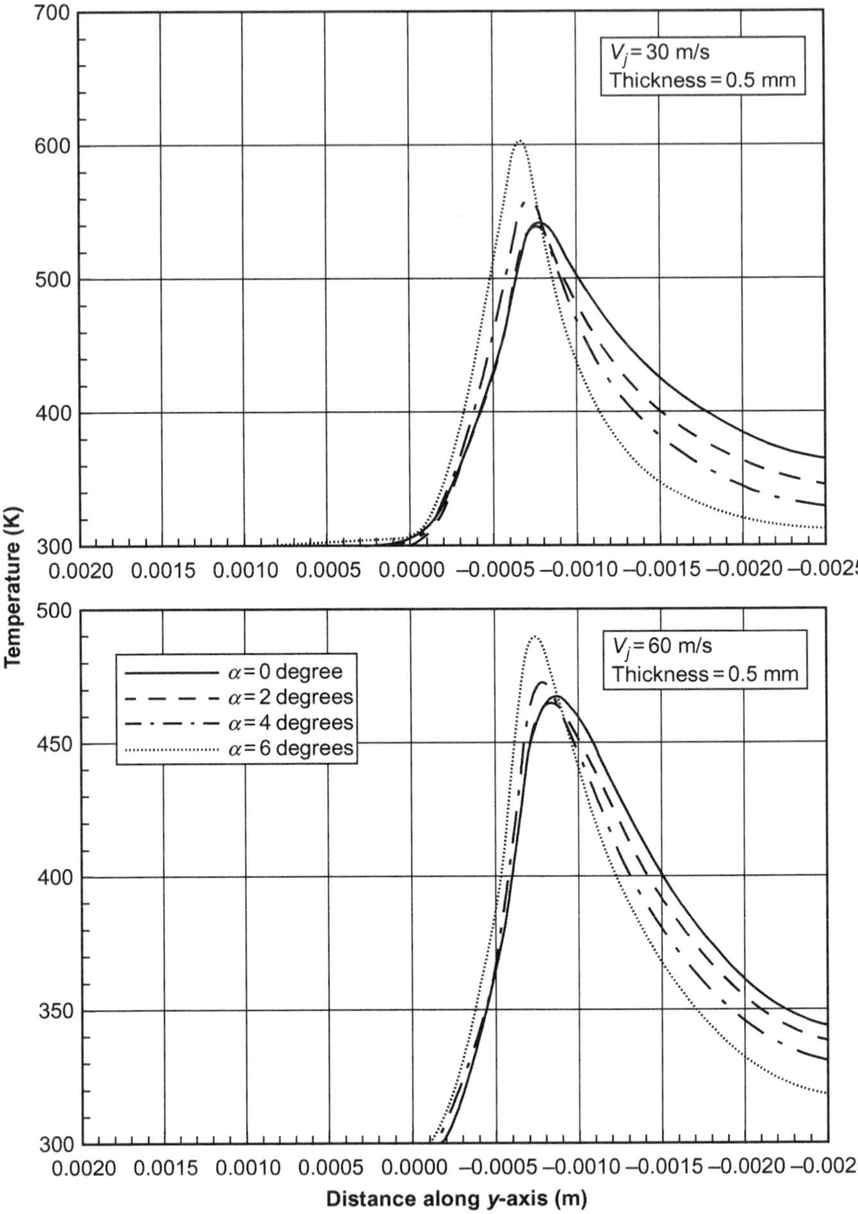

Fig. 4.14 Temperature distribution along laser beam axis for two different averaged jet velocities, kerf thickness of 0.5 mm, and four kerf wedge angles [9].

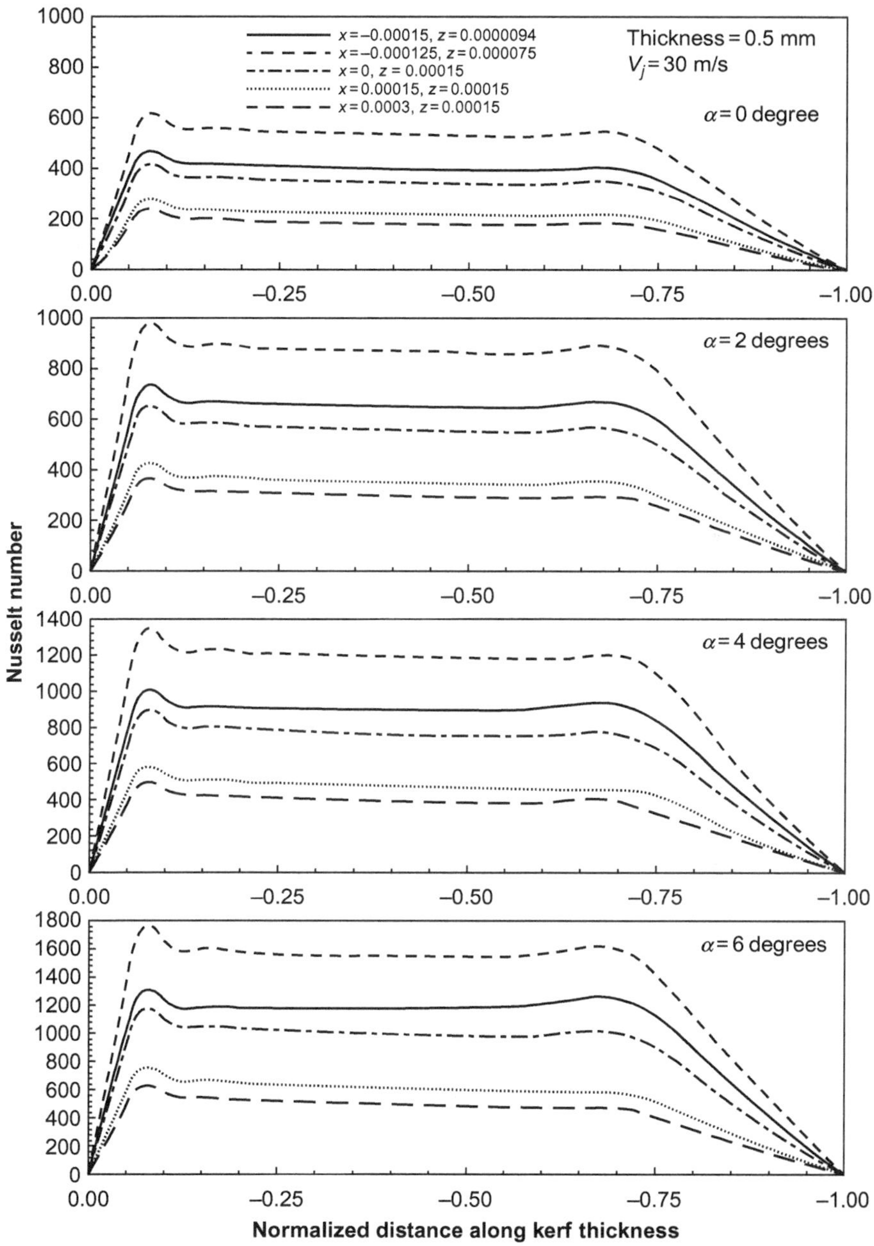

Fig. 4.15 Nusselt number distribution at five locations along the kerf wall for kerf thickness = 0.5 mm, averaged jet velocity = 30 m/s, and four kerf wedge angles [9].

The Nusselt number distribution is shown at five different locations along the kerf wall. Location A ($x=-0.00015$, $z=0.0000094$ m) is almost at the kerf edge (where the kerf wall intersects the symmetry plane). The other four locations (B, C, D, and E) are distributed along the kerf wall (see Fig. 4.11). The location B ($x=-0.000125$, $z=0.000075$ m) corresponds to the location where the Nusselt number is the highest. The Nusselt number increases as the distance increases along the kerf wall from location A ($x=-0.00015$, $z=0.000094$) in the direction of the positive x-axis. The Nusselt number attains the maximum value at location B ($x=-0.000125$, $z=0.000075$ m). Moreover, the Nusselt number decreases as the distance increases further in the direction of the positive x-axis from location B. It is evident that the Nusselt number increases sharply in the region of the kerf top edge due to the convective cooling of the jet in the kerf. The gradual increase in the Nusselt number along the kerf thickness is observed at some locations. The Nusselt number decays along the wall toward the kerf exit after attaining the second peak. The decay in the Nusselt number becomes sharp as the kerf wall wedge angle increases. When comparing the Nusselt number distribution at location B, where the Nusselt number attains highest values; for kerf thickness = 0.5 mm, two averaged jet velocities, and four kerf wedge angles. As the wedge angle increases, Nusselt number increases. The blockage effect of the kerf leads to pressure buildup that increases the density of the fluid in the kerf while enhancing the amount of heat transfer in the kerf.

Fig. 4.16 shows the skin friction distribution along the normalized kerf thickness for kerf thickness = 0.5, two averaged jet velocities, and four kerf wall wedge angles. The location ($x=-0.000125$, $z=0.000075$ m) corresponds to the location of the maximum Nusselt number. Skin friction increases sharply at the inlet of the kerf and decreases with increasing kerf wall wedge angle. The skin friction attains its maximum before the sharp decay along the kerf thickness. However, the skin friction decreases monotonically as the kerf depth increases for the kerf wall wedge angle = 0 degree. This is true for both averaged jet velocities. The gradual increase in skin friction occurs for the kerf wall wedge angles = 4 and 6 degrees up to the point where the second peak of the skin friction is attained. The skin friction, then, decreases toward the kerf exit. Moreover, the skin friction decreases in this region as the kerf wedge angle decreases. The skin friction is large for the averaged jet velocity = 60 m/s, which is more pronounced for the kerf wedge angles = 2 and 4 degrees.

4.5.3.2 Influence of kerf thickness on heat transfer and skin friction

The second case is involved with a jet impingement onto a heated slot (kerf) in relation to the laser cutting process. The flow field in the kerf is simulated using the practical assisting gas parameters. The heat transfer rates form the kerf surface and the skin friction at the kerf wall is predicted. The data in Table 4.4 are used in the simulations and findings are presented in line with the previous study [9–12]. Fig. 4.17 shows the schematic view of the kerf and the coordinate systems.

Fig. 4.18 shows the pressure coefficient contours in the symmetry plane for the kerf thickness of 0.5 and 1.0 mm. The negative values of the pressure coefficient are

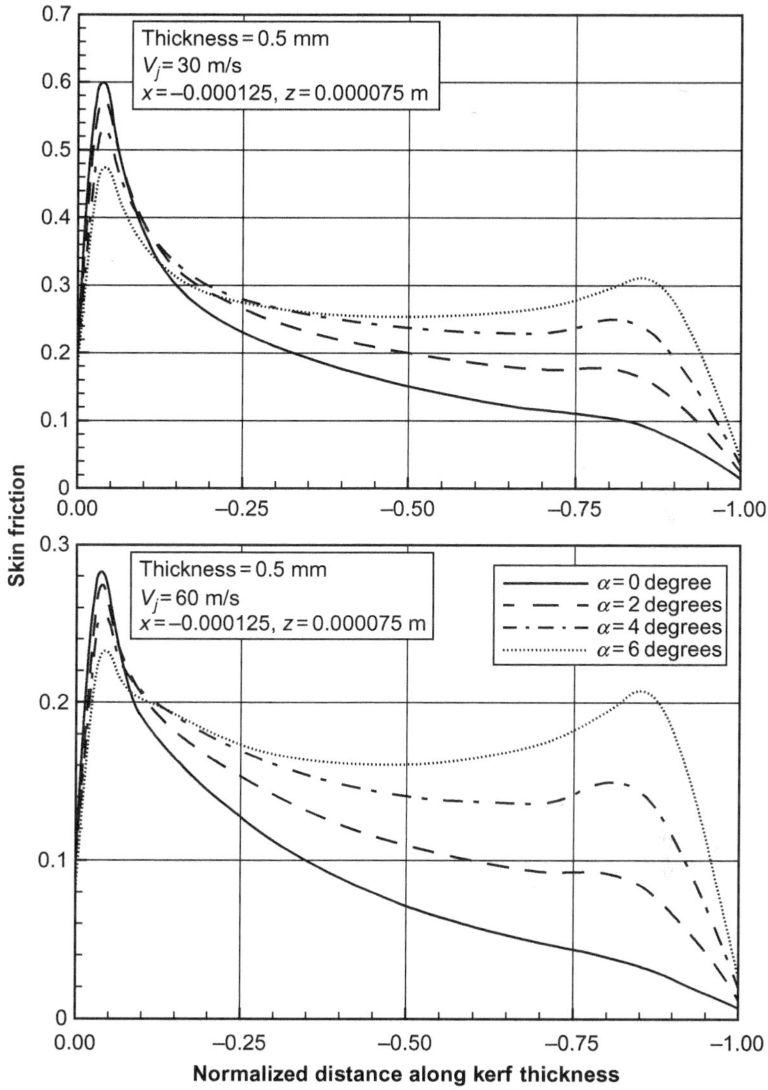

Fig. 4.16 Skin friction distribution at location B ($x = -0.000125$, $z = 0.000075$ m) along kerf wall for kerf thickness = 0.5 mm, two averaged jet velocities, and four kerf wedge angles [9].

observed in the jet outer boundary. This indicates the reverse flow and the possible formation of circulation cell in this region. This situation is observed for all the kerf thicknesses. The decrease in the flow velocity as the flow approaches the kerf top surface is associated with pressure increase in this region. Also, the kerf thickness has a negligible effect on the pressure field in this region. As the kerf thickness increases from 1.0 to 2.0 mm, a slight rise in the pressure coefficient is observed in the kerf. This is because of the radial deceleration of the flow after the initial flow expansion

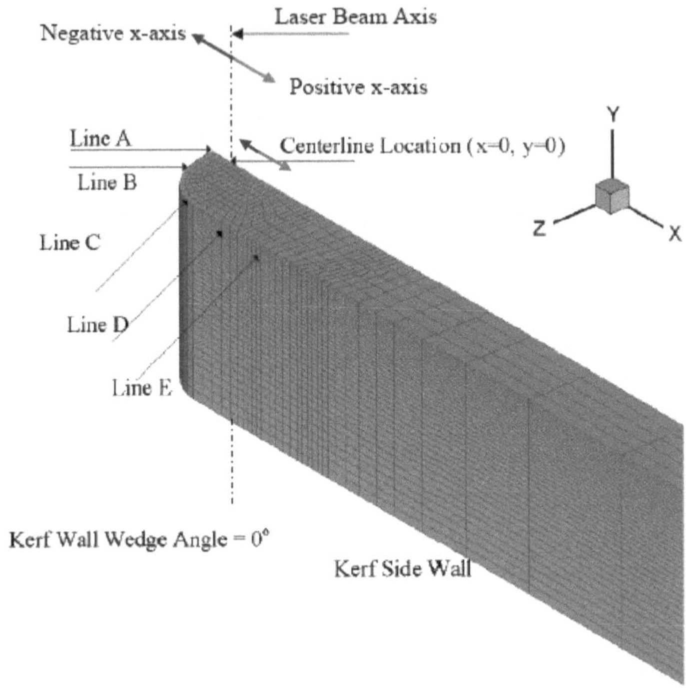

Fig. 4.17 Coordinate system and location of lines around the kerf wall [10].

in the kerf. Moreover, the slight pressure rise in the kerf can also be attributed to the increase in thermodynamic pressure in the region close to the kerf walls, which remain at an elevated temperature (1500°K). Static pressure reduces after exiting the nozzle, because of the jet expansion, and then increases to the maximum at a location close to the kerf top surface because of the formation of a stagnation zone and blockage effect of the kerf inlet. The kerf thickness has no significant effect on pressure distribution in this region. A sharp pressure drop is observed in the entry region of the kerf for all thicknesses. Moreover, the sharp pressure drop is followed by a gradual drop near the kerf exit in the case of the kerf thickness of 2.0 mm. The location of the minimum pressure is close to the kerf bottom exit for all the kerf thicknesses considered in the simulations. Pressure increases slightly after the flow exits the kerf before decreasing to the atmospheric pressure.

Fig. 4.19 shows the Nusselt number distribution along the normalized kerf thickness. The kerf thickness is normalized with the maximum thickness; that is, the normalized kerf thickness of 0 corresponds to the kerf inlet, and the normalized kerf thickness of -1 corresponds to the kerf exit. The Nusselt number distribution is shown at 5 locations along the kerf wall. Location A ($x=-0.00015$, $z=0.0000094$ m) is almost the kerf edge (where the kerf wall intersects the symmetry plane, Fig. 4.17). The other four locations (B, C, D, and E) are distributed along the kerf wall (Fig. 4.17). Location B ($x=-0.000125$, $z=0.000075$ m) corresponds to the location where the Nusselt number attains the highest value. The trends of the curves show that

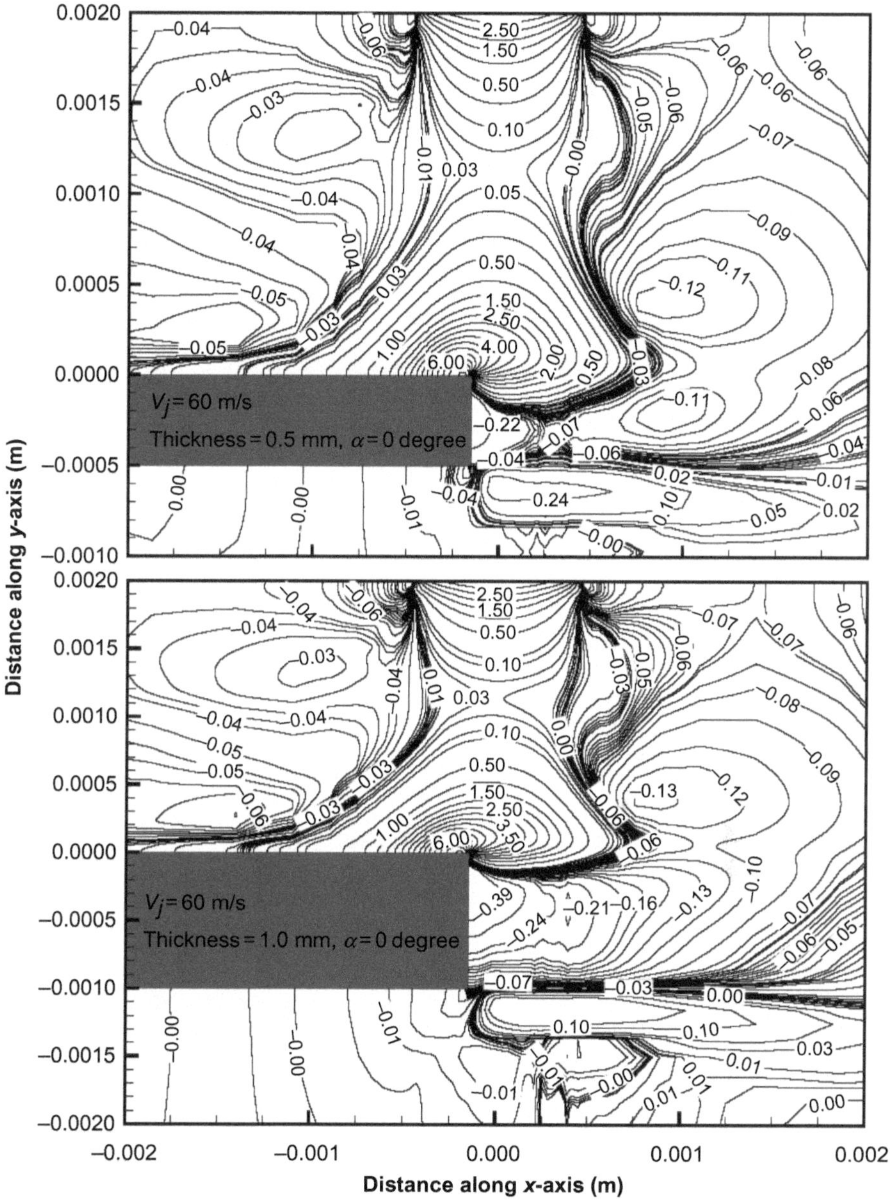

Fig. 4.18 Pressure coefficient in the symmetry plane for jet-averaged velocity of 60 m/s, kerf thickness of 0.5 and 1.0 mm, and kerf wedge angle of 0 degree [10].

the Nusselt number increases as one moves along the kerf wall from location A ($x = -0.00015$, $z = 0.000094$) in the direction of the positive x-axis to attain the maximum value at location B ($x = -0.000125$, $z = 0.000075$ m). The Nusselt number, then, decreases as one moves from location B in the direction of positive x-axis (see Fig. 4.17). The trend shows that the Nusselt number increases sharply in the region of

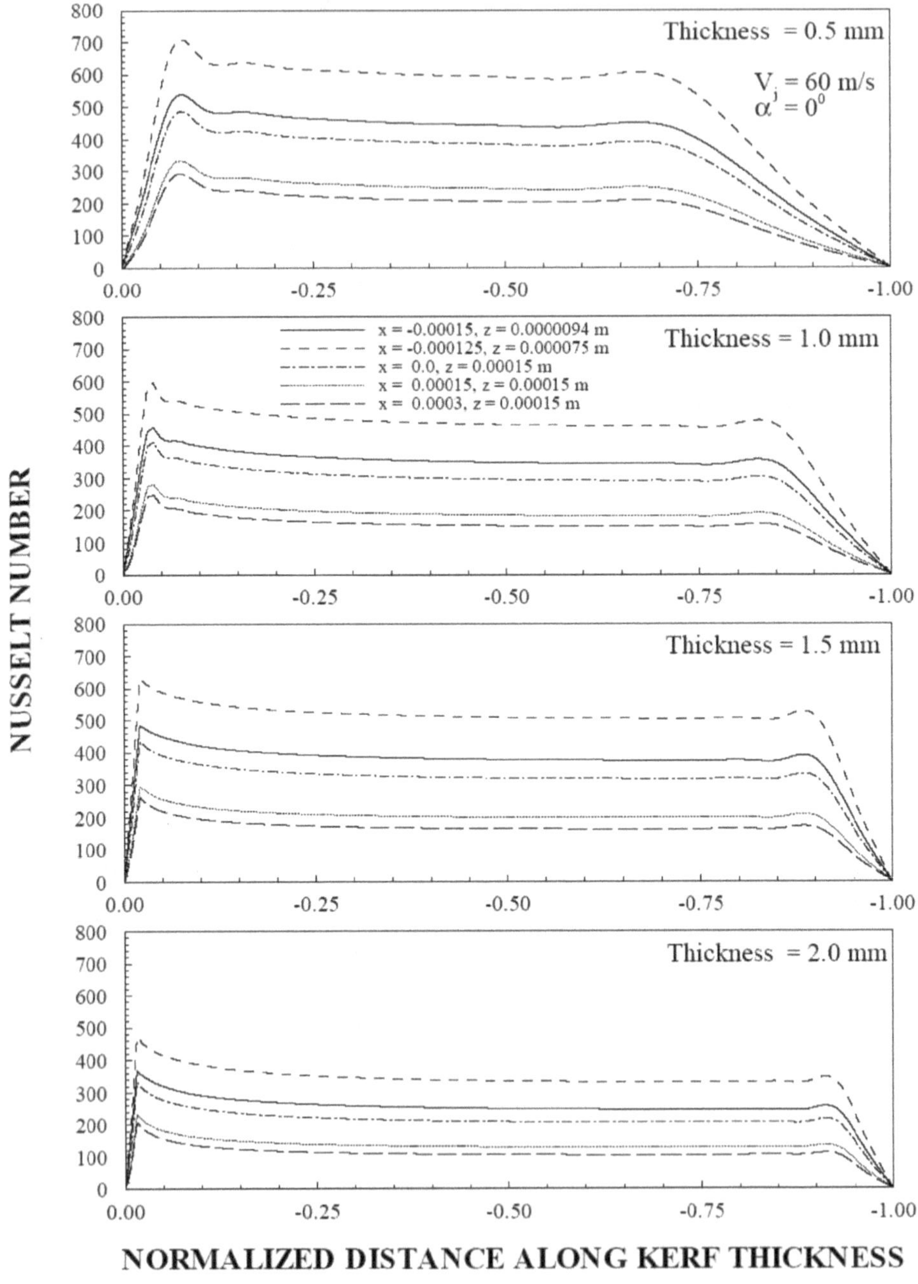

Fig. 4.19 Nusselt number distribution at five locations along the kerf wall for kerf thickness = 0.5, 1.0, 1.5, and 2.0 mm, averaged jet velocity = 60 m/s, and kerf wedge angle = 0 degree [10].

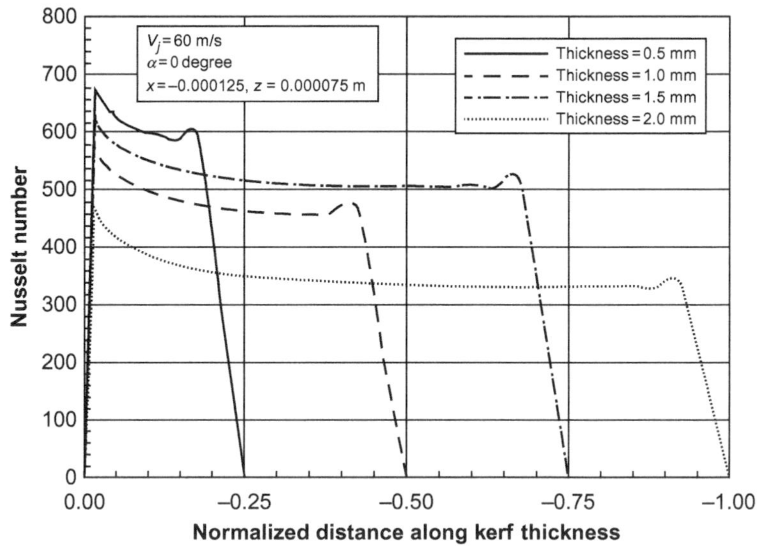

Fig. 4.20 Nusselt number distribution at location B ($x = -0.000125$, $z = 0.000075$ m) along kerf wall for kerf thickness = 0.5, 1.0, 1.5, and 2.0 mm, averaged jet velocity = 60 m/s, and kerf wedge angle = 0 degree [10].

the kerf top edge due to the convection cooling of affect of the jet in this region. Moreover, the appearance of the secondary peak in the Nusselt number distribution is attributed to the sharp rise in the turbulence levels, which accompany transition from an accelerating region flow to a decelerating jet in the kerf.

Fig. 4.20 shows the Nusselt number distribution at location B, where the Nusselt number attains the maximum values, for kerf thickness of 0.5, 1.0, 1.5, and 2.0 mm. The kerf thickness is normalized with the maximum thickness; in which case, the normalized kerf thickness of 0 corresponds to the kerf inlet, and the normalized kerf thickness of −1 corresponds to the kerf exit for the thickness of 2.0 mm. The normalized kerf thickness of −0.25 corresponds to the exit of the kerf for the thickness of 0.5 mm; normalized kerf thickness of −0.5 corresponds to the exit of the kerf for the thickness of 1.0 mm; and normalized kerf thickness of −0.75 corresponds to the exit of the kerf for the thickness of 1.5 mm. The Nusselt number attains the highest values at location B in the kerf for the thickness of 0.5 mm. The values of the Nusselt number attained at location B in the kerf with the thickness of 1.5 mm are higher than those corresponding to the thickness of 1.0 mm. These are also higher than those attained in the kerf with the thickness of 2.0 mm. This is attributed to the heat-transfer rates from the kerf wall surface; in which case, increasing the kerf wall size while increasing the kerf thickness results in an increased thermal boundary layer thickness around the kerf wall. This, in turn, suppresses the value of the Nusselt number for large-thickness.

Fig. 4.21 shows the skin friction distribution along the normalized kerf thickness for four different kerf thicknesses. The kerf thickness is normalized with the

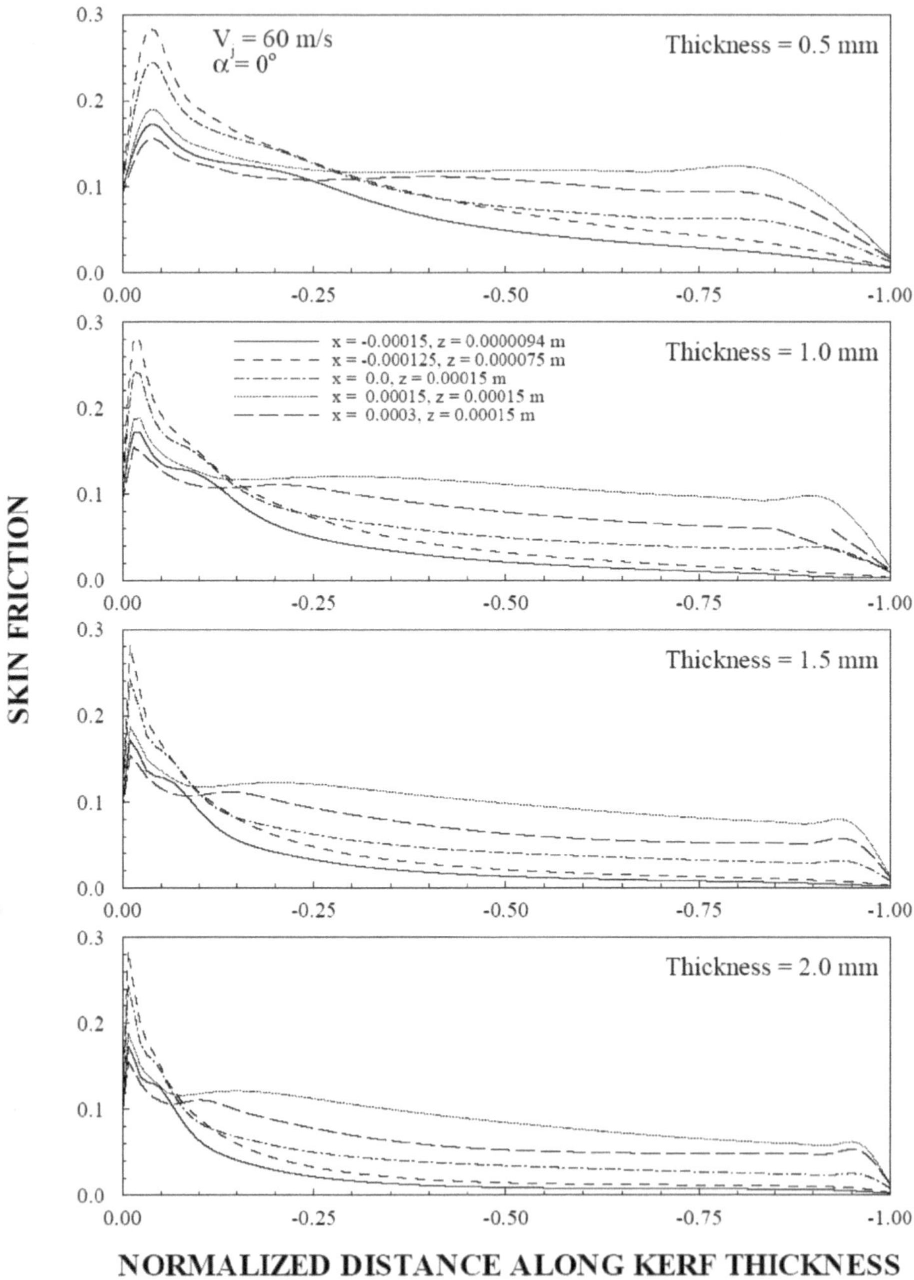

Fig. 4.21 Skin friction distribution at five locations along the kerf wall for kerf thickness = 0.5, 1.0, 1.5, and 2.0 mm, averaged jet velocity = 60 m/s, and kerf wedge angle = 0 degree [10].

maximum thickness; that is, the normalized kerf thickness 0 corresponds to the kerf inlet, and normalized kerf thickness −1 corresponds to the kerf exit. The skin friction distribution is shown at five locations along the kerf wall. Location A ($x = -0.00015$, $z = 0.0000094$ m) is almost the kerf edge (where the kerf wall intersects the symmetry plane). The other 4 locations (B, C, D, and E) are distributed along the kerf wall and are the same locations (Fig. 4.17) for which the Nusselt number distribution is shown in Fig. 4.20. The skin friction increases at the inlet of the kerf due to flow acceleration in this region. The skin friction, then, reduces as the distance along the kerf depth increases. This is attributed to the shear layer developed within the vicinity of the kerf wall, which is influenced by the high wall temperature leading to high thermodynamic pressure in this region. A slight increase in the skin friction over a short distance is also observed. The second peak in the skin friction is observed in the kerf, where the rate of fluid strain becomes high due to the increase in the flow velocity because of the pressure drop in the kerf. As the distance along the kerf wall increases further toward the kerf exit, the skin friction decreases.

4.5.3.3 Influence of assisting gas type on the Nusselt number and skin friction

The effect of the assisting gas type on the heat transfer coefficient and skin friction at the kerf surface is presented in line with the previous study [9–12]. The data used in the simulations are included in Table 4.4. A schematic view of the gas jet impingement and kerf geometry are shown in Figs. 4.22 and 4.23, respectively.

Fig. 4.24 shows the pressure variation along the centerline for three assisting gases. Upon emerging from the nozzle exit, pressure decays in the fluid due to flow expansion. However, the presence of a slot in the flow path results in a sharp rise of pressure before the jet enters into the kerf, which is associated with the blockage effect of the slot in the region next to the kerf inlet. Because the kerf inlet corresponds to $y = 0.0$ m, a pressure drop is observed in the region next to the kerf inlet. This may occur because of the flow path, which forms a streamline curvature, and radial acceleration of the flow develops in a radial jet above the kerf inlet. Due to the velocity increase in the entry region of the kerf, pressure drops sharply and then becomes gradual toward the kerf exit. This is more pronounced for a helium jet because it has a higher velocity in this region as compared to those of air and nitrogen. The presence of a thick shear layer and an increase in thermodynamic pressure in the fluid toward the kerf exit contributes to a gradual decay in pressure in this region. Moreover, pressure increases slightly once the flow exits the kerf, which is attributed to the flow mixing in this region.

Fig. 4.25 shows Nusselt number variation along the kerf length (Fig. 4.1) for three different assisting gas types. It should be noted that the x-axis location is $x = -0.000125$ m and the z-axis location is $z = 0.000075$ m (Fig. 4.23). The Nusselt number increases sharply at the kerf inlet and reduces slightly with the increasing distance from the kerf inlet. As the distance increases further, the Nusselt number remains almost constant along the kerf length. Moreover, the Nusselt number decreases toward the kerf exit. The sharp rise of the Nusselt number is associated with

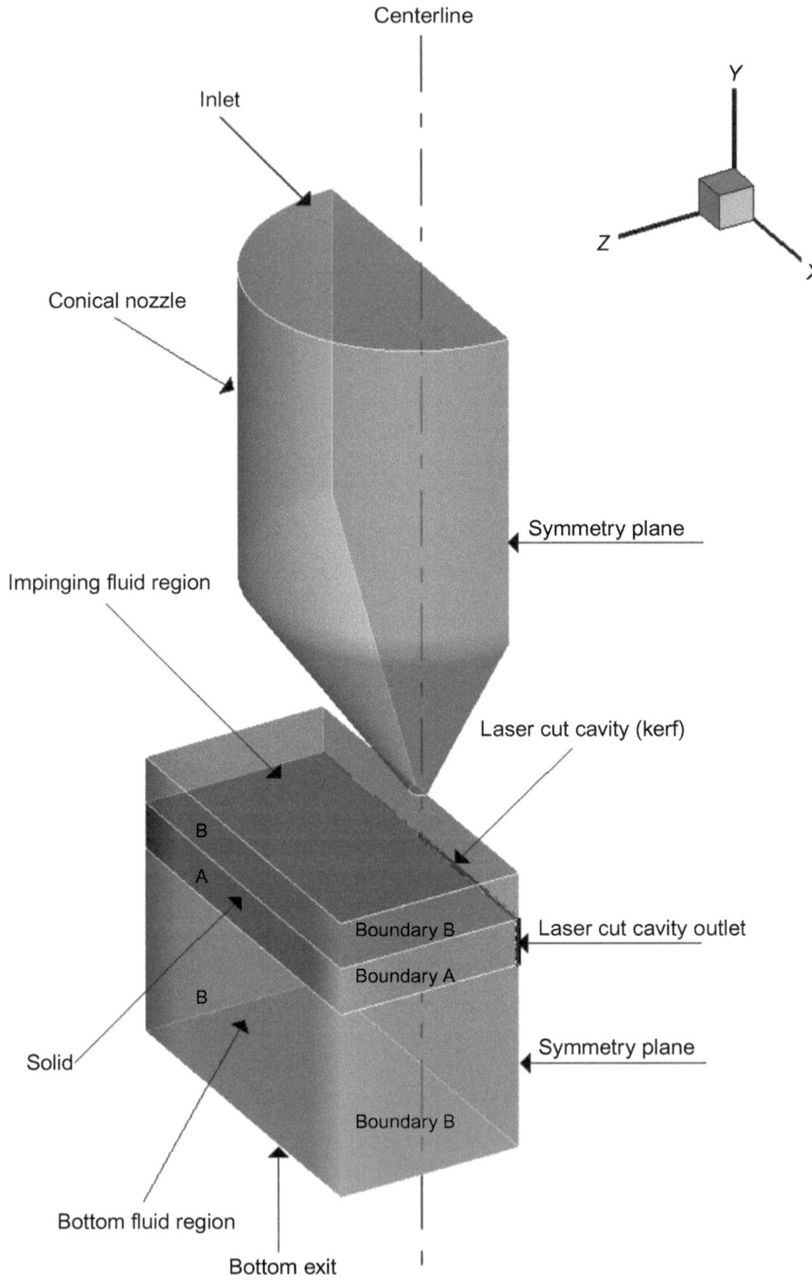

Fig. 4.22 A three-dimensional view of jet impingement onto the kerf [11].

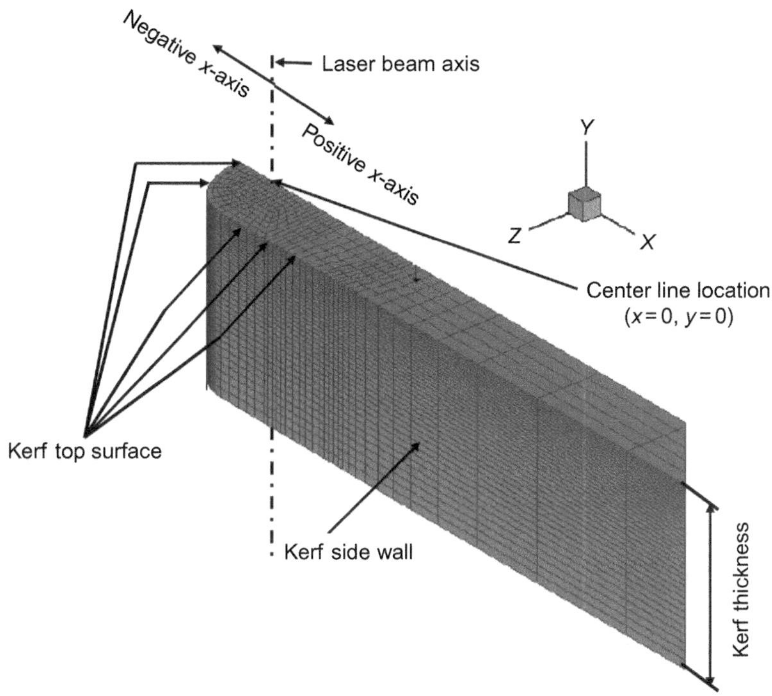

Fig. 4.23 Laser cut kerf geometry [11].

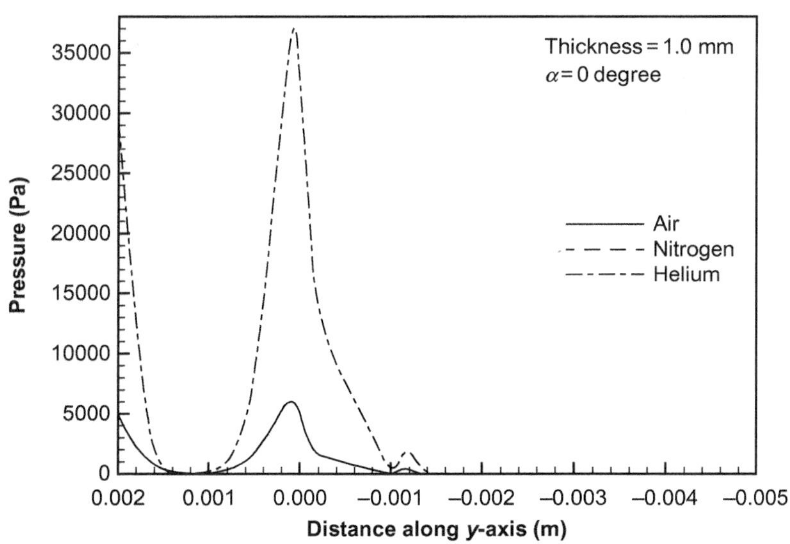

Fig. 4.24 Pressure distributions along the centerline for three different gases [11].

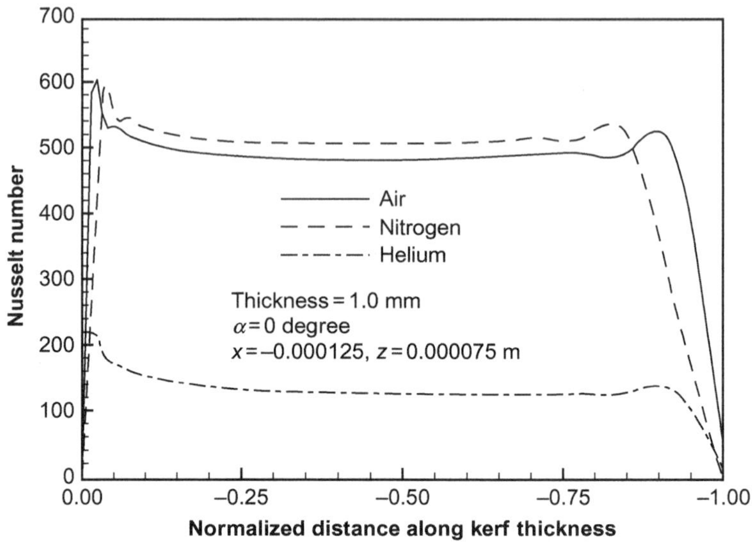

Fig. 4.25 Nusselt number distributions along the kerf thickness for three different gases [11].

heat transfer from the kerf wall to the fluid at the kerf inlet. Because the fluid temperature is low and the kerf wall temperature is high, the amount of heat transfer is high at the kerf inlet. Once the fluid temperature increases at some distance along the kerf surface, the amount of heat transfer reduces, which results in a slight decay of the Nusselt number in this region. The flow velocity and temperature increase as the distance increases in the kerf. Although temperature increase in the fluid has an adverse effect on the amount of heat transfer to the fluid, increasing velocity enhances the convection heat transfer from the kerf wall to the fluid. Consequently, restraint of heat transfer due to a temperature increase balances the enhancement of the heat transfer due to convection. This results in steady-like behavior of the Nusselt number in the kerf. However, the presence of a high temperature shear layer toward the kerf exit enhances the temperature increase and lowers the convection heat transfer in this region. Consequently, the Nusselt number reduces in this region. When comparing the influence of the assisting gas type on the Nusselt number, it is evident that helium results in the lowest Nusselt number, followed by air and nitrogen. This is associated with the density, specific heat capacity, and thermal conductivity of the gases used; in which case, helium has the lowest density, the highest specific heat capacity, and low thermal conductivity [17].

Fig. 4.26 shows the skin friction along the kerf length (Fig. 4.23) for different assisting gases. The skin friction increases sharply at the kerf inlet due to the sharp drop of pressure in this region. As the distance increases further in the kerf, the skin friction reduces gradually toward the kerf exit, which, then, decays sharply. Thermodynamic pressure at the kerf wall vicinity due to the kerf wall high temperature alters the thermal boundary layer thickness and fluid density in the region close to the kerf wall. Consequently, once the fluid temperature increases in the kerf, the rate of fluid

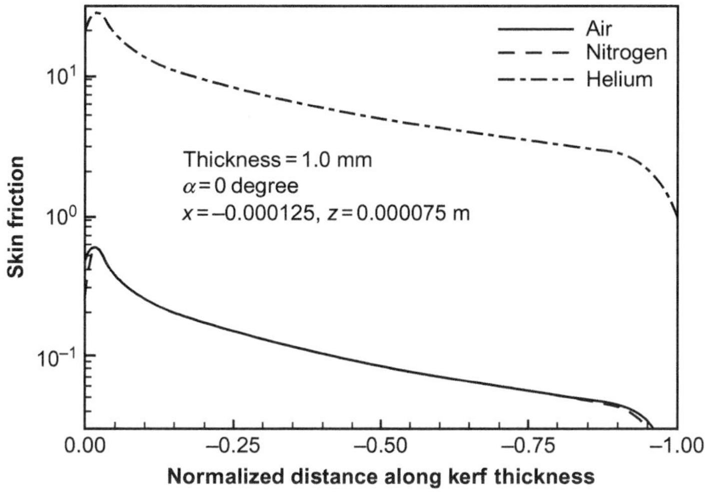

Fig. 4.26 Skin friction distributions along the kerf thickness for three different gases [11].

strain in the kerf vicinity changes notably because of the density variation. This results in the gradual decay of the skin friction in the kerf. However, the shear layer thickening toward the kerf exit modifies the rate of fluid strain in this region significantly. In addition, attainment of the high fluid temperature alters both the density and viscosity of the fluid. This influences the behavior of the skin friction at the kerf exit. Consequently, the skin friction decays sharply in this region. The influence of the assisting gas type on the skin friction is notable; in which case, helium results in the highest skin friction because of its relatively high velocity due to its high rate of fluid strain in the kerf wall surface region. Air and nitrogen assisting gases show similar behavior due to their velocity behavior in the kerf and the close values of their viscosities.

4.5.3.4 Effect of jet velocity on heat transfer rates

The variation of the Nusselt number and the skin friction at the kerf surfaces are presented for four average jet velocities at the nozzle exit and two kerf wall wedge angles. The findings are discussed in line with the previous study [9–12]. The data used in the simulations are given in Table 4.4. Fig. 4.27A and B shows the distribution of the meshes used in the simulations and kerf site.

Fig. 4.28 shows the pressure distribution along the laser beam axis for different averaged jet velocities at the nozzle exit and two different kerf wedge angles. Static pressure reduces as the jet expands toward the kerf top surface ($y = 0.001$ m). Pressure then increases toward the kerf because of the stagnation formation and because of the blockage effect of the kerf. However, a sudden drop in static pressure is observed close to the kerf surface. This pressure drop continues within the kerf. This is because of the jet expansion in this region. This is more pronounced for a high-averaged jet velocity at the nozzle exit. The influence of the kerf wedge angle on pressure distribution is

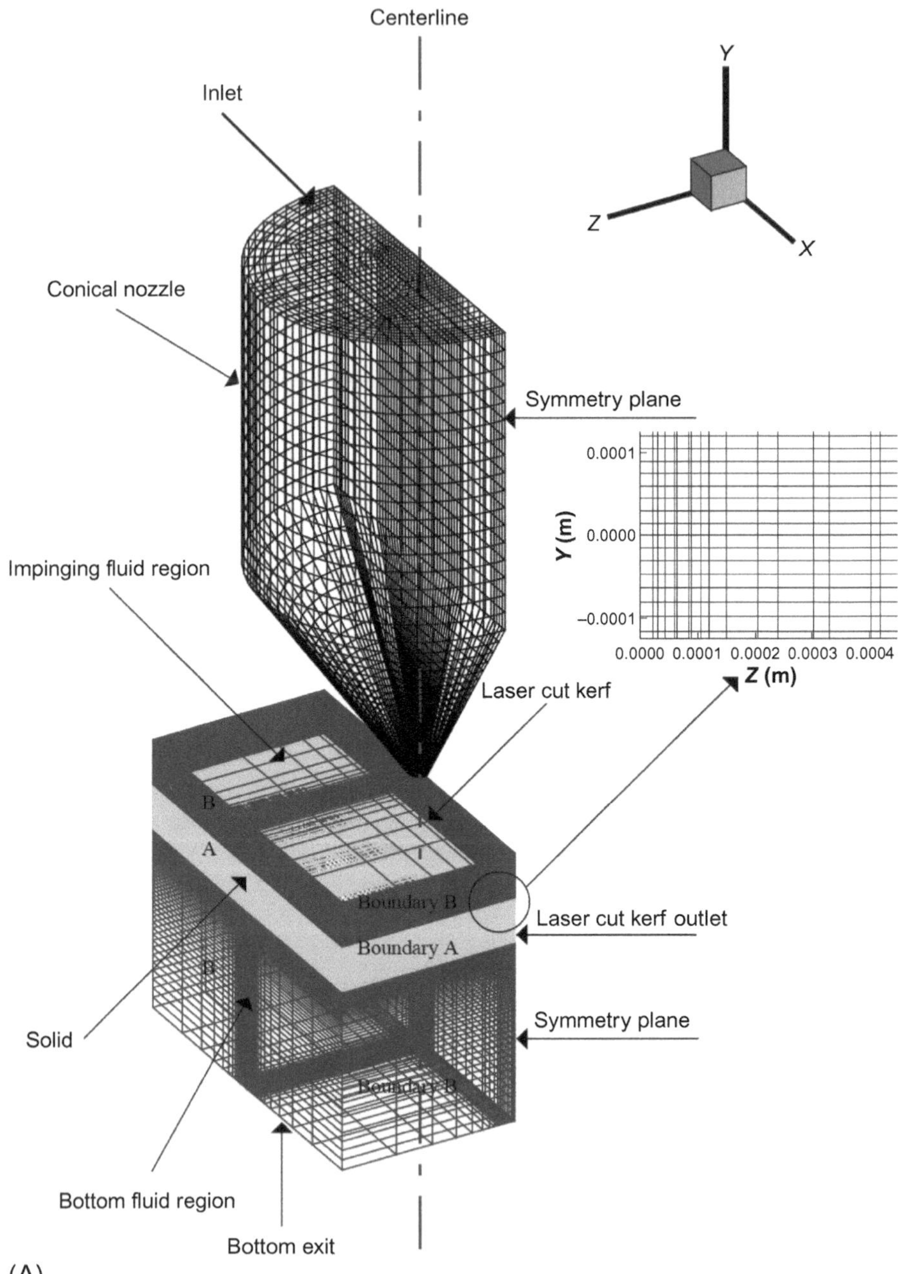

Fig. 4.27 (A) Mesh generated in the solution domain [12]. (B) Close view mesh generated in the kerf [12].

(Continued)

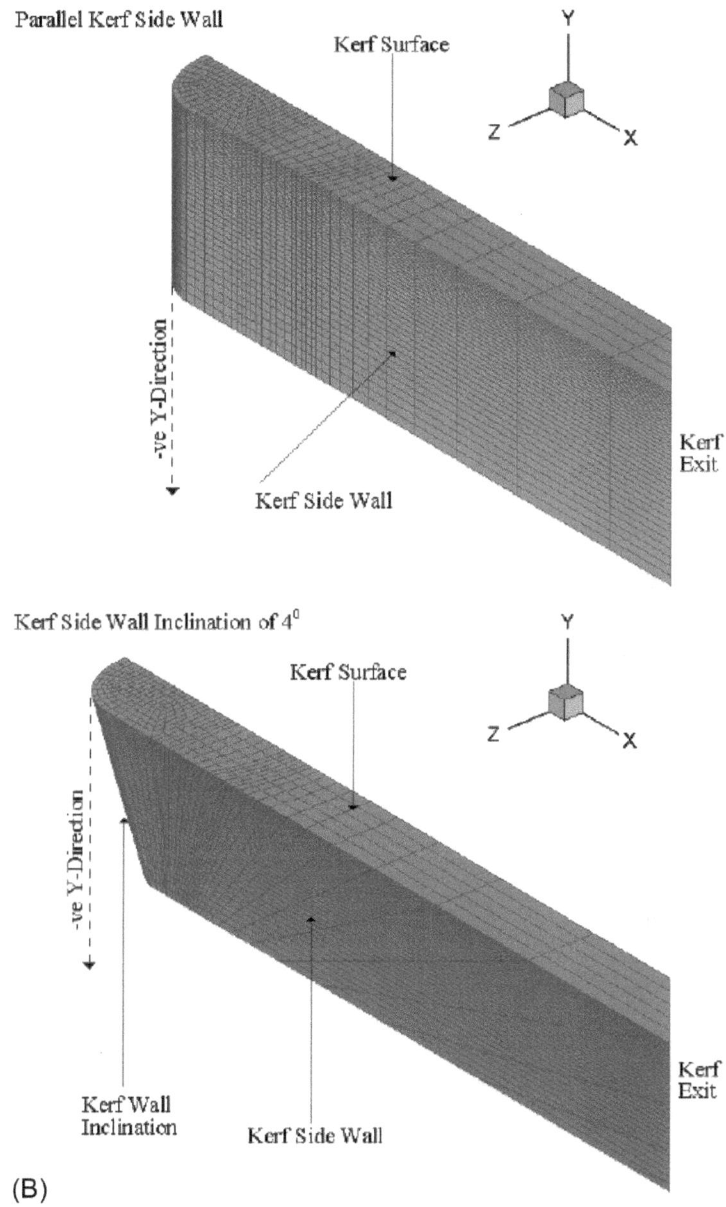

Fig. 4.27—Cont'd

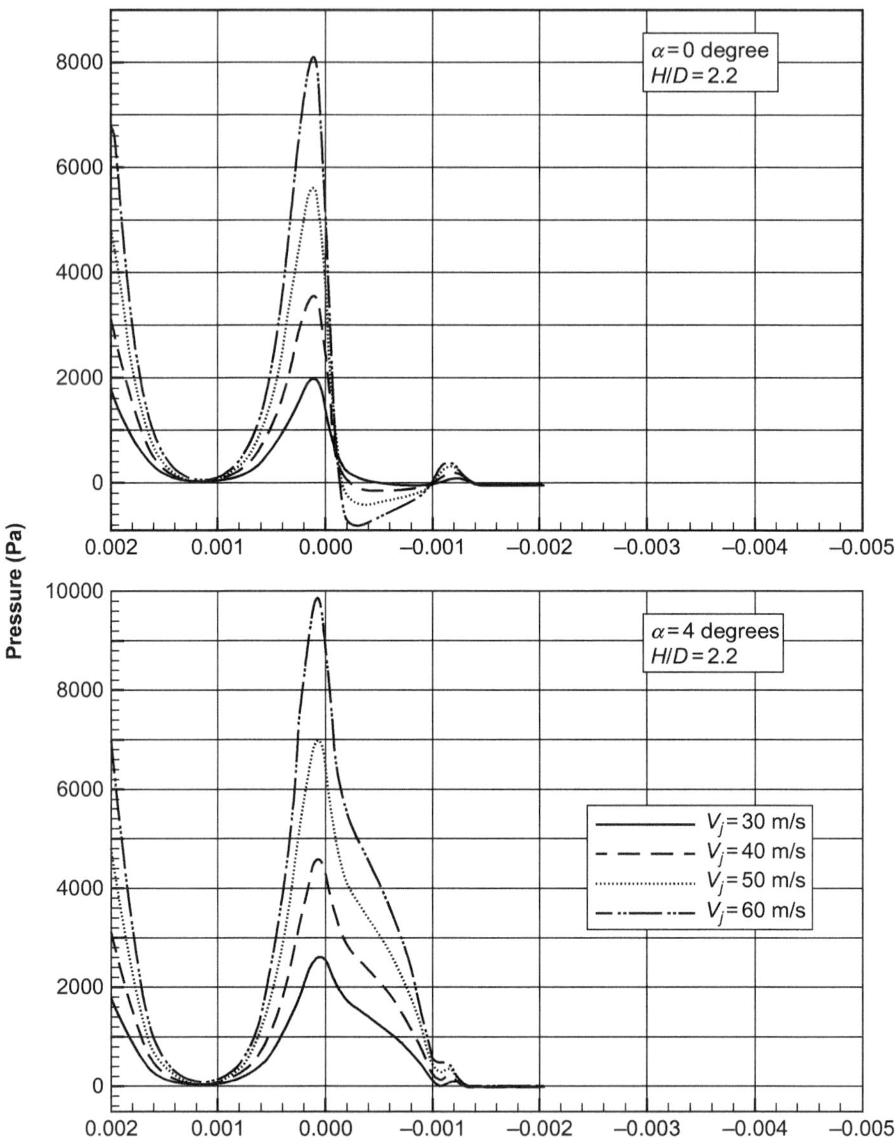

Fig. 4.28 Pressure variation along the laser beam axis for different averaged jet velocities of kerf wedge angles of 0 and 4 degrees [12].

significant; in which case, a pressure rise toward the kerf inlet ($y=0.001$ m) becomes considerably high because of the enhanced blockage effect of the kerf wedge. Moreover, a pressure drop is sharp for the kerf wedge angle = 0 degree in the neighborhood of the kerf inlet while it is gradual for the kerf wedge angle = 4 degrees. The attainment of negative pressure in the kerf for the kerf wedge angle = 0 degree indicates the overexpansion of the jet in this region (from $y = -0.0001$ m toward $y = -0.001$ m). It should be noted that the gradual drop in pressure for the higher kerf wedge angle can be associated with the temperature rise in the kerf due to heat transfer from the kerf wall to the jet.

Fig. 4.29 shows the Nusselt number distribution along the normalized kerf wall thickness for different averaged jet velocities at the nozzle exit and two kerf wedge angles. The kerf wall thickness is normalized with the maximum thickness; that is, a normalized kerf wall thickness = 0 represents the kerf inlet and normalized kerf

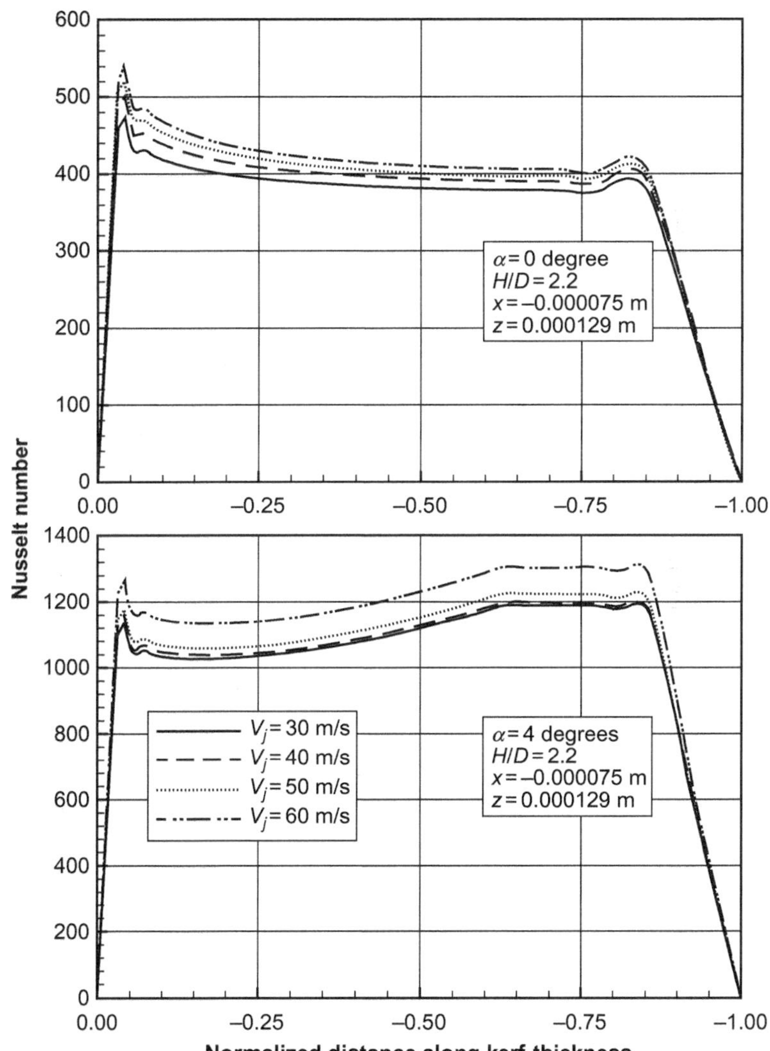

Fig. 4.29 Nusselt number distribution along the normalized kerf wall thickness for different averaged jet velocities at the nozzle exit and kerf wedge angles of 0 and 4 degrees [12].

wall thickness = −1 is the kerf exit. The curves are plotted at a location ($x = -0.000075$ m; $z = 0.000129$ m) where the Nusselt number attained the highest value. In the case of the kerf wall wedge angle $\alpha = 0$ degree, the Nusselt number increases sharply from the kerf edge due to convective cooling of the jet in the kerf and reaches its maximum ($Nu \asymp 520$). It should be noted that the kerf wall temperature was kept at 1500°K to resemble the laser cutting process. The Nusselt number decreases gradually in the kerf, and then reduces sharply toward the kerf exit. The radial acceleration of the flow in the kerf influences the Nusselt number; in which case, the Nusselt number increases with increasing jet velocity. However, this increase is not significant in the regions close to the kerf inlet and kerf exit. As the kerf wedge angle increases ($\alpha = 4$ degrees), the Nusselt number increases gradually toward the kerf exit ($Nu \asymp 1300$). This gradual increase is associated with the blockage effect of the kerf. In this case, pressure buildup increases the density of the fluid in the kerf while enhancing the amount of heat transfer in the kerf. Moreover, the Nusselt number reduces sharply toward the exit of the kerf, similar to the case for which the kerf wedge angle = 0 degree. The influence of averaged jet velocity at the nozzle exit on the Nusselt number is more pronounced for the kerf wedge angle = 4 degrees.

Fig. 4.30 shows the skin friction distribution along the normalized kerf thickness for different averaged jet velocities at the nozzle exit and two kerf wedge angles. The location ($x = -0.000075$ m; $z = 0.000129$ m) correspond to the location of the maximum Nusselt number. The skin friction increases sharply at the inlet of the kerf due to the flow acceleration within the kerf. As the kerf depth increases, the skin friction reduces. This is attributed to the shear layer developed within the vicinity of the kerf wall, which is influenced by the high wall temperature because of the formation of high thermodynamic pressure in this region. The influence of the mean jet velocity at the nozzle exit is notable in such a way that increasing the mean jet velocity reduces the skin friction. This is because of the thermal boundary layer formed in the vicinity of the kerf wall; that is, reducing the mean jet velocity enhances the thermal boundary layer thickness because of the low axial momentum of the flow in the surface region. Consequently, thickening of the thermal boundary layer increases the shear stress at the wall. In addition, reducing the averaged jet velocity reduces the velocity head ($1/2\ \rho V_j^2$) while increasing the skin friction. The influence of kerf wedge angle on the skin friction is considerable; in which case, skin friction first drops, and later increases with increasing depth along the kerf wall. The increase in the skin friction toward the kerf exit is because of the blockage effect of the wedge angle. Therefore, pressure buildup toward the kerf exit modifies the boundary layer thickness and the rate of fluid strain in this region. Consequently, the skin friction increases. This is more pronounced for the low-averaged jet velocity ($V_j = 30$ m/s).

4.6 Entropy analysis of the laser cutting process

In this study, laser cutting of thick sections is considered and modeling of the laser cutting process is introduced. Thermal efficiency of the cutting process and the specific energy required for the cutting are formulated. In the analysis, a lump parameter

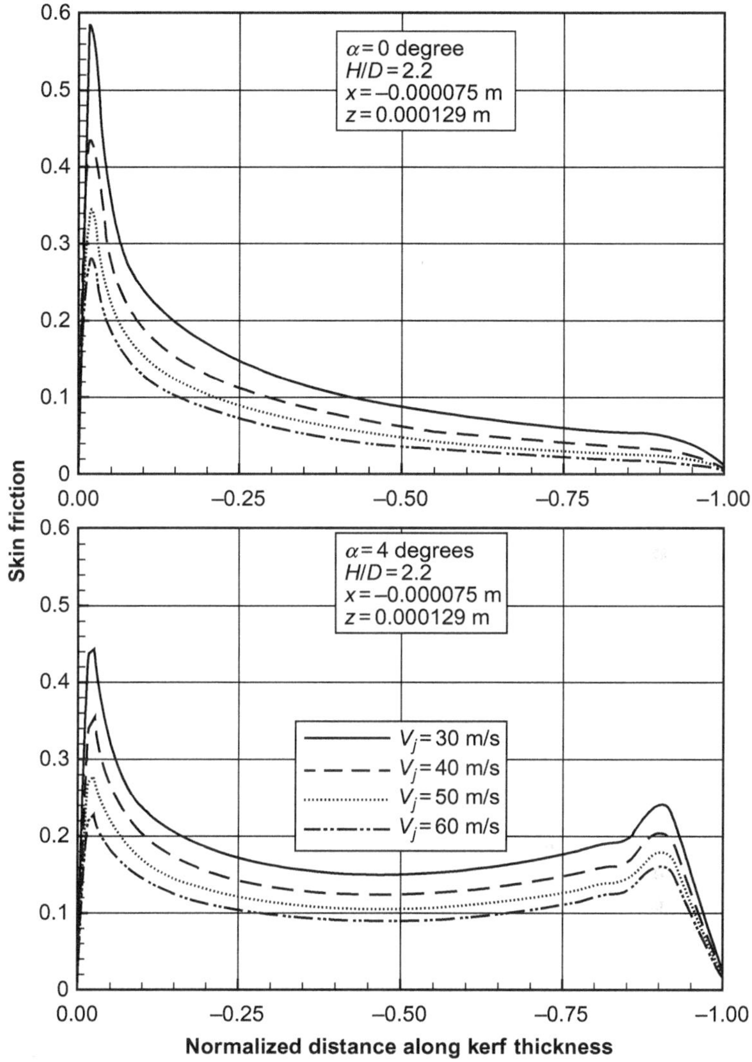

Fig. 4.30 Skin friction distribution along the normalized kerf wall thickness for different averaged jet velocities at the nozzle exit and kerf wedge angles of 0 and 4 degrees [12].

approach is introduced to formulate energy gain by the substrate material during the laser cutting process. The analysis and findings are presented in line with the previous study [18].

The specific energy required for a laser cutting of a metallic substrate is formulated using a lump parameter technique. In this case, the rate of energy balance for cutting of sheet metal is considered in the formulations. The cutting situation and geometric view of the cut section is shown in Fig. 4.31.

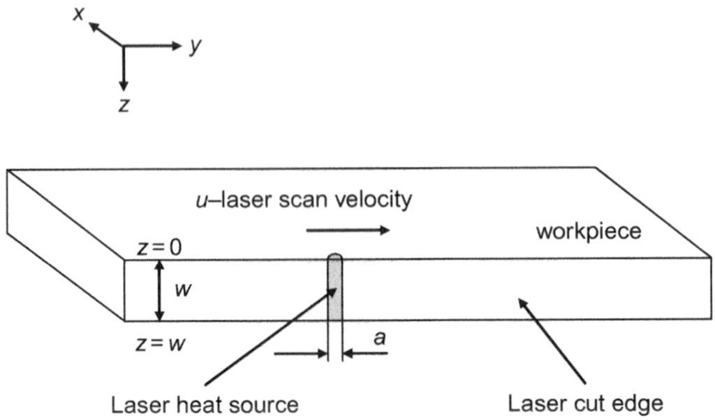

Fig. 4.31 A laser heating situation and coordinate system.

The rate of energy required for cutting of a section can be written as:

$$\dot{E}_r = \dot{E}_l + \dot{m}[\Delta u + L_m + \beta L_{en}] \quad (4.50)$$

where $\dot{E}_l$, $\dot{m}$, Δu, L_m, β, L_{ev} are the rate of total losses during the laser cutting, the rate of total mass of material removed during laser cutting, internal energy gain of the substrate material removed, latent heat of melting, fraction of mass evaporated, and latent heat of evaporation, respectively.

The total material removed during laser cutting, after assuming constant density of the substrate material, is:

$$\dot{m} = \rho\frac{d\forall}{dt} = \rho\frac{d(w \cdot L \cdot Th)}{dt} = w \cdot Th\frac{dL}{dt} \quad (4.51)$$

where $\forall$ is the volume of the laser cut hole, ρ is the density of the substrate material, w and L are the width and length of the cut section, and Th is the thickness of the sheet metal. The term dL/dt represents the cutting velocity, which is the laser scanning velocity during the cutting process.

In addition, the rate of total loss includes the rate of conduction and convection losses during the cutting; that is:

$$\dot{E}_{loss} = \dot{E}_{conduction} + \dot{E}_{convection} \quad (4.52)$$

where $\dot{E}_{conduction}$ and $\dot{E}_{convection}$ are the rate of conduction and convection losses.

The rate of conduction loss from the cutting wall site can be written as:

$$\dot{E}_{conduction} = \frac{d}{dt}\left[\int_0^{Th}\left(\int_0^{50w} \rho \cdot C_pT \cdot L \cdot dy\right)dx\right] \quad (4.53)$$

where 40w represents the distance away from the solution domain in the y-direction (radial direction).

It is assumed that the natural convection occurs at the top and bottom surfaces of the cut section. The heat transfer coefficients across both surfaces are the same. In this case, the convection losses can be written as:

$$\dot{E}_{convection} = A_s[h_1(T_s - T_o) + h_2(T_s - T_o)] = w \cdot L \cdot (T_s - T_o)(h_1 + h_2) \tag{4.54}$$

where A_s is the surface area of the cut section, h_1 and h_2 are the heat transfer coefficient at top and bottom surfaces of the cut section.

The specific energy required for the cutting is determined from:

$$E_{spec} = \frac{1}{\forall}\int_o^{t_c} \dot{E}_r dt \tag{4.55}$$

where t_c is the time taken to cut the edges in the sheet metal.

Thermal efficiency of the laser drilling process is the ratio of the rate of energy required for through cutting over the power used for the cutting process. In this case, thermal efficiency becomes:

$$\eta = \frac{\dot{E}_{req}}{P} \tag{4.56}$$

where P is the laser power on the workpiece surface during the cutting process. The solutions of Eqs. (4.50)–(4.56) are obtained by using Mathematica software. The data used in the solution are given in Table 4.5.

Fig. 4.32 shows thermal efficiency with laser output power for different cutting velocities. Thermal efficiency increases with reduced laser power. This is more pronounced for high cutting velocities. In this case, power required for cutting reduces with increased velocity, which is more pronounced for high-cutting velocities. This is because the energy required for cutting per unit time reduces with increased cutting speed. Moreover, as the laser power increases, thermal efficiency remains the same, provided that thermal efficiency attains high values with increased cutting velocity.

Table 4.5 Properties used in the simulations

Source of variation	Value	Units
Fraction of evaporation contribution (β)	0.1	–
Latent heat of melting of workpiece	247,112	J/kg
Thermal conductivity of workpiece	63	W/mK
Specific heat capacity of workpiece	460	J/kgK
Thermal diffusivity of workpiece	1.6134×10^{-5}	m^2/s
Melting temperature of workpiece	1700	°K

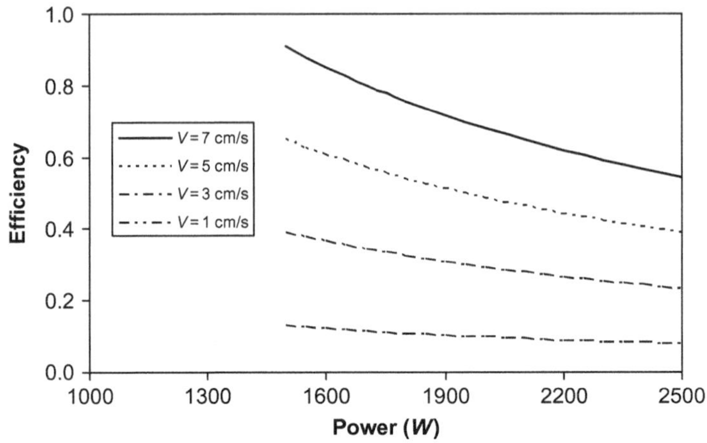

Fig. 4.32 Thermal efficiency of laser cutting with laser power for different scanning speeds [18].

In the case of the effect of cutting speed on thermal efficiency, increasing cutting speed improves thermal efficiency significantly because the energy required per unit time increases so that the ratio of energy required per unit time for cutting to laser power intensity increases. It should be noted that although increasing cutting speed while keeping the laser power constant enhances thermal efficiency, the cutting process may not be completed due to insufficient laser power for complete cutting.

References

[1] Z. Yilbas, Classification of striation patterns using a neural network in CO_2 laser cutting, Lasers Eng. 7 (1988) 25–37.

[2] O.L. Davies, Design and Analysis of Industrial Experiments, first ed., Longman Group Ltd., London, 1978.

[3] B.S. Yilbas, M. Rashid, CO_2 laser cutting of Incoloy 800 HT alloy and quality assessment, Lasers Eng. 12 (2002) 135–145.

[4] S.Z. Shuja, B.S. Yilbas, Heating of a moving slab with two laser spots: influence of relative spacing of spots on temperature field, Lasers Eng. 26 (1-2) (2013) 59–70.

[5] S.Z. Shuja, B.S. Yilbas, The influence of gas jet velocity in laser heating-a moving workpiece case, Proc. Inst. Mech. Eng. C: J. Mech. Eng. Sci. 214 (2000) 1059–1078.

[6] Y. Ootao, Y. Tanigawa, Three-dimensional solution for transient thermal stresses of functionally graded rectangular plate due to nonuniform heat supply, Int. J. Mech. Sci. 47 (2005) 1769–1788.

[7] COMSOL, http://www.comsol.com/, 2015.

[8] F.P. Incorpera, D.P. Dewitt, T.L. Bergman, A.S. Lavine, Fundamentals of Heat and Mass Transfer, sixth ed., John Wiley and Sons, Hoboken, NJ, 2007.

[9] O. Melhem, B.S. Yilbas, S.Z. Shuja, Jet impingement onto kerf: effect of kerf wedge angle on heat transfer rates and skin friction, Optics Laser Technol. 56 (2014) 76–87.

[10] O. Al-Melhem, B.S. Yilbas, S.Z. Shuja, Jet Impingement onto a laser produced kerf: influence of kerf thickness on heat transfer and skin friction, Int. J. Numer. Meth. Heat Fluid Flow 21 (6) (2011) 754–778.

[11] O. Melhem, B.S. Yilbas, S.Z. Shuja, A.Z. Sahin, Influence of assisting gas type on the Nusselt number and the skin friction on slots in relation to laser cutting, Heat Transfer Eng. 34 (10) (2013) 852–862.

[12] O. Melhem, B.S. Yilbas, S.Z. Shuja, Three-dimensional consideration of jet impingement onto kerf in relation to laser cutting process: effect of jet velocity on heat transfer rates, Opt. Lasers Eng. 49 (3) (2011) 384–395.

[13] H.K. Versteeg, W. Malalasekera, An Introduction to Computational Fluid Dynamics: The Finite Volume Method, second ed., Pearson Education Ltd., England, 2007.

[14] www.fluent.com, 2015.

[15] D. Sahoo, M.A.R. Sharif, Mixed-convective cooling of an isothermal hot surface by confined slot jet impingement, Numer. Heat Transfer Part A: Appl. 45 (2004) 887–909.

[16] S.V. Patankar, Numerical Heat Transfer, McGraw-Hill, New York, 1980.

[17] B. Sagot, G. Antonini, A. Christgen, F. Buron, Jet impingement heat transfer on a flat plate at a constant wall temperature, Int. J. Thermal Sci. 47 (2008) 1610–1619.

[18] B.S. Yilbas, Laser cutting of thick sheet metals: effects of cutting parameters on kerf size variations, J. Mater. Process. Technol. 201 (1–3) (2008) 285–290.

Some applications of laser cutting 5

5.1 Introduction

Laser cutting of sheet metals has several advantages over conventional cutting methods. Some of these advantages include high-speed and local processing, precision of operation, low cutting waste, and net shaping. However, laser cutting faces problems when processing difficult-to-cut materials. This is because of the fact that localized heating, due to the laser irradiation at the surface, can be suppressed by the reflectivity of the substrates or heat dissipation because of high thermal conductivity. In addition, laser cutting of ceramics is challenging because of the thermally-induced cracking generated during the laser-cutting process. In the laser-cutting process, the end product is highly-dependent on the proper selection of the laser parameters and the workpiece material properties such that the high quality of the desired end product is ensured. One of the high-thermal conductivity materials of the metallic substrate is bronze, which is used in bearing applications due to its low friction coefficient. Because laser cutting is involved with high-temperature processing, temperature gradients developed in the cutting region result in high-thermal stresses. Depending on the mechanical properties of the substrate material used, cracks can be formed at the cut surfaces, limiting the practical application of the laser cut components. Consequently, the investigation into laser cutting of difficult-to-cut materials becomes essential. This, in turn, provides useful information on the machining capability lasers in practical applications.

In this chapter, laser cutting of bronze, alumina tiles, Kevlar, carbon composite, and wedges are presented. In addition, the examples of straight and circular cutting into these materials are included. The formulation of thermal stress and temperature fields are provided, and findings are discussed in line with the previous studies [1–11].

5.2 Heating and stress analysis

The high-thermal conductivity materials are difficult to cut by a laser because of heat dissipation from the localized irradiated region toward the solid bulk. Therefore, when examining the cutting process, analysis of temperature and stress fields is fruitful. In addition, morphological and metallurgical changes in the cut section are important because of high rate of heat flux from the kerf surface toward the solid bulk. Two different cutting configurations are considered, namely straight and circular cuttings. In this case, straight cutting is associated with laser movement of straight lines during the cutting process. However, laser movement completes a circular shape in the circular cutting arrangement. This requires different formulation of the laser power intensity in the analysis. Consequently, the analysis related to the formulation of temperature and stress fields are given incorporating the straight and circular laser cuttings, and findings are presented following the mathematical formulas.

The Laser Cutting Process. https://doi.org/10.1016/B978-0-12-812982-1.00005-9

5.2.1 Thermal analysis

Fig. 5.1 shows the schematic view of the straight laser cutting and the coordinate system. In the analysis, the solid-body heat conduction with temperature-dependent conductivity, internal energy (including latent heat effects) and convection and radiation boundary conditions are considered. The Fourier heat transfer equation, pertinent to the laser heating process, can be written as:

$$\rho\frac{\partial H(T)}{\partial t} = (\nabla(k(T)\nabla T)) + S_o \tag{5.1}$$

where H is the temperature-dependent enthalpy including the latent heat of solidification, ρ is the density, k is the temperature-dependent thermal conductivity, and S_o is the heat source term resembling the laser beam.

In the case of a straight laser cutting, the laser source moves along the x-axis, with the laser beam intensity distribution remaining at the Gaussian distribution at the surface; that is:

$$S_o = I_o \delta e^{-\delta z}(1 - r_f)e^{\left(-\frac{x^2+y^2}{a^2}\right)} \tag{5.2}$$

I_o is laser power peak density, δ is the absorption coefficient, a is the Gaussian parameter, r_f is the surface reflectivity, and x and y are the axes while the laser beam scans the surface along the x-axis.

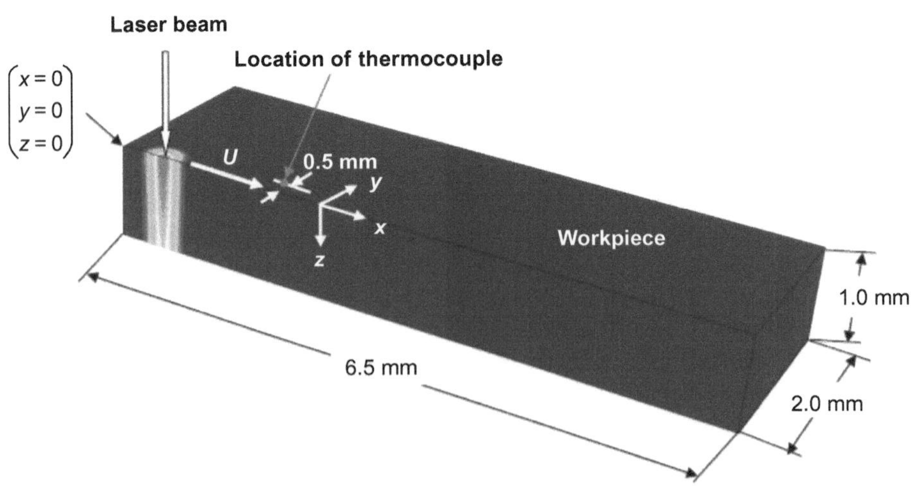

Fig. 5.1 A shows a schematic view of laser straight cutting process and coordinate system. Half of the workpiece is shown in order to demonstrate the laser beam location in the workpiece.

In the case of a circular laser cutting, the laser beam completes a circular shape during the cutting process. In this case, power intensity distribution remains the same at the workpiece surface; however, power intensity can be formulated as:

$$S_o = I_o \delta e^{-\delta z}\left(1 - r_f\right) e^{\left(-\frac{(x - rSinwt)^2 + (y - rCoswt)^2}{a^2}\right)} \tag{5.3}$$

I_o is the laser power peak density, δ is the absorption coefficient, a is the Gaussian parameter of the laser beam, r_f is the surface reflectivity, ρ is the density, and x and y are the axes while the laser beam scans the surface along the circumference of the cut hole, then, $r = \sqrt{x^2 + y^2}$, that is, circular cutting is carried out and the laser beam motion follows the circumference of the laser cut hole with a constant angular velocity ω. Fig. 5.2 shows a typical laser circular cutting arrangement.

The effective penetration depth of the laser beam after the key-hole formation is estimated through considering the multiple reflections from the cut edges and incorporating the exponential decay of the laser intensity along the absorption depth. It can be simplified as:

$$\delta = \frac{1}{L} \ln\left(\frac{I_o}{I_L}\right) \tag{5.4}$$

where L is the thickness of the workpiece, I_o is the peak power intensity at the workpiece surface, and I_L is the depth below the surface where the laser power intensity

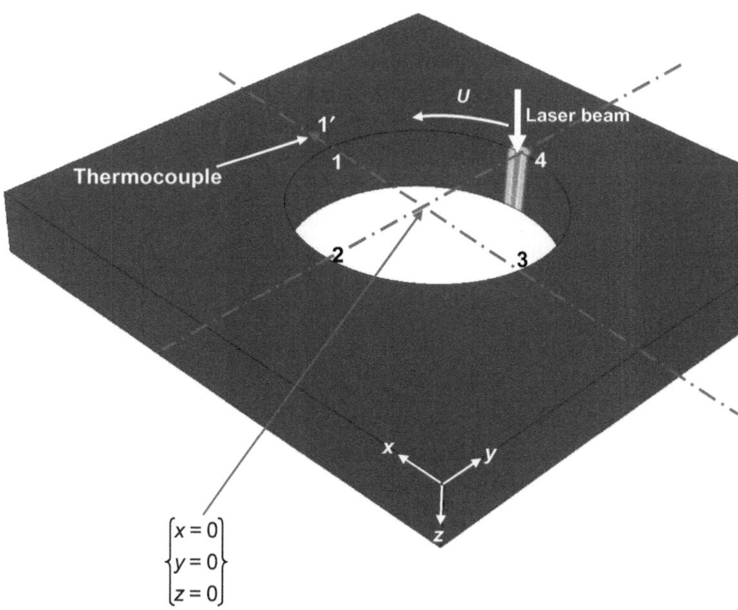

Fig. 5.2 Schematic view of the laser circular cutting process and the location of thermocouple.

reduces $1/e$ value. It should be noted that the absorption of the incident beam by the substrate material is assumed to behave similar to the Lambert-Beer's law, that is, exponential decay of the intensity along the workpiece thickness. The laser beam axis is parallel to the z-axis (Fig. 5.1). It should be noted that the laser beam intensity distribution is assumed to be Gaussian at the irradiated surface. It is assumed that once temperature in the irradiated region exceeds the melting temperature of the substrate material, superheating in the liquid phase occurs. In this case, the pressure in the liquid phase may rise. Because the irradiated surface is freely exposed to atmosphere, the liquid phase expands freely in the melted zone. Therefore, the pressure increase in the liquid phase can be neglected. In addition, the specific heat capacity of the liquid phase is kept the same in the simulation during the superheating of the liquid phase. However, the evaporation at the surface does not take place during the heating cycle, because surface temperature remains below the boiling temperature of the substrate material in the present study.

The convective and radiation boundary conditions are considered at the free surface of the workpiece. Therefore, the corresponding boundary condition is:

At the irradiated surface (top surface):

$$\frac{\partial T}{\partial z} = \frac{h_f}{k}(T_s - T_{amb}) + \frac{\varepsilon\sigma}{k}\left(T_s^4 - T_{amb}^4\right) \tag{5.5}$$

where h_f (=3000 W/m^2 K [12]) is the forced-convection heat transfer coefficient due to the assisting gas.

At the bottom surface:

$$\frac{\partial T}{\partial y} = \frac{h}{k}(T_s - T_{amb}) + \frac{\varepsilon\sigma}{k}\left(T_s^4 - T_{amb}^4\right) : \frac{\partial T}{\partial z} = \frac{h}{k}(T_s - T_{amb}) + \frac{\varepsilon\sigma}{k}\left(T_s^4 - T_{amb}^4\right) \tag{5.6}$$

where h (=20 W/m^2 K) is the heat transfer coefficient due to natural convection, and T_s and T_{amb} are the surface and ambient temperatures, respectively, ε is the emissivity (ε=0.9 is considered), and σ is the Stefan-Boltzmann constant ($\sigma = 5.67 \times 10^{-8}$ W/m^2 K). At a far away boundary (at edges of the solution domain), a constant temperature boundary is assumed (T=293 K); that is,

$$x = \infty;\ y = \infty;\ z = t_h\ \rightarrow T = 293\text{K} \tag{5.7}$$

where t_h is the thickness of the workpiece. Initially (prior to laser cutting), the substrate material is assumed to be at a constant ambient temperature, that is, $T = T_{amb}$, which is considered as constant (T_{amb}=293 K).

Eq. (5.1) is solved numerically with the appropriate boundary conditions to predict the temperature field in the substrate material. However, to analyze the phase-change problem, the enthalpy method is used [13]. The specific heat capacity is associated with the internal energy gain of the substrate material; that is, $Cp(T) = \frac{\partial U}{\partial T}$. However, the internal energy gain during the phase change is associated with the latent heat of fusion, which is given separately in terms of solidus and liquidus temperatures (the lower and upper temperature bounds of the phase-change range) and the total internal energy associated with the phase change, called the latent heat [13].

5.2.2 Thermal stress analysis

Thermal stress developed in the laser cutting region is involved with elasto-plastic material behavior under the high-temperature gradients. The formula for the thermal-stress problem is given below.

Solidification involves a small strain, so the assumption of the small strain is adopted in this work. The thermal strains which dominate thermo-mechanical behavior during solidification are on the order of only a few percent. Several previous solidification models [14,15] confirm that the solidified metal part indeed undergoes only a small deformation during solidification. The displacement spatial gradient is small $\nabla \mathbf{u} = \partial \mathbf{u}/\partial \mathbf{x}$ so $\nabla \mathbf{u} : \nabla \mathbf{u} \approx 1$ and the linearized strain tensor becomes [16]:

$$\varepsilon = \frac{1}{2}\left[\nabla \mathbf{u} + (\nabla \mathbf{u})^T\right] \tag{5.8}$$

The small strain formulation can be used, where Cauchy stress tensor is identified with the nominal stress tensor $\boldsymbol{\sigma}$, and $\mathbf{b}$ is the body force density with respect to initial configuration.

$$\nabla . \boldsymbol{\sigma}(\mathbf{x}) + \mathbf{b} = 0 \tag{5.9}$$

The rate representation of total strain in this elastic-viscoplastic model is given by [13]:

$$\dot{\varepsilon} = \dot{\varepsilon}_{el} + \dot{\varepsilon}_{ie} + \dot{\varepsilon}_{th} \tag{5.10}$$

where $\dot{\varepsilon}_{el}, \dot{\varepsilon}_{ie}, \dot{\varepsilon}_{th}$ are the elastic, inelastic (plastic + creep), and thermal-strain rate tensors, respectively. The stress rate $\dot{\sigma}$ depends on elastic strain rate, and in this case of linear isotropic material and negligible large rotations, it is given by Eq. (5.11) in which "**:**" represents the inner tensor product.

$$\dot{\sigma} = \underline{\underline{\mathbf{D}}} : (\dot{\varepsilon} - \dot{\varepsilon}_{ie} - \dot{\varepsilon}_{th}) \tag{5.11}$$

$\underline{\underline{D}}$ is the forth order isotropic elasticity tensor given by Eq. (5.12).

$$\underline{\underline{\mathbf{D}}} = 2\mu \underline{\underline{\mathbf{I}}} + \left(K_B - \frac{2}{3}\mu\right) \mathbf{I} \otimes \mathbf{I} \tag{5.12}$$

Here μ, K_B are the shear modulus and bulk modulus, respectively, and are in general functions of temperature, while $\underline{\mathbf{I}}$, and $\mathbf{I}$ are fourth and second order-identity tensors, and "$\otimes$" is the notation for outer tensor product.

Inelastic strain includes both strain-rate independent plasticity and time-dependent creep. Creep is significant at the high temperatures of the solidification processes and is indistinguishable from plastic strain [17]. The inelastic strain-rate is defined here with a unified formulation using a single internal variable [18,19], equivalent inelastic strain $\overline{\varepsilon}_{el}$ to characterize the microstructure. The equivalent inelastic strain-rate $\dot{\overline{\varepsilon}}_{el}$ is a function of equivalent stress $\overline{\sigma}$, temperature T, equivalent to inelastic strain $\overline{\varepsilon}_{el}$.

$$\dot{\overline{\varepsilon}}_{el} = f(\overline{\sigma}, T, \overline{\varepsilon}_{el}) \tag{5.13}$$

$$\overline{\sigma} = \sqrt{\frac{2}{3}\sigma'_{ij}\sigma'_{ij}} \tag{5.14}$$

σ' is a deviatoric stress tensor defined in Eq. (5.15).

$$\sigma'_{ij} = \sigma_{ij} - \frac{1}{3}\sigma_{kk}\delta_{ij} \tag{5.15}$$

The workpiece is assumed to harden isotropically, so the von Mises loading surface, associated plasticity, and normality hypothesis in the Prandtl-Reuss flow law is applied [20]:

$$(\dot{\varepsilon}_{ie})_{ij} = \frac{2}{3}\dot{\overline{\varepsilon}}_{el}\frac{\sigma'_{ij}}{\overline{\sigma}} \tag{5.16}$$

Thermal strains arise due to volume changes caused by both temperature differences and phase transformations, including solidification and solid-state phase changes; that is,

$$(\varepsilon_{th})_{ij} = \int_{T_o}^{T} \alpha(T)dT\delta_{ij} \tag{5.17}$$

where α is the temperature-dependent coefficient of thermal expansion, T_o is the reference temperature, and δ_{ij} is Kronecker's delta. It is normally accounted for in stress analyses through a temperature-dependent differential thermal expansion coefficient, $\alpha(T)$. Numerical finite element codes, such as ABAQUS, can be used to simulate the laser cutting process. ABAQUS/Standard analysis, a spatially varying thermal expansion, can be defined for homogeneous solid continuum elements by using a distribution, which includes the tabulated values for the thermal expansion [12]. ABAQUS uses an implicit backward-difference scheme for time integration of both temperature and displacements at every material integration point [13].

5.3 Numerical simulations

Finite element discretization can be carried out using the ABAQUS software [13]. The simulation can be performed in ABAQUS/Standard and consists of sequential thermal-stress analysis. In the sequential thermal-stress analysis, a large number of elements are used to create the model using two element types; for the heat transfer analysis, the mesh elements of type DC3D4 (4-node Linear heat transfer tetrahedron) are used, and the stress analysis C3D4 (4-node Linear 3D stress tetrahedron) is used. The fixed-boundary conditions can be applied on both ends of the work piece, resembling the experimental laser cutting situation.

In the stress analysis, displacements are stored by ABAQUS at the nodal positions as a solution variable, and loads are defined as prescribed displacements and forces. Employing the interpolation functions, it is possible to calculate the strain and stress increments at any point within the element using the compatibility and constitutive equations. ABAQUS transforms the mechanical equilibrium equations into a set of simultaneous equations, such that the nodal displacements and forces are related to each other through the elemental stiffness matrix.

A typical example of laser cutting conditions may include the following: Laser heat flux, with a Gaussian distribution and prescribed velocity along the hole circumference through user subroutine DFLUX, is applied to the thermal model. The Gaussian parameter "a" is $a = 0.0002$ m, in accordance with the experimental power-intensity distribution. The peak laser intensity is 0.5×10^{11} W/m^2 and the reflectivity is 0.62 set in the simulations in accordance with the experimental tests. The temperature-time history resulting from the thermal analysis is used as input to the thermal stress analysis. The workpiece is considered to be an elastic body, which is modeled as a von Mises elastic-plastic material with isotropic hardening and with a yield stress that changes with temperature.

5.4 Laser high-thermal conductive materials

Straight laser cutting of bronze is considered because bronze has a high-thermal conductivity and is difficult to cut by using the thermal process.

5.4.1 *Straight laser cutting of bronze*

The findings of simulations and experimental data for laser cutting of bronze are presented below in light of the previous study [1]. Tables 5.1 and 5.2 give the data used in the simulations.

Table 5.1 Thermal properties of bronze used in the simulations

Density (kg/m^3)	8860
Latent heat of fusion (J/kg)	209,000
Solidus temperature (K)	1227
Liquidus temperature (K)	1367
Specific heat capacity (J/KgK)	377
Thermal conductivity (W/mK)	69.2

Table 5.2 **Properties of bronze used in the simulations [21]**

Temperature (°K)	Modulus of elasticity, *E* (Pa)	Yield stress (Pa)	Poisson's ratio	Expansion coefficient (1/K)
294	9.65E+10	2.10E+08	0.34	1.54E−05
366	9.44E+10	2.05E+08	0.34	1.58E−05
422	9.24E+10	1.97E+08	0.34	1.62E−05
477	9.10E+10	1.95E+08	0.34	1.66E−05
533	8.89E+10	1.57E+08	0.34	1.73E−05
589	8.61E+10	1.50E+08	0.34	1.77E−05
644	8.27E+10	9.48E+07	0.34	1.81E−05
866	7.58E+10	1.00E+07	0.34	2.00E−05
977	6.89E+10	1.00E+07	0.34	2.10E−05
1088	4.82E+10		0.34	2.16E−05

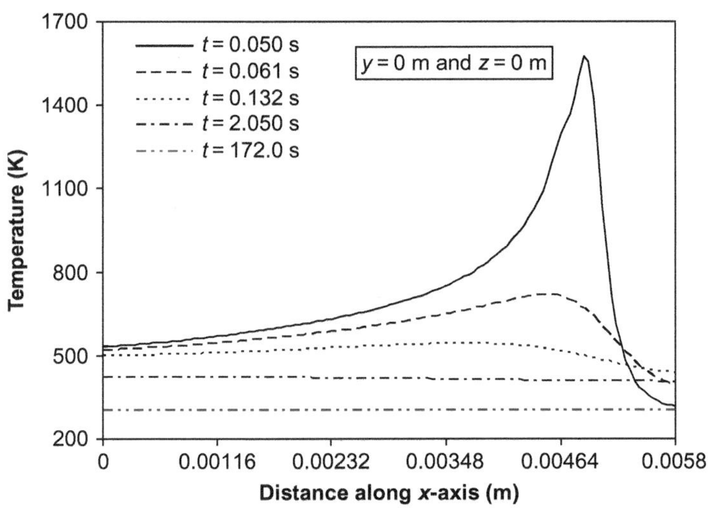

Fig. 5.3 Temperature variation along the x-axis for different cooling periods. The cooling period starts at $t = 0.05$ s [1].

Fig. 5.3 shows the temperature variation predicted along the x-axis for different cooling periods while Fig. 5.4 shows temperature contours. The laser cutting takes place along the x-axis. It should be noted that the y-axis location is at the edge of the cut ($y = 0$ m, Fig. 5.1) and the z-axis location is at the top surface ($z = 0$ m, Fig. 5.1). The cooling cycle starts immediately after the cutting process is completed and the laser beam power ceases. The cooling cycle ends when the temperature at the cutting edges reduces to the initial temperature. The temperature is at the maximum at

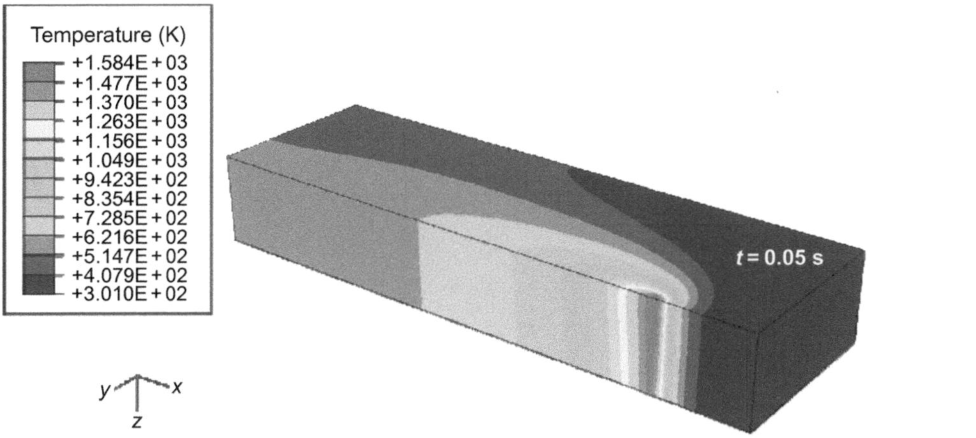

Fig. 5.4 Temperature contours in the workpiece immediately after the cooling period starts [1].

the cut edge immediately after the laser power ceases; in which case, the x-axis location is at $x=0.005$ m and the cooling period is $t=0.05$ s. The maximum temperature is higher than the melting temperature of the substrate material. In this case, liquid-phase heating takes place at the kerf edge. The temperature rise is slow at $0\leq x\leq 0.004$ m and becomes sharp along the x-axis to reach the melting point. The temperature rise becomes gradual during the phase change prior to liquid-phase heating taking place. The slow rise of temperature corresponds to the initial laser cut edge along $0\leq x\leq 0.004$ m, and convection, radiation, and conduction from the kerf edge results in temperature decay in this region. This appears as a gradual rise in temperature along the x-axis. However, once the melting temperature is reached due to the differences in the liquidus and solidus temperatures, small and gradual rise of temperature is observed during the melting process. Moreover, temperature decays sharply in the frontal region of the laser beam ($x=0.005$ m). This results in the development of the high-temperature gradient in this region. As the cooling period progresses, the peak temperature reduces significantly. This is because of the high thermal conductivity of the substrate material, which lowers the peak temperature due to the high conduction heat transfer from the cut edge to the solid bulk. In addition, temperature decay from its peak value along the x-axis becomes gradual. As the cooling period progresses further, the temperature reduces almost to the initial temperature at $t=172$ s when the cooling cycle ends.

Fig. 5.5 shows the von Mises stress distribution along the x-axis for the cooling periods similar to those shown in Fig. 5.4, while Fig. 5.6 shows the von Mises stress contours along the cut edge. The von Mises stress remains low along the cut edge $0\leq x\leq 0.0035$ m and reduces significantly when the axial distance is increased. The occurrence of a low von Mises stress is attributed to a higher temperature in the early cooling period ($t=0.05$ s); in which case, the elastic modulus becomes

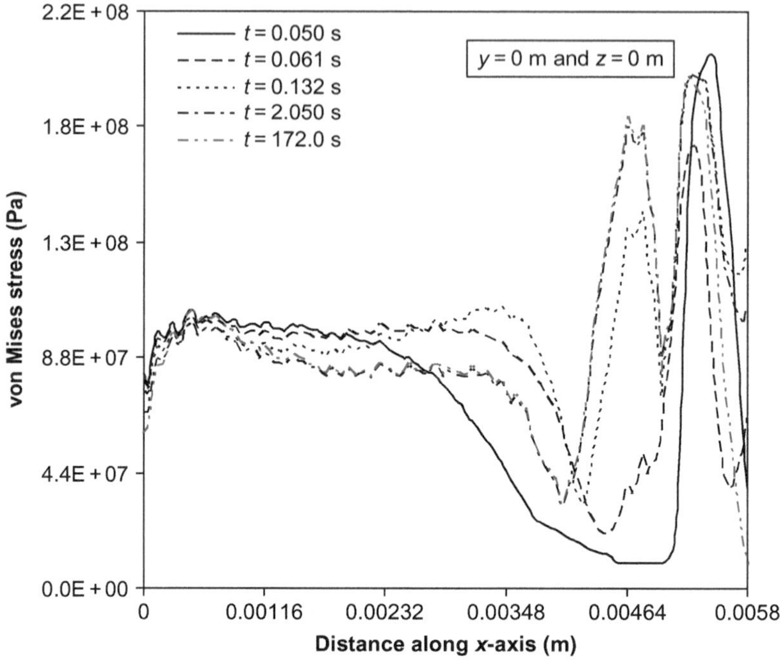

Fig. 5.5 von Mises stress variation along the x-axis for different cooling periods. The cooling period starts at $t = 0.05$ s [1].

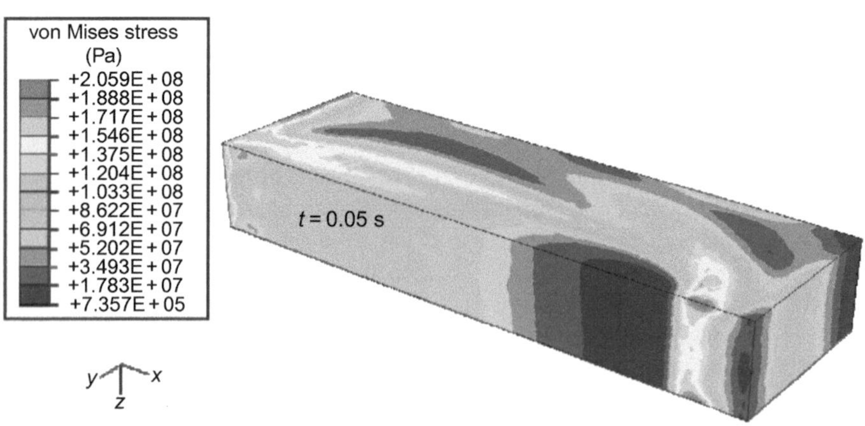

Fig. 5.6 von Mises stress contours in the workpiece immediately after the cooling period starts [1].

low with increasing temperature (Table 5.2). However, the von Mises stress increases to reach its peak value in the frontal region of the laser beam during the early cooling period. The high-temperature gradient developed in this region is responsible for the attainment of a high von Mises stress in the early cooling period. As the cooling period increases, the von Mises stress increases in the resolidified region, forming a second stress peak along the *x*-axis. Consequently, two stress peaks occur: (i) one in the resolidified edge, and (ii) in the frontal region of the laser beam where the laser beam power ceases. The magnitude of the stress peak is higher at a location in the frontal region of the laser beam than that corresponding to a resolidified region at the cutting edge. Once the cooling period ends, the stress field becomes the residual stress. The maximum von Mises stress is on the order of 180 MPa after the cooling period ends. The maximum von Mises stress is higher than the yield stress of the substrate material indicating a plastic deformation taking place in the region of the cut edge. It should be noted that elasto-plastic material behavior is incorporated in the analysis to predict the residual stress in the cut edges.

Fig. 5.7 shows the temperature distribution along the *y*-axis for different cooling periods at the *x*-axis location $x = 0.005$ m where the laser beam power ceases and the *z*-axis location is $z = 0$ m at the top surface of the workpiece. It should be noted that $y = 0$ m represents the cut-edge surface. Temperature decay is gradual in the vicinity of the cut-edge surface and it becomes sharp as the distance increases away from the surface during the early cooling period. The gradual decay of temperature in the surface region is associated with the melt zone generated in this region where the liquid phase remains almost at the melting temperature in this region. However, the

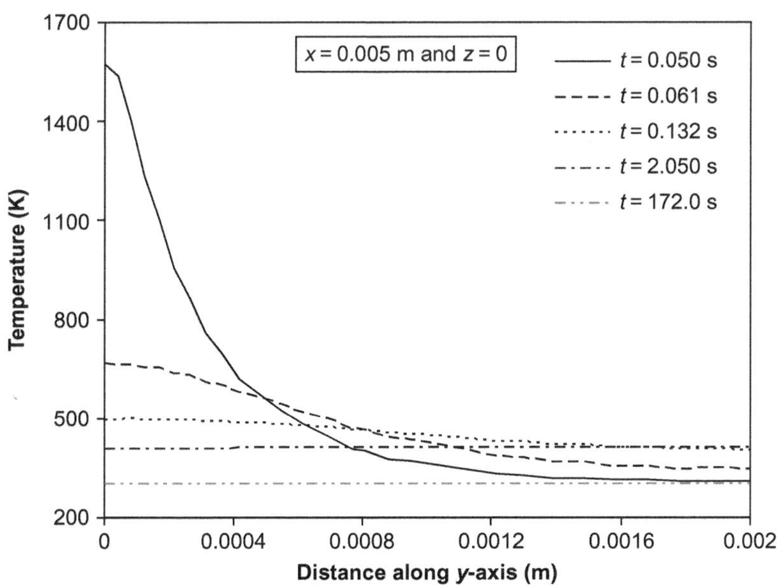

Fig. 5.7 Temperature variation along the *y*-axis for different cooling periods [1].

sharp decay in temperature is attributed to the conduction losses from the cut edge surface vicinity to the solid bulk. As the cooling period increases, the temperature at the cut edge surface region reduces significantly due to high thermal conductivity of the substrate material. This lowers the temperature gradient in this region with a progressing cooling period. In addition, as the cooling period progresses further, temperature along the *y*-axis reduces to its initial temperature. This is again because of the high rate of heat conduction from the cut-edge surface to the solid bulk due to high thermal conductivity of bronze.

Fig. 5.8 shows the von Mises stress distribution along the *y*-axis for different cooling periods similar to those shown in Fig. 5.7. The von Mises stress attains low values in the vicinity of the cut edge in the early cooling period because of the attainment of a high temperature in this region; it becomes high at some distance away from the surface in the early cooling period. This is attributed to the high-temperature gradient developed in this region. It should be noted that increasing the temperature reduces the elastic modulus (Table 5.2). As the cooling period increases, the von Mises stress in the surface region of the cut edge increases, and at the location some distance away from the edge surface, the maximum von Mises stress occurs. This is associated with the free expansion of the cut edge surface during the cooling period while resulting in a relatively lower value of the von Mises stress in the surface vicinity as compared to that some distance away from the surface.

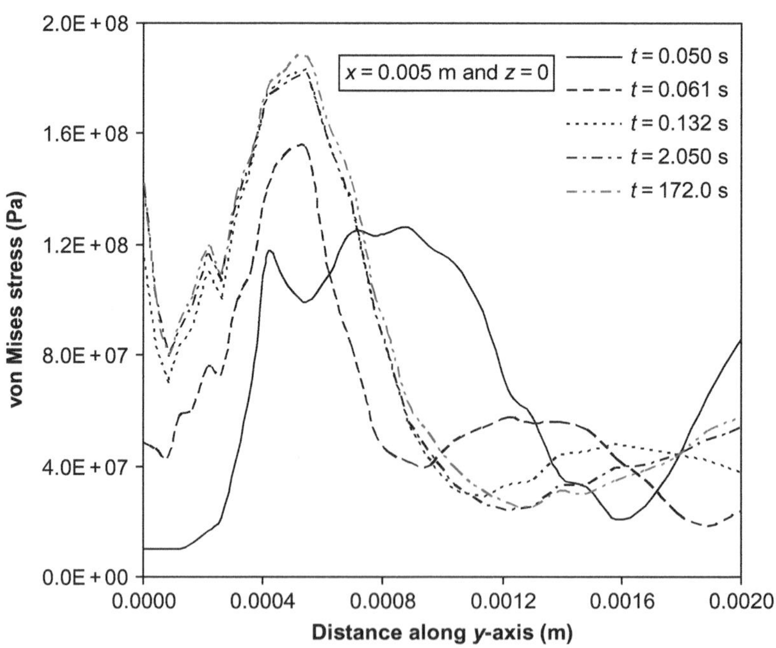

Fig. 5.8 von Mises stress variation along the *y*-axis for different cooling periods [1].

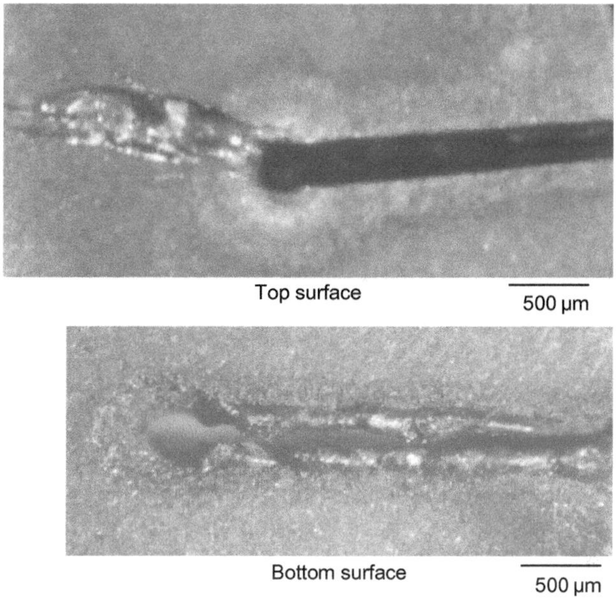

Fig. 5.9 Optical photographs of the top and bottom surfaces of the laser cut section [1].

Fig. 5.9 shows the optical photograph of the top view of the laser cuts. It can be observed that the top surface of the laser is free from sideways burning and the striation depth is shallow. This is because of the nitrogen assisting gas, which is used at high pressure (600 MPa) during the laser-cutting process. However, some resolidified material around the key-hole is observed. This is due to the material ejected from the key-hole onset of cutting initiation. On the other hand, the dross attachment is evident from the bottom surface of the cut edge. The dross attachment forms irregular patterns at the cut edge indicating the transient nature of the molten flow without regular behavior. The dross attachment is associated with the solidification of the molten metal leaving the kerf around the cut edges at the kerf exit. Although the nitrogen assisting gas at high pressure is used to purge the molten metal out of the kerf exit, the drag force generated at the molten metal surface due to the assisting gas is not sufficiently high enough to purge out the molten metal from the kerf. In addition, high conduction heat loss from the molten metal to the kerf edge lowers the molten metal temperature while increasing the molten metal viscosity. This contributes to the dross attachment at the kerf exit. Fig. 5.10 shows the SEM micrograph of the side view of the kerf surface. Although the depth of the striations is small, the striation pattern changes toward the kef exit; in which case, the striations become straight. This is attributed to one or both of the following reasons: (i) the temperature around the hole exit reduces due to the conduction and convection losses while modifying the viscosity of the molten flow and changing the molten flow direction toward the kerf exit, and (ii) the assisting gas loses its momentum toward the kerf exit, due to frictional losses in

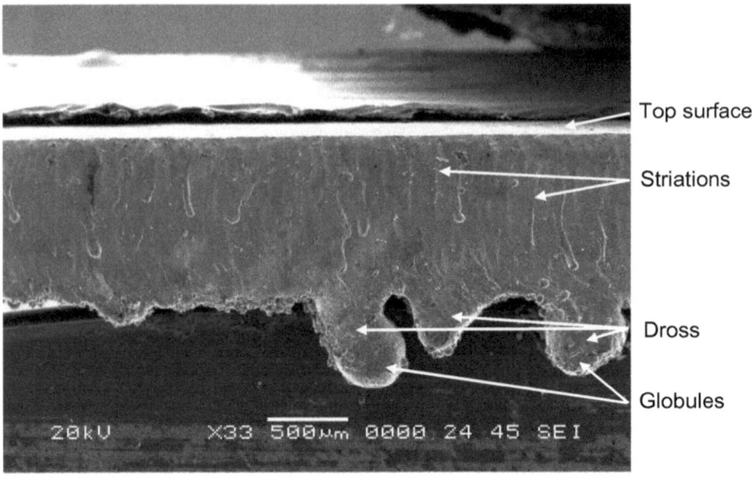

Fig. 5.10 SEM micrograph of the kerf surface [1].

the kerf, altering the drag force acting on the molten flow while influencing the flow direction. The close examination of the SEM micrograph reveals the formation of globule-like drosses attaching to the kerf exit, which is associated with the surface tension force. The cast layer forms at the cut edge surface due to the solidification of the molten metal at the kerf surface. The thickness of the cast layer is on the order of 2.5 μm. The affected zone, which is observed below the cast layer, is on the order of 10 μm. The small depth of the heat-affected zone is attributed to the high rate of heat conduction from the kerf surface to the solid bulk.

Table 5.3 gives the elemental composition of the as-received material and the laser cut section obtained from the EDS analysis. The depletion of the phosphorous in the kerf surface, which is due to the evaporation of phosphorous from the kerf surface during the cutting process, is observed. Oxygen is also present, as shown in Table 5.3, indicating the formation of the oxide compounds at the kerf surface despite the fact that nitrogen gas at high pressure is used as the assisting gas during the cutting process. This indicates that nitrogen gas pressure is not enough to totally prevent the oxidation reactions taking place in the kerf. Fig. 5.11 shows the XRD diffractogram corresponding to the laser cutting section. The XRD diffractogram reveals that the Cu_3N compound is formed in the laser-cutting region, which is associated with the

Table 5.3 EDS data obtained as received and the laser cutting region (wt%)

	N	O	Sn	P	Cu
Laser cutting region	6.74	24.40	5.21	0.0	Balance
As received	0.0	0.0	6.01	0.1	Balance

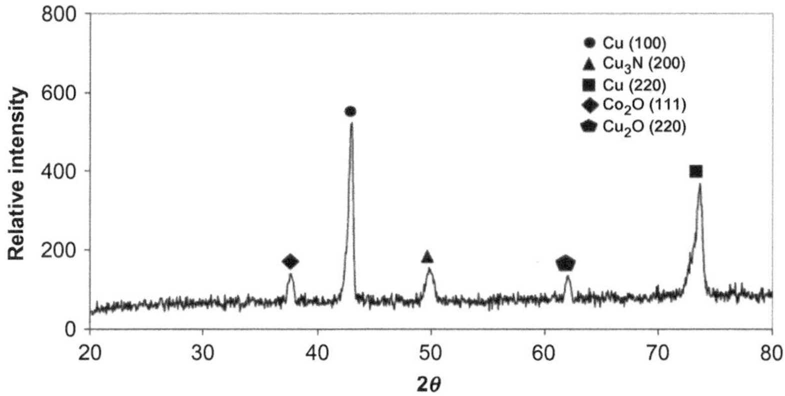

Fig. 5.11 XRD diffractogram for the laser-cut section [1].

presence of high-pressure nitrogen assisting gas. In addition, formation of Cu_2O is also evident from the XRD peaks, indicating some oxidation reactions taking place in the cutting section, despite the fact that assisting gas at high pressure is used during the cutting process.

5.4.2 Laser circular cutting of bronze

The simulations and experimental data for laser circular cutting of bronze are presented below in light of the previous study [2].

Fig. 5.12 shows the temperature distribution along the top and bottom circumference of the hole for different cutting durations while Fig. 5.13 shows temperature contours in the cutting region. The temperature attains its maximum at a location of the laser beam during the cutting process. Consequently, the location of the maximum temperature changes as the distance increases along the circumference of the hole. Temperature decays sharply from its maximum value to the initial temperature in the region next to the laser-irradiated spot. This is because of the low workpiece temperature in the region next to the laser-cutting section.

However, temperature decays gradually along the circumference of the section being cut. This is because of the laser heating of this region during the cutting process. Consequently, convective and conduction losses from the cutting section gradually lower the temperature in this region. The maximum temperature exceeds the melting temperature of the substrate material. Therefore superheating of the liquid phase takes place in the cutting region. In addition, a phase change is observed from the temperature curves; in which case, the slow rise of temperature occurs between solidus and liquidus temperatures. As the laser beam moves along the hole circumference, temperature decay in the region behind the laser spot becomes sharper. For example, this occurs particularly in the region 0.02 m $\leq$ distance $\leq$ 0.026 m along the circumference at 0.1425 s of the cutting duration because the laser beam location is

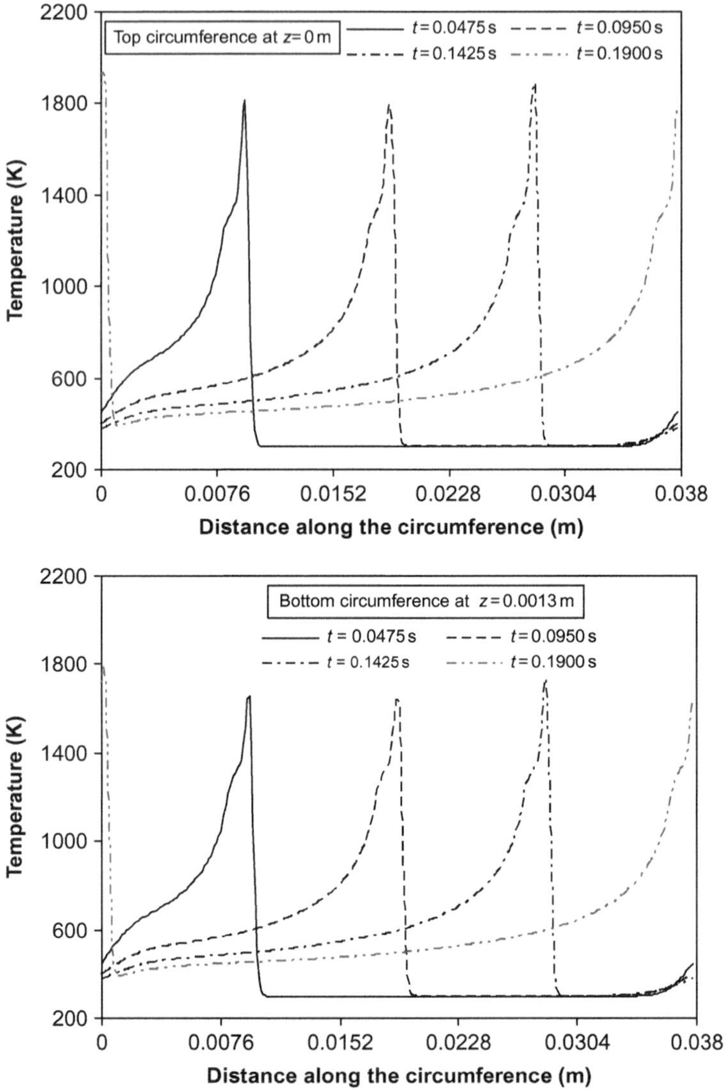

Fig. 5.12 Temperature distribution along the top and bottom circumferences for different cutting periods [2].

just 0.026 m in front of the circumference at 0.1425 s. The sharp decay of temperature results in the formation of high temperature gradients in this region. In the case of the bottom circumference, the maximum temperature remains slightly lower than that corresponding to the top circumference. Moreover, temperature remains well above the melting temperature of workpiece material. The superheating of the liquid phase in the irradiated spot results in a sharp decay of temperature in the

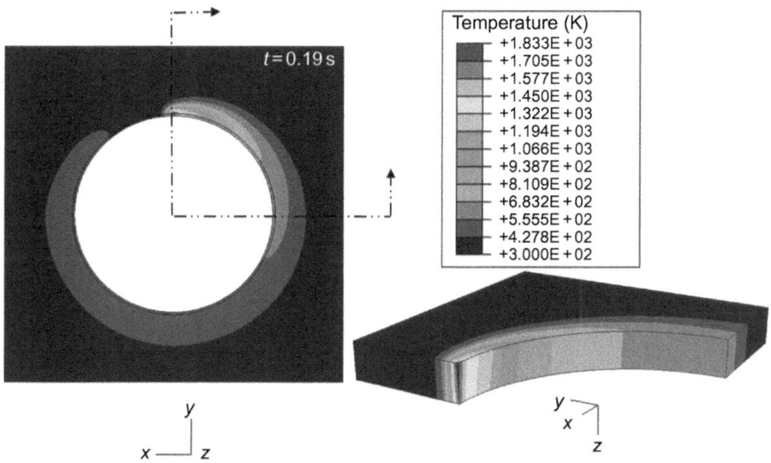

Fig. 5.13 Temperature contours in the cutting section at $t=0.19$ s of the cutting period (end of cutting) [2].

neighborhood of the cutting section, which is similar to those corresponding to the top circumference.

Fig. 5.14 shows the von Mises stress distribution along the top and bottom circumference of the hole for different cutting durations, while Fig. 5.15 shows the von Mises stress contours in the cutting section. The von Mises stress attains low values in the region where the temperature is high at the initiation of the cooling period. This is associated with the low elastic modulus of the workpiece at elevated temperatures (Table 5.2). However, the von Mises stress attains high values in the region next to the irradiated spot edge, particularly in the frontal region of the laser-irradiated spot during the cutting process. This is because of the large temperature gradient developed in this region through the attainment of the large temperature difference (Fig. 5.12). The variation of the von Mises stress along the top and bottom circumferences of the hole is almost the same. This is attributed to the similar temperature distributions along the top and bottom circumferences during the cutting process.

Fig. 5.16 shows the temperature variation along the circumference of the hole after the cutting process is completed. It should be noted that the time corresponding to the completion of cutting is $t=0.19$ s. In this case, the temperature remains high at the point where the cutting is terminated. This corresponds to the distance 0.038 m and 0 m around the circumference. However, as the cooling period progresses after the completion of the cutting ($t>0.19$ s), temperature reduces sharply around the circumference of the hole, except at a location where the cutting process has just ceased. This is associated with the occurrence of the superheating in the liquid phase; that is, heat transfer via conduction and convection does not reduce temperature sharply in the melting pool. It should be noted that the influence of the drag force due to the assisting gas in the vicinity of the cut surface is considered to be small. This effect is significant

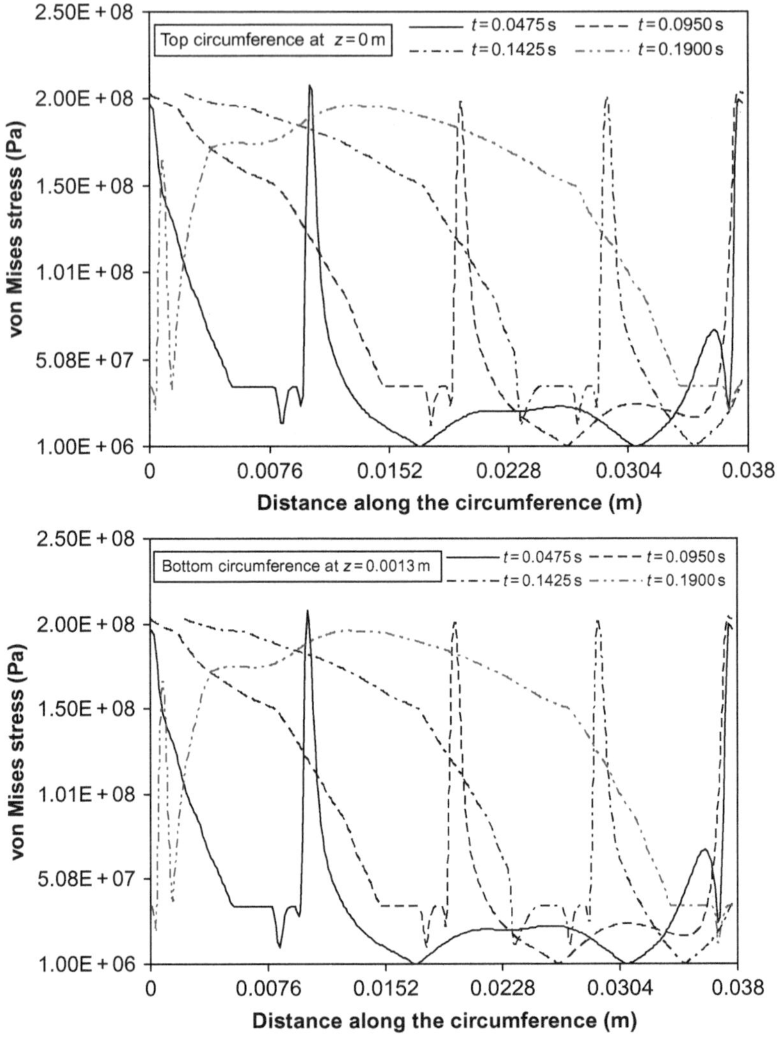

Fig. 5.14 von Mises stress distribution along the top and bottom circumferences during different cutting periods [2].

in the kerf, because the assisting gas jet considerably contributes to the removal of the molten metal from the kerf. However, the cooling accelerates in this region once the temperature reduces below the melting temperature of the substrate material. Temperature reduces to the initial temperature after 100 s. Moreover, the temperature gradient variation with time along the circumference of the hole is important for the thermal stress formation in the cutting section. Although the peak temperatures around the circumference at the laser beam locations are the same, temperature decay at each location around the circumference is not the same due to the laser movement during the cutting.

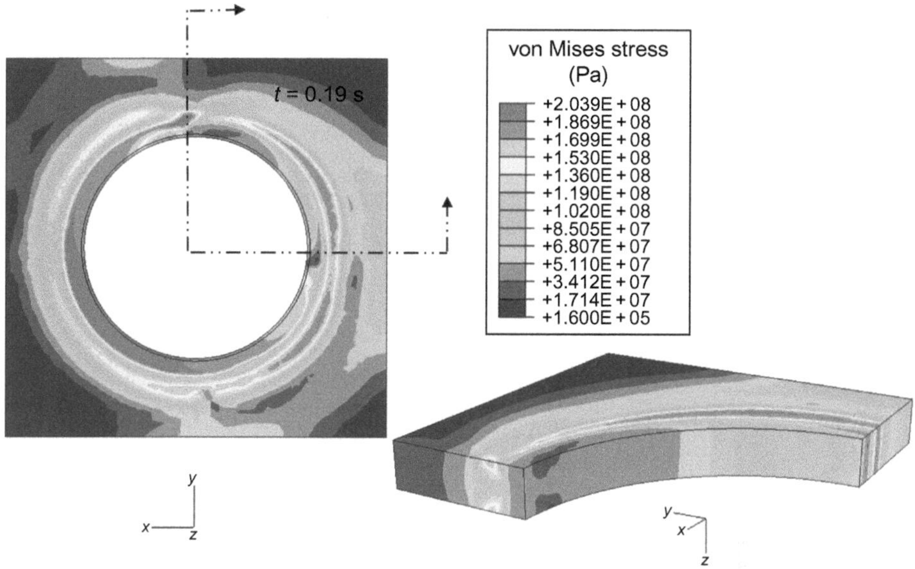

Fig. 5.15 von Mises stress contours in the cutting section at $t = 0.19$ s of the cutting period (end of cutting) [2].

This results in a variation of the temperature gradient in the region at the end of the laser cut, which is different from those points around the circumference of the hole.

Fig. 5.17 shows the von Mises stress distribution along the top and bottom circumferences of the hole during the cooling period, while Fig. 5.18 shows the residual stress contours in the cutting region. The cooling period corresponds to the time after the completion of the cutting process. The von Mises stress attains low values in the region where the laser cutting ceases. Although the hole circumference undergoes cooling, and the temperature reduces around the circumference immediately after the cutting process ends, the temperature remains high at the point where the laser beam terminates. This, in turn, lowers the von Mises stress at the onset of cooling in this region. As the cooling period increases, the von Mises stress increases around the hole circumference, which is more pronounced in the region away from the location where the cutting terminates. Moreover, as the cooling period progresses further, the von Mises stress becomes the residual stress and it stays almost the same along the hole circumference. This is true for the top and bottom circumferences. The maximum residual stress at the end of the cooling period is on the order of 200 MPa, which is almost uniformly distributed along the hole surface.

Fig. 5.19 shows optical photographs of the front and back surfaces of the hole, which is cut in the bronze workpiece. Dross attachment at the bottom surface of the cut edges is observed. Fig. 5.20 shows SEM micrographs of the kerf wall. The striation formation is evident around the kerf wall and the striation length is of the order of 100 μm. The averaged roughness (Ra) of the cut surface is on the order of

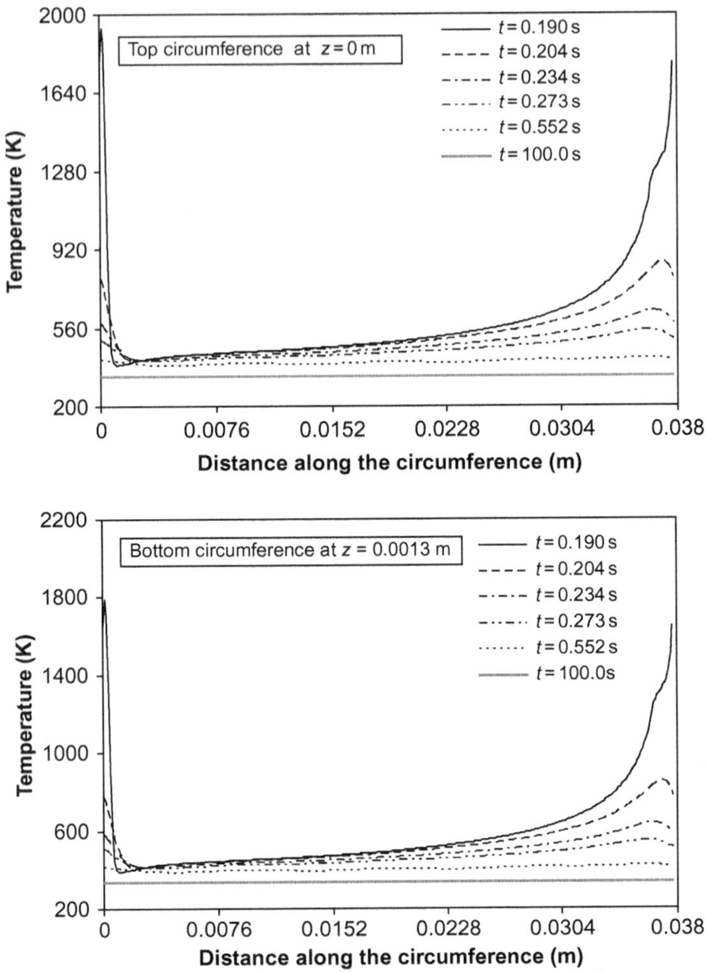

Fig. 5.16 Temperature distribution along the top and bottom circumferences during different cooling periods (after cutting ends) [2].

8 μm. Moreover, the striation patterns change toward the hole exit; where they are partially bent and closely spaced. This is caused by one or all of the following reasons: (i) the temperature around the hole exit reduces, which in turn, reduces the viscosity of melt flow (consequently, the shear force generated in the molten flow modifies the flow pattern and striation formation in this region), and (ii) the assisting gas impinging into the kerf loses its momentum toward the kerf exit, resulting in less gas pressure and reduced drag force at the hole exit. As a consequence of this, the striation pattern changes at the hole exit. Moreover, the dross attachment is also observed at the hole exit. This is associated with the solidification of the molten metal before exiting the

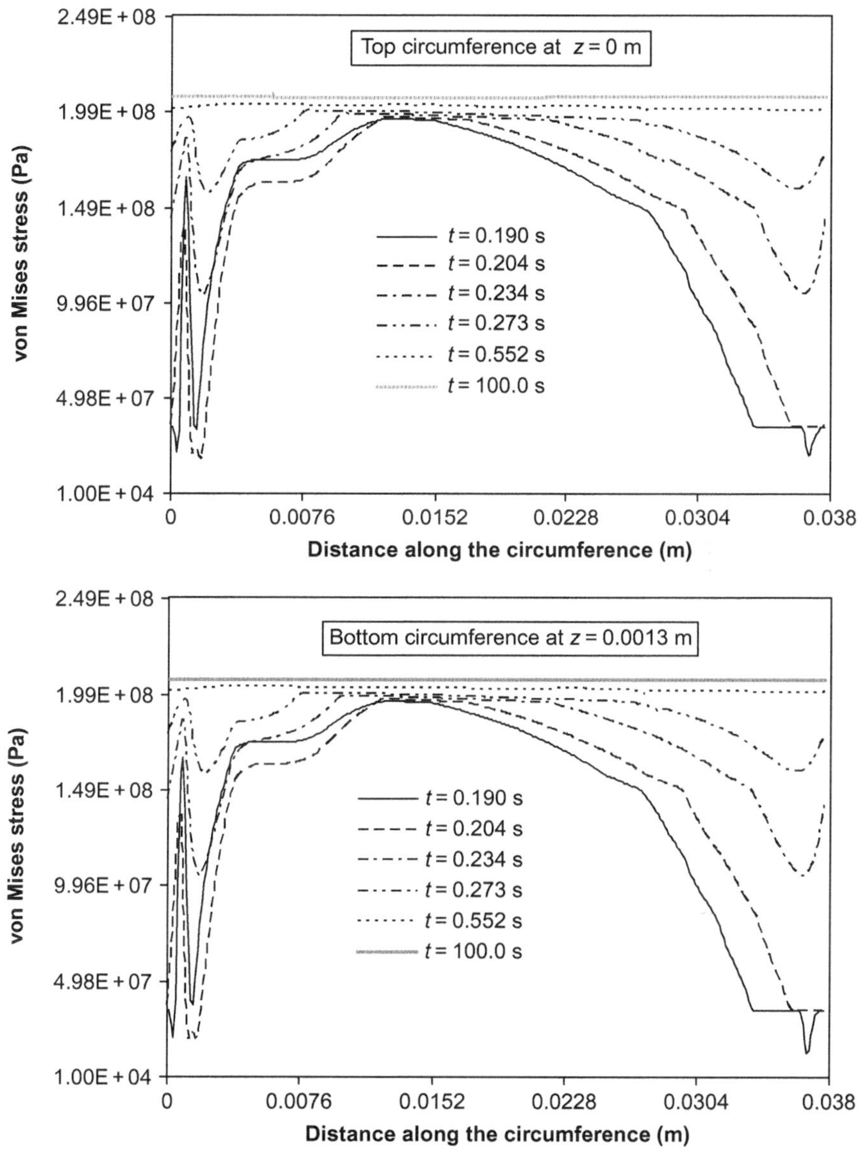

Fig. 5.17 von Mises stress distribution along the top and bottom circumferences during different cooling periods (after cutting ending) [2].

hole. This occurs because of the high thermal conductivity of bronze, which results in a temperature drop in the molten flow through heat transfer between the base substrate and the molten metal. Although the dross attachment at the hole exit influences the thermal stress development at the hole exit, due to the transient complex nature of the phenomena, it is omitted in the simulations.

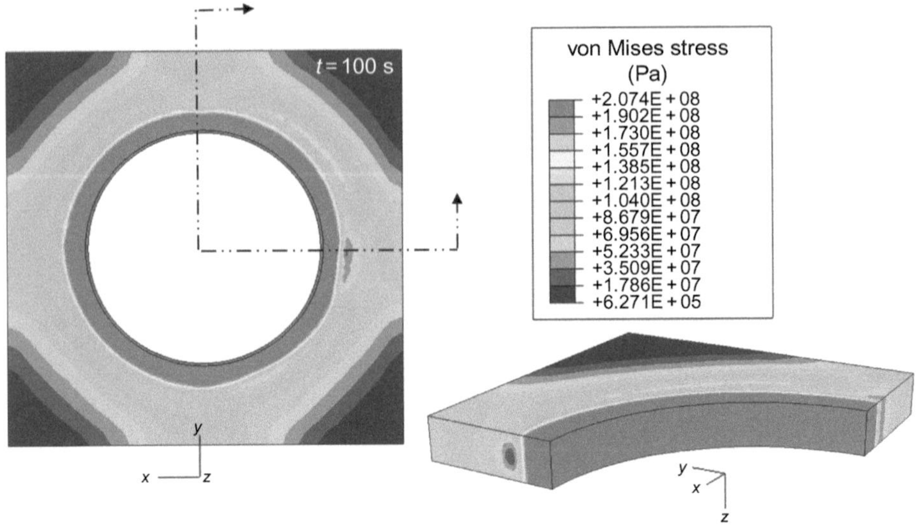

Fig. 5.18 Residual stress contours in the cutting section at $t = 100$ s of the cutting period (end of cooling period) [2].

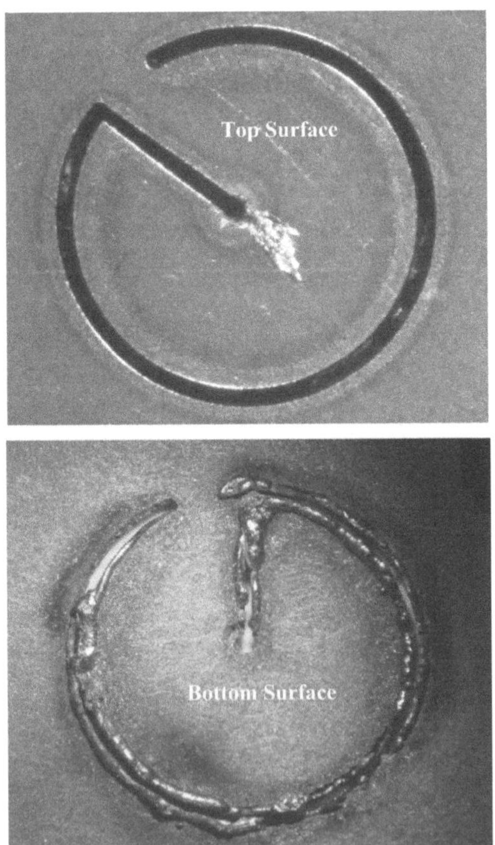

Fig. 5.19 Optical photograph of laser cut hole [2].

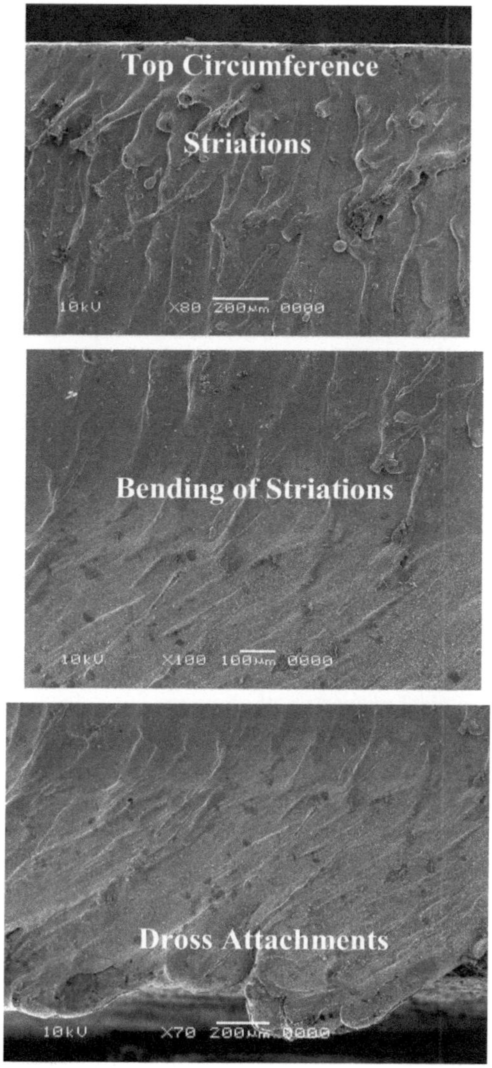

Fig. 5.20 SEM micrographs of kerf surface of the hole [2].

5.5 Laser cutting of high-elastic modules and materials

Alumina is considered to be one of the high-elastic modulus materials with low thermal conductivity. During laser cutting, high-temperature gradients causes the formation of thermal stress fields around the cutting section. This, in turn, results in high residual stresses in the close region of the cutting edges and possible formation of micro-cracks in the cut section. Consequently, heating and thermal stress simulations provide useful information on the laser cutting process. The simulation and experimental data in relation to laser cutting of alumina tiles are presented in light of the previous studies [3–6].

5.5.1 Laser straight cutting of alumina tiles

The predictions of temperature and stress fields, together with the experimental results in relation to straight laser cutting alumina tiles, are presented in line with the previous study [3]. The data used in the simulations are given in Table 5.4A–C.

Fig. 5.21 shows the temperature distribution along the cut edge corresponding to the top and bottom of the cut edges for different cooling times while Fig. 5.22 shows the temperature contours in the cutting section. The cooling cycle begins when the laser power is turned off at the end of cutting process; therefore, the cooling cycle begins at $t=0.05$ s soon after the cutting cycle ends. The temperature increases to

Tables 5.4 (A)–(C) Properties of alumina used in the simulations [22] and simulation conditions. L is the thickness, I_o is the peak power intensity, α is the thermal expansion coefficient; $T_{solidus}$ and $T_{liquidus}$ are the liquidus and solidus temperatures of alumina, respectively, and L_m is the latent heat of alumina tiles:

(A) Simulation conditions

L (m)	I_o (W/m^2)	α (1/T)	$T_{solidus}$ (K)	$T_{liquidus}$ (K)	L_m (kJ/kg)
0.0015	8.3×10^{12}	7.5×10^{-6}	2260	2323	900

(B) Thermal and mechanical properties of alumina

T (K)	Cp (J/kg K)	k (W/m K)	E (GPa)	Poisson's ratio
300	780	37.06	380	0.270
400	950	28.19	375	0.272
500	1000	21.81	371	0.274
600	1100	17.23	366	0.276
700	1110	13.93	362	0.277
800	1140	11.56	357	0.279
900	1165	9.86	353	0.281
1000	1190	8.63	348	0.283
1100	1210	7.75	343	0.285
1200	1235	7.12	339	0.287
1300	1255	6.66	334	0.289

(C) Flow stress of alumina at different temperatures

Temp. (°K)	Flow stress (GPa)
300	4.70
473	4.20
673	3.60
873	3.00
1073	1.80
1273	1.40
1473	1.00

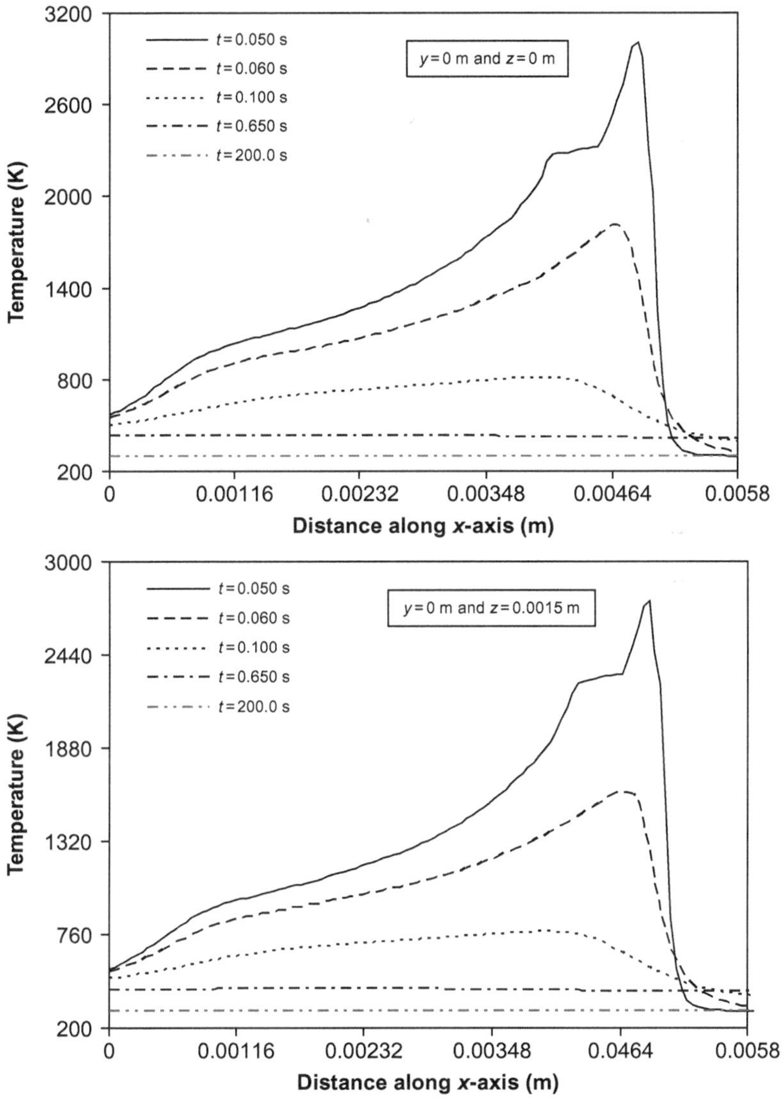

Fig. 5.21 Temperature variation at the top and bottom edges of the kerf along the x-axis during the cooling cycle. The cooling cycle begins immediately after the cutting ends [3].

reach maximum at the irradiated spot. Temperature decay is low along the initially-cut section ($0\ \mathrm{m} \leq x \leq 0.005\ \mathrm{m}$) because of the convective, radiative, and conductive heat loss from this region. The temperature remains almost the same in the region next to the peak temperature at the onset of cooling cycle. This indicates the phase change, which takes place almost at a constant temperature due to the difference between the solidus and liquidus temperature is small (Table 5.4A). The temperature exceeds the melting temperature of the substrate material before reaching its peak in the irradiated

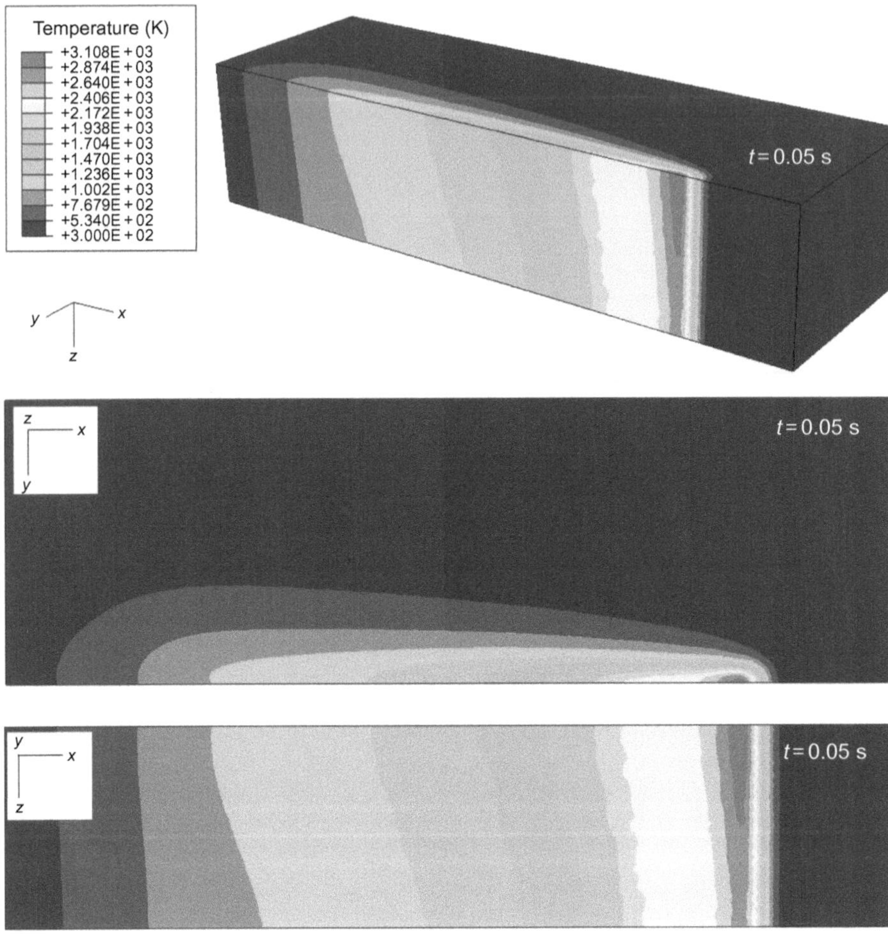

Fig. 5.22 Temperature contours inside the cutting section at the cooling cycle initiation [3].

region. This shows the superheating of the liquid phase taking place in the irradiated region. As the cooling progresses, the peak temperature reduces and the temperature decay along the x-axis becomes gradual except in the region of the superheated liquid where the decay remains high. As the cooling progresses further, temperature decays almost uniformly along the x-axis, reducing the initial temperature of the substrate material when the cooling cycle ends. In the case of temperature distribution at the bottom surface of the cut edge, temperature behavior is similar to that corresponding to the top surface. However, the peak temperature reduces slightly at the surface of the cut edge because of: (i) the laser energy reaching the bottom region of the cutting section is lower than that at the surface, and (ii) the assisting gas cools the bottom edge prior to exiting the kerf. Nevertheless, the reduction in the peak temperature is not significantly high.

Fig. 5.23 shows the von Mises stress distribution along the x-axis for different cooling times while Fig. 5.24 shows the von Mises stress contours in the cut section. The von Mises stress attains low values in the region where temperature is high (Fig. 5.21), particularly in the irradiated region during the early cooling periods. This is associated with the low-elastic modulus of the substrate material at high

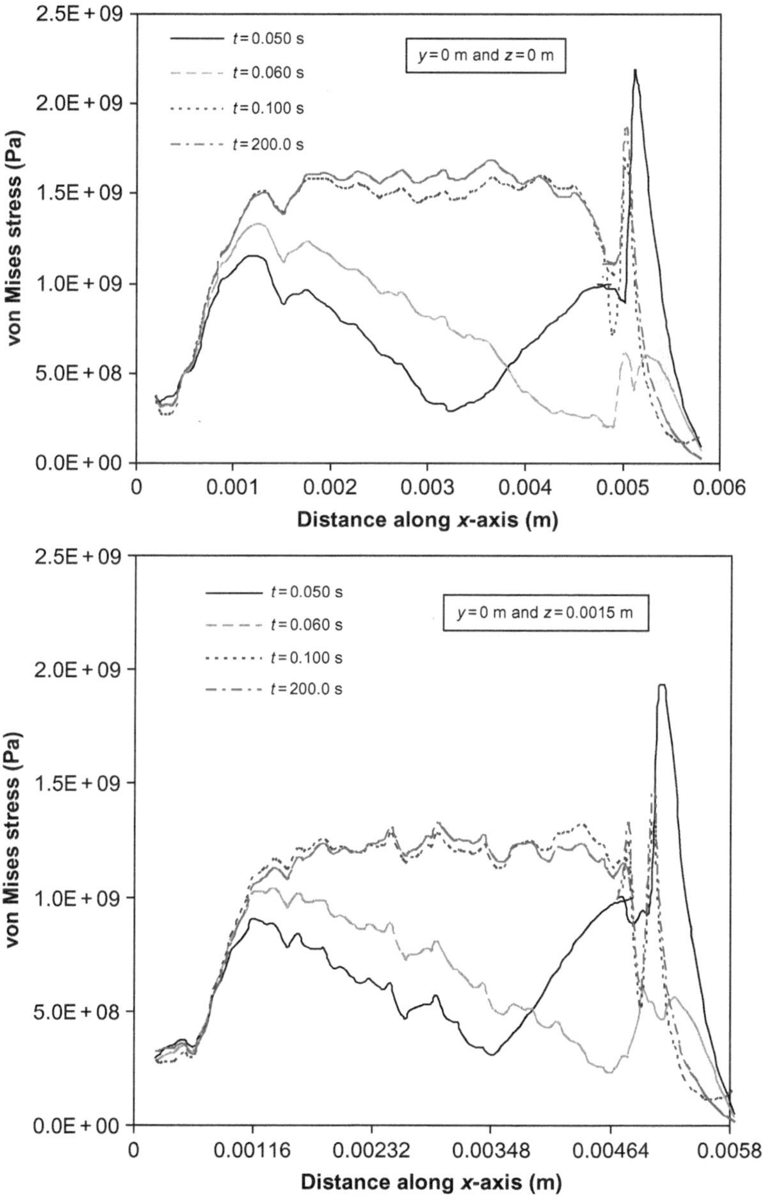

Fig. 5.23 The von Mises stress distribution at the top and bottom edges of the kerf along the x-axis during the cooling cycle. The cooling cycle begins immediately after the cutting ends [3].

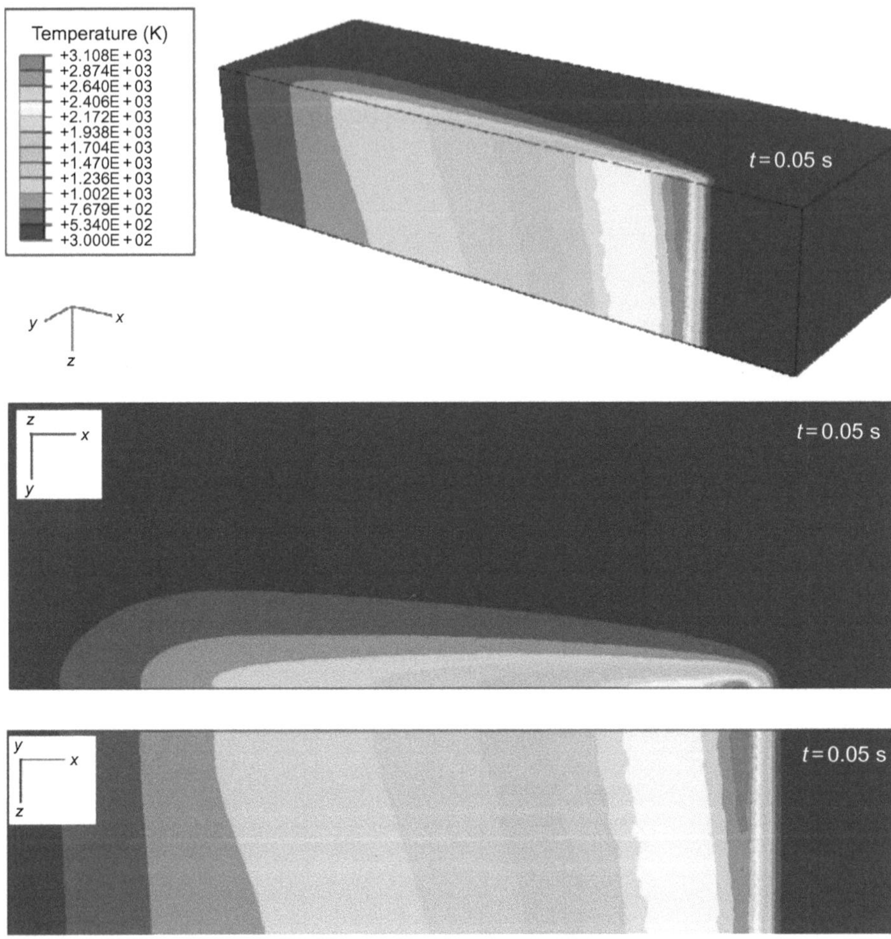

Fig. 5.24 The von Mises stress contours inside the cutting section at the cooling cycle initiation [3].

temperatures (Table 5.4). Consequently, thermal softening of the irradiated region results in attainment of low von Mises stress in the high temperature region. The von Mises stress attains high values in the region where temperature decay is sharp; that is, $x > 0.005$ m. This is because of high thermal strain developed in this region despite the fact that elastic modulus is relatively lower than that at room temperature. As the cooling period progresses, the von Mises stress increases along the x-axis, which is more pronounced in the initially-cut edges along the kerf. As the cooling period progresses further, temperature reduces to initial temperature of the substrate material and the cooling cycle ends. In this case, the stress field becomes the residual stress. When comparing the von Mises stress distribution to the top and bottom surfaces of the cut edges, both stress curves behave the same; however, the maximum von Mises stress magnitude reduces slightly at the bottom surface.

Fig. 5.25 shows temperature variation along the y-axis for different cooling periods. It should be noted that the y-axis location $y = 0$ represents the cut edge surface (Fig. 5.1). Temperature decay in the surface vicinity $0 \leq y \leq 5 \times 10^{-6}$ m is gradual and it becomes sharp for $5 \times 10^{-6} \leq y \leq 200 \times 10^{-6}$ m. As the distance increases further away from the cut edge surface, temperature decay becomes gradual during the early cooling period ($t = 0.05$ s). The gradual temperature decay in the surface vicinity is associated with the internal energy gain and the occurrence of the superheated liquid in this region. As the distance increases in the region next to the surface vicinity, conduction losses from this region to the solid bulk lowers the temperature sharply, resulting in high-temperature gradients in this region. As the cooling period increases, temperature decay becomes gradual, which is more pronounced in the surface vicinity. In this case, the superheated liquid cools at a relatively slower rate as compared to that of the solid phase in the neighborhood of the surface vicinity. Consequently, the slow cooling rate in the surface vicinity causes self-annealing of the solidified material in this region. As the cooling period progresses, temperature reduces to initial temperature along the y-axis in the cutting region.

Fig. 5.26 shows the von Mises stress distribution along the y-axis for different cooling periods. The von Mises stress remains high in the early cooling periods in the region where temperature decay is sharp, while the temperature is relatively lower than the melting temperature of the substrate material. The attainment of a high von Mises stress is attributed to the temperature dependent on elastic modulus, which is high at low temperatures. Consequently, balance between the high elastic modulus

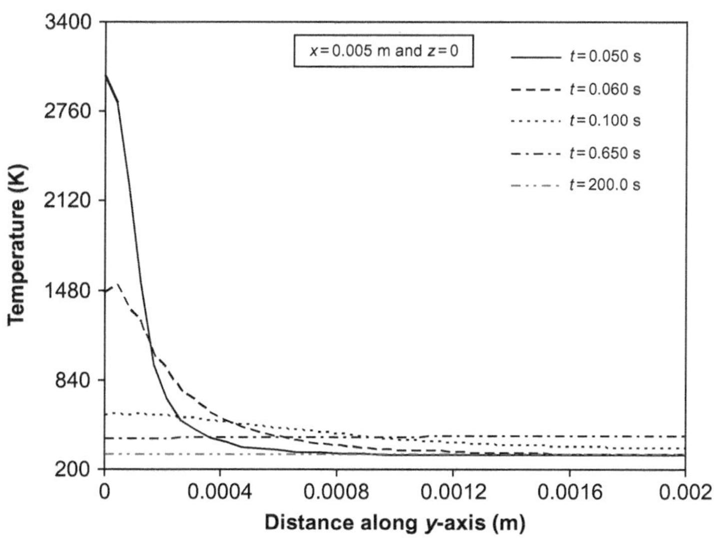

Fig. 5.25 Temperature variation along the y-axis during the cooling cycle. The cooling cycle begins immediately after the cutting ends [3].

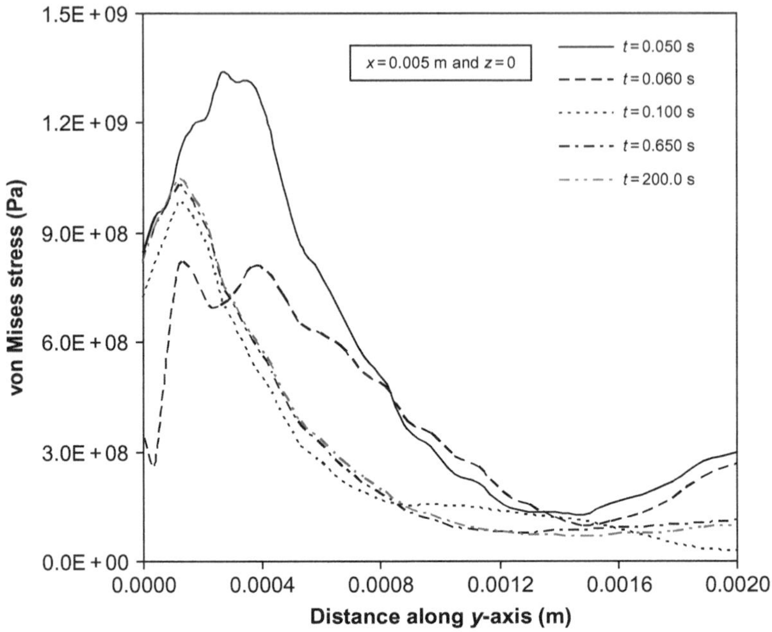

Fig. 5.26 The von Mises stress distribution along the *y*-axis during the cooling cycle. The cooling cycle begins immediately after the cutting ends [3].

and a low temperature gradient results in a high-stress field in the region next to the surface vicinity, and a low-stress field in the surface vicinity. In addition, the annealing effect due to slow cooling rate in the cut edge surface vicinity contributes to the attainment of von Mises stresses in the surface vicinity. As the cooling progresses further, the temperature reduces to the initial temperature and the stress field becomes the residual stress. The maximum von Mises stress at the end of the cooling cycle occurs in the region next to the surface vicinity and its value is on the order of 1 GPa.

Fig. 5.27 shows the von Mises stress distribution along the *z*-axis for different cooling periods. The von Mises stress remains almost the same along the workpiece thickness in the early cooling periods except in the region of the top and bottom surfaces of the cut edges. In this case, the von Mises stress reduces because of the free expansion of the surface along the *z*-axis in this region. As the cooling period increases, the von Mises stress increases in the mid-thickness of the kerf surface. This behavior can be attributed to the compressive stress behavior in this region because there is no free surface except at the top and bottom of the cut sections. The peak value of the von Mises stress does not change toward the end of the cooling cycle. The maximum von Mises stress occurs at the mid-height of the kerf surface, which corresponds to the mid-height of the workpiece. The magnitude of the von Mises stress is on the order of 2.5 GPa, which is higher than that of the free surface. Fig. 5.28 shows the

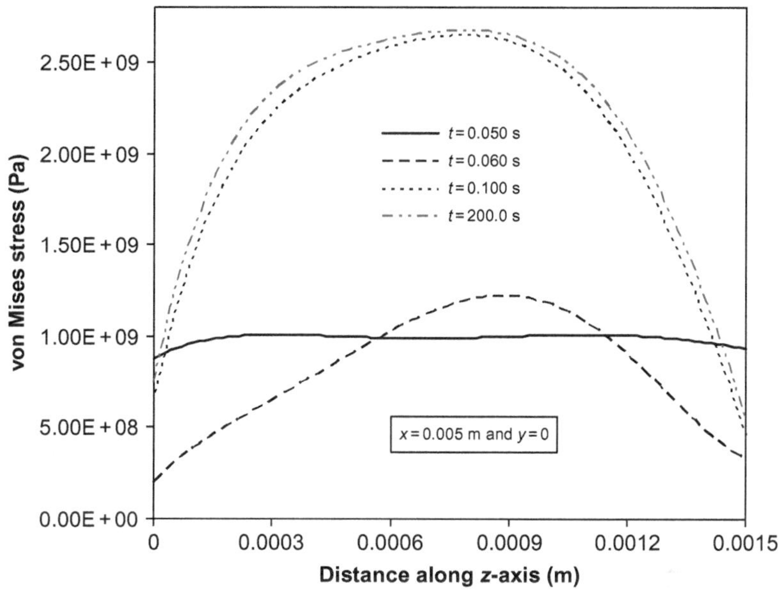

Fig. 5.27 The von Mises stress distribution along the *z*-axis during the cooling cycle. The cooling cycle begins immediately after the cutting ends [3].

Fig. 5.28 Residual stress contours in the cutting section at the end of the cooling period [3].

residual stress contours predicted from the simulations. The residual stress remains high at the mid-height of the cut section.

Fig. 5.29 shows an optical photograph of the top and bottom surfaces of the laser cut section. The cuts with parallel edges are observed and thermal agitation resulting in sideways burning around the cut edges does not occur. The dross attachment at the

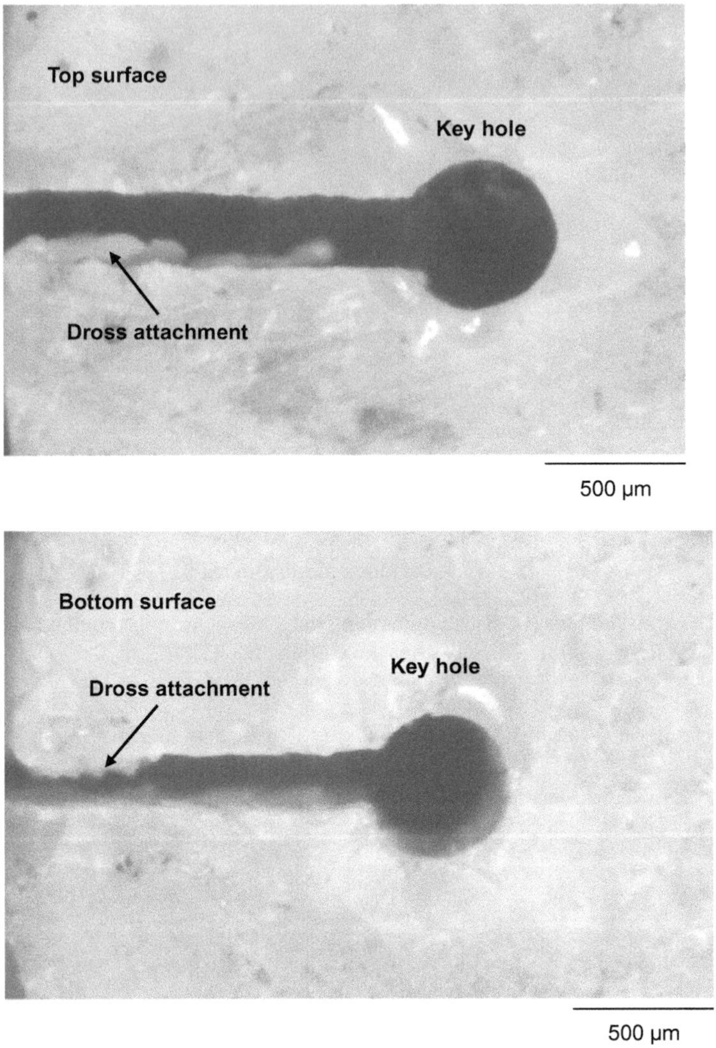

Fig. 5.29 Optical photographs of top and bottom surfaces of laser cut sections [3].

bottom surface of the cut section is notable. However, the amount of the dross attachment is not considerably large to block the kerf exit. The large crack formation around the cut edges is not visible, indicating that the fracturing and delaminating of the cut edge at the top and bottom surfaces are unlikely. Fig. 5.30 shows SEM micrographs of kerf surfaces. The formation of striation pattern is evident at the kerf surface. However, the depth of the striation pattern is shallow, which can be associated with the low thermal conductivity and high melting temperature of alumina. In this case, the melting zone is limited with the surface vicinity of the kerf. The micro-crack formation at the kerf surface is evident, which is attributed to the high-cooling rates resulting in

Fig. 5.30 SEM micrograph of kerf surfaces [3].

rapid solidification of the recast layer at the kef surface. The multicrack formation forms the crack network in the recast layer at the surface. This relaxes the stresses in the recast layer while minimizing the delamination and peels off the recast layer from the kerf surface. The dross attachment at the bottom surface of the cut edge is visible. This occurs because of one or both of the following reasons: (i) the liquid temperature in the molten flow along the kerf surface reduces toward the kerf exit lowering the fluid viscosity in this region. This, in turn, enhances the shear stress while enhancing the amount of material attachments at the kerf exit, and (ii) the assisting gas jet velocity reduces in the kerf due to a blockage effect which lowers the jet momentum at the kerf exit; consequently, the amount of molten material purged out from the kerf exit is reduced. Nevertheless, the dross attachment appears to be local at the cut edges, and the formation of the continuous film of dross attachment is less likely at the kerf exit.

Fig. 5.31 shows XRD diffractogram of the laser cut section. The formation of metastable γ-Al_2O_3 phase into α-Al_2O_3 takes place through thermal relaxation of non-equilibrium γ-Al_2O_3 phase [3]. The presence of AlON peak in the diffractogram reveals that Al_2O_3 undergoes a reduced reaction forming AlO ($Al_2O_3 \rightarrow 2AlO + 1/2\ O_2$). Because nitrogen at high pressure is used as an assisting gas, it enables

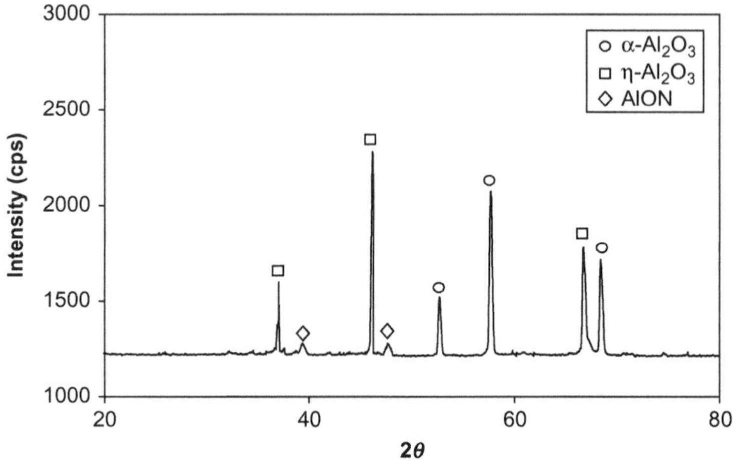

Fig. 5.31 XRD diffractogram obtained from laser cut section [3].

the formation of AlON through the reaction $2AlO+N_2 \rightarrow 2AlON$. Moreover, the residual stress obtained from the XRD technique agrees well with the predictions. In this case, the measured residual stress is on the order of 850 ± 30 MPa while its counterpart predicted from the simulations is on the order of 920 MPa at the surface vicinity.

5.5.2 Circular laser cutting of alumina tiles

Circular laser cutting of alumina tiles are challenging because of the high-stress levels developed in the close region of the hole circumference. The simulation results and data are given below in line with the previous study [4].

Fig. 5.32 shows the temperature variation along the top and bottom circumferences of the holes for different cooling periods, while Fig. 5.33 shows temperature contours around the hole. The temperature reduces sharply around the hole circumference once the cooling period progresses. It should be noted that the cooling period initiates at different times for two hole diameters; in which case, the cooling period starts at $t=0.21$ s for a hole diameter of 10 mm, and 0.314 s for a 15 mm hole diameter. Temperature decay from the melting temperature of the substrate material is at a higher rate for a 15 mm hole diameter than that of a 10 mm hole diameter. This is associated with the heat conduction taking place around the hole circumference. In this case, the internal energy gain of the substrate material around the hole circumference causes a temperature increase during the cutting process.

Because the cutting speed is high enough, the rate of cooling is low due to the conduction heat transfer from the hole edges to solid bulk. This, in turn, results in a high temperature region remaining along the hole wall at the onset of cooling initiation. This behavior is true for the top and bottom circumferences of the hole. In addition, the temperature attains significantly higher values at a location around the hole circumference where the laser beam is located, and the region in front of the laser beam

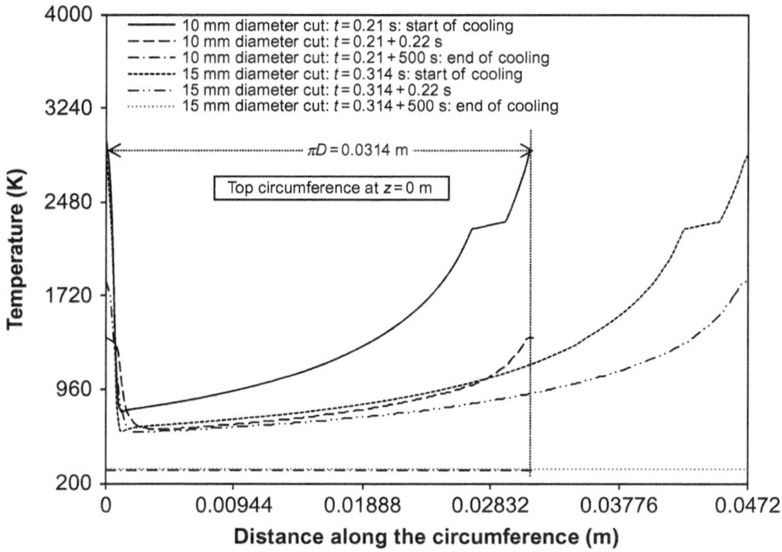

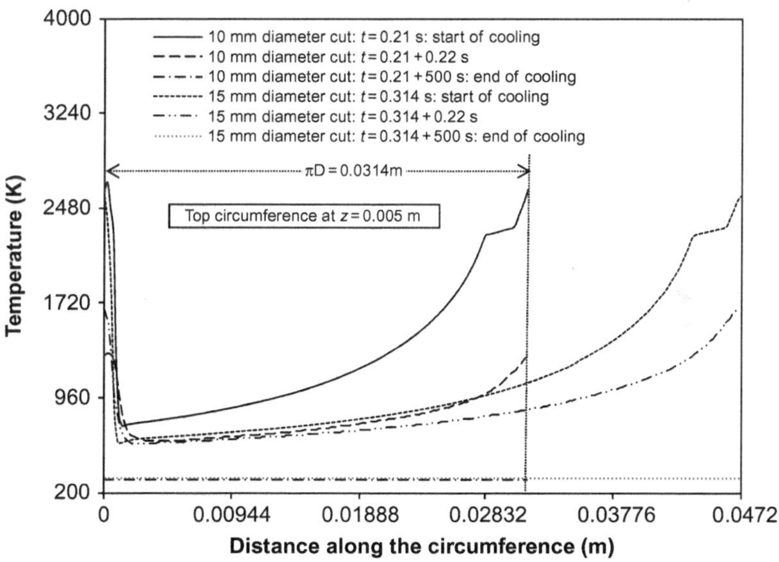

Fig. 5.32 The temperature distribution along the top and bottom circumferences of the hole for different cooling periods. The circumference of a 10 mm hole diameter is 0.0314 m, and a 15 mm hole diameter is 0.0471 m [4].

temperature remains low prior to the initiated cooling cycle. This appears as a sharp decay in temperature in Fig. 5.32. Once the cooling cycle progresses further, the temperature reduces significantly around the hole circumference for both hole diameters. The convection, conduction, and radiation heat transfer from the hole wall is responsible for the rapid decay of temperature. Moreover, the presence of a mushy zone due

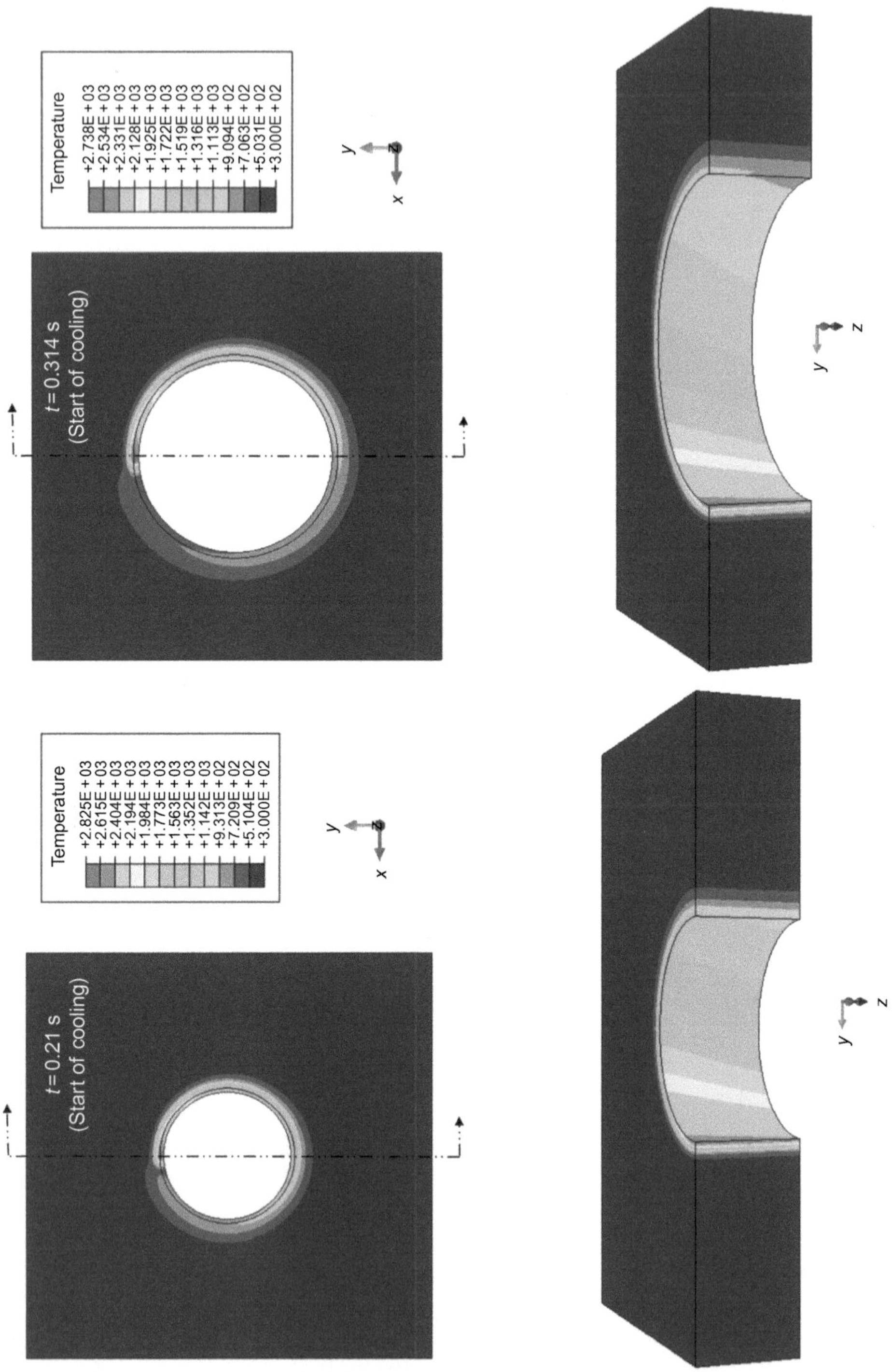

Fig. 5.33 The temperature contours inside the workpiece for two hole diameters: (A) 10 mm and (B) 15 mm [4].

to a phase change between the liquidus and solidus temperatures are evident from temperature curves. However, the temperature increases above the melting temperature of the substrate material in the central region of the irradiated spot, provided that the temperature remains below the evaporation temperature of the substrate material. This causes superheating of the liquid phase in this region. This phenomenon disappears and replaces with solid-phase cooling once the cooling cycle progresses in this region. The maximum temperature is slightly lower at around the bottom circumference of the hole as compared to that of the top circumference. This is attributed to modeling of the absorption of the incident beam, which is governed with the Lambert's Beer law; in which case, the laser beam intensity reduces exponentially with reducing depth.

Fig. 5.34 shows the von Mises stress distribution along the top and bottom of the hole circumferences for different cooling periods while Fig. 5.35 shows the contour plots of the von Mises stress around the hole. The von Mises stress remains low in the early cooling period when temperature is high. This is associated with the temperature dependent on the elastic modulus of the substrate material. It should be noted that in the simulations the temperature dependent on elastic modulus is incorporated and as temperature increases, elastic modulus reduces. The von Mises stress becomes small in the region where the cutting started at the end of the cooling periods. This is associated with the self-annealing effect of the previously cut sections. Heat transfer from the location of the laser beam toward the initially-cut regions is responsible for low-cooling rates around the initially-cut region. This, in turn, generates a self-annealing effect while lowering the stress levels in this region. The von Mises stress attains high values in the region where the temperature gradient is high. This corresponds to the region of the irradiated spot when the cutting ceases. When comparing the von Mises stress distribution along the hole circumferences due to two different diameters, it behaves almost the same way for two diameters at the end of the cooling period. However, this behavior differs during the cooling period because of the differences in cooling rates for each hole diameter. This indicates that the stress levels developed around the hole circumference differ considerably during the cutting of each hole; however, this behavior changes once the cooling period ends after the cutting process. In this case, stress behavior and its magnitude becomes almost identical for both hole diameters. When comparing the von Mises stress distribution around top and bottom circumferences, the magnitude of it becomes lower at the bottom circumference as compared to the top circumference. The difference between the stress distributions is associated with the cooling rates and temperature distribution at both circumferences, which differ slightly as observed from Fig. 5.32.

Fig. 5.36 shows the von Mises stress along the *z*-axis for two hole diameters and different cooling periods. The von Mises stress attains lower values in the region close to the top and bottom surfaces of the hole. This is attributed to free expansion of the surfaces. In this case, the surface thermally expands freely at the surface because there is no constraint presents at the surface. This lowers the strain and stress levels in this region. This situation changes in the region of the mid-thickness of the workpiece. In this case, the von Mises stress attains high values because of the high strain formed in this region. The heated material is not free to expand in this region and the high strain formed due to the high-temperature heating during the cutting in this region. The maximum value of

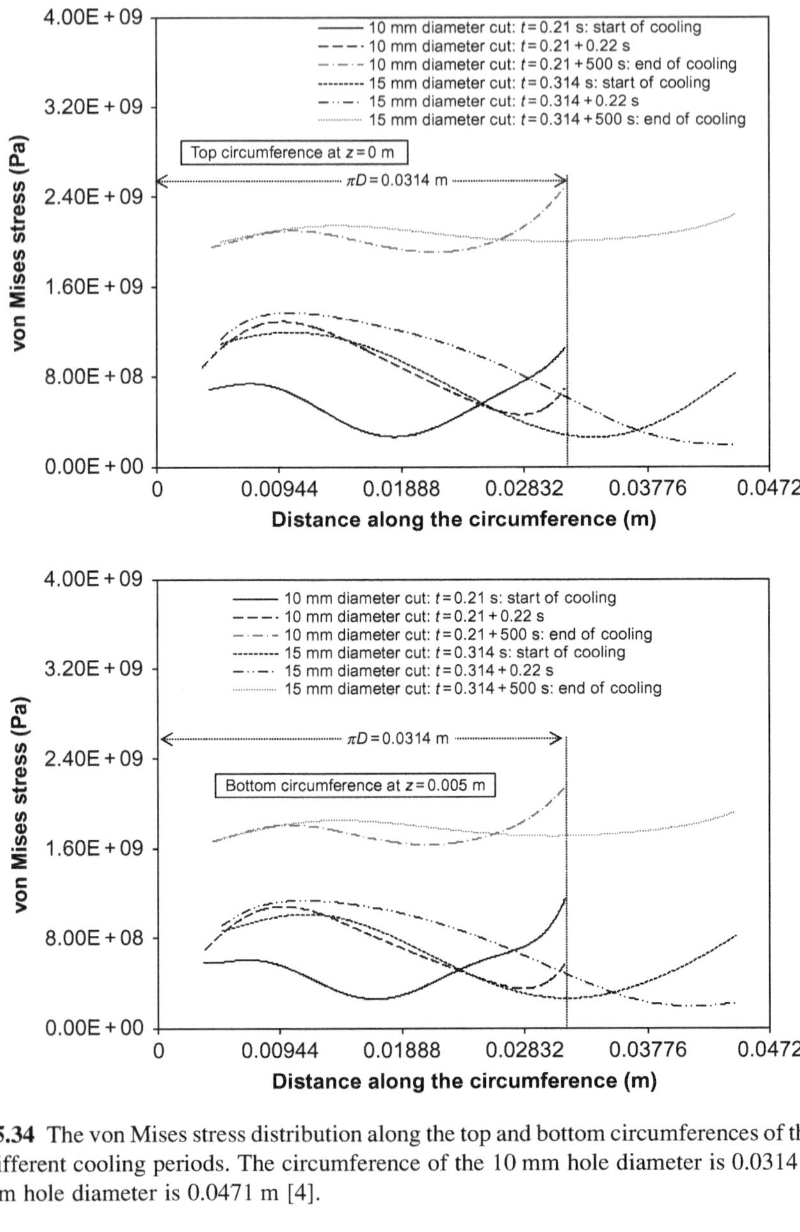

Fig. 5.34 The von Mises stress distribution along the top and bottom circumferences of the hole for different cooling periods. The circumference of the 10 mm hole diameter is 0.0314 m and 15 mm hole diameter is 0.0471 m [4].

the von Mises stress is on the order of 3.5 GPa, which is considerably high as compared to that predicted for metals. The influence of the hole diameter on the von Mises stress is not considerable, provided that von Mises corresponds to the 10 mm diameter, which is slightly higher than the 15 m diameter.

Fig. 5.37 shows SEM micrographs of the laser-cut section. It is evident from the micrographs that the cut edges are free from large defect sides and no major cracks

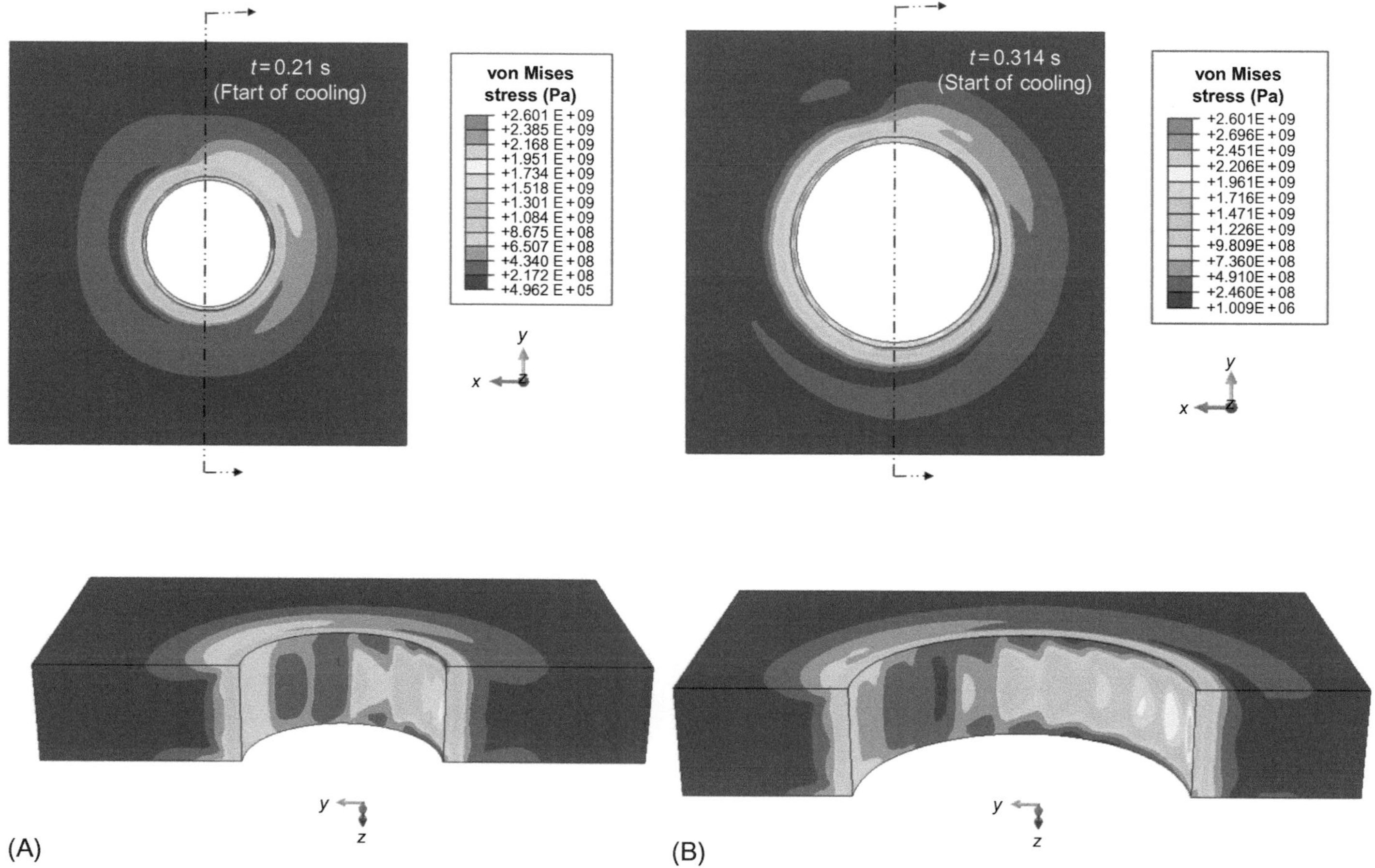

Fig. 5.35 The von Mises stress contours inside the workpiece for two hole diameters: (A) 10 mm and (B) 15 mm [4].

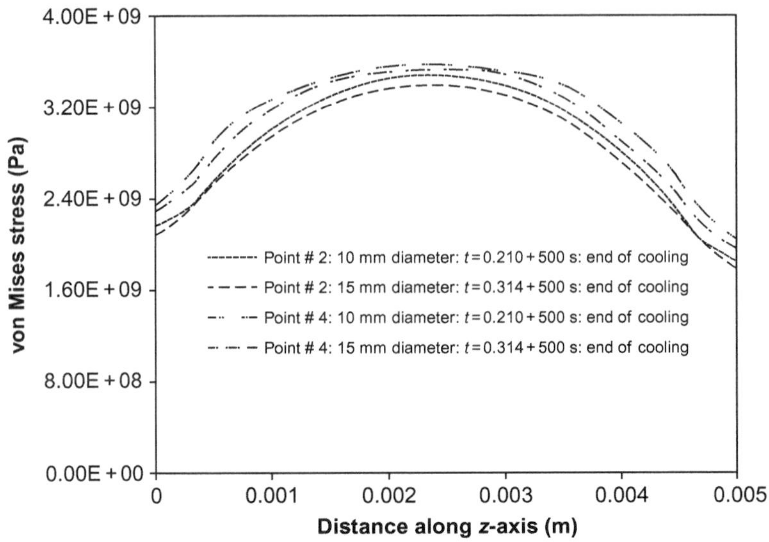

Fig. 5.36 The von Mises stress distributions along the z-axis at the end of the cooling periods for two locations along the hole circumference (Fig. 1). $z = 0$ m corresponds to the hole top surface, and $z = 0.005$ m is the hole bottom surface [4].

are observed. Unlike metals, the sideways burnings around the cut edges are not visible. The striation pattern is evident from the SEM micrographs. The formation of striations occurs because of instabilities associated with the melt flow at the kerf surface during cutting. However, the depth of the striations is shallow due to the low thermal conductivity of aluminum; in which case thermal erosion caused by the heat diffusion is not substantial as it is the case for metals. The striations are slightly bent toward the kerf exit and they are closely spaced. This behavior is attributed to the change of the viscosity of the molten flow, which reduces in temperature at the kerf exit region. Although high-pressure assisting gas is used to purge the molten flow out of the kerf width during the cutting, the shear force developed in the molten flow increases significantly due to the high flow. This, in turn, modifies the flow pattern toward the kerf exit and striation patterns changes in this region. In addition, the momentum of the assisting gas reduces toward the kerf exit which, in turn, results in reduced drag force in this region. The dross attachment at the kerf exit is evident from the SEM micrograph. This occurs because of the attainment of a relatively low temperature (slightly above the liquidus temperature) at the circumference of the hole exit because of the convection cooling of the assisting gas. The reduced temperature increases the viscosity of the molten flow at the kerf exit. In addition, the assisting gas reduces its axial momentum due to the flow friction in the narrow kerf width. Consequently, the combination of high-viscosity molten flow and low assisting gas pressure reduces the rate of molten material ejected from the kerf exit. The remaining molten material, which are not ejected from the kerf exit, attaches to the kerf while forming the burrs at the hole exit. The cracks are formed at the kerf surface, which is associated with

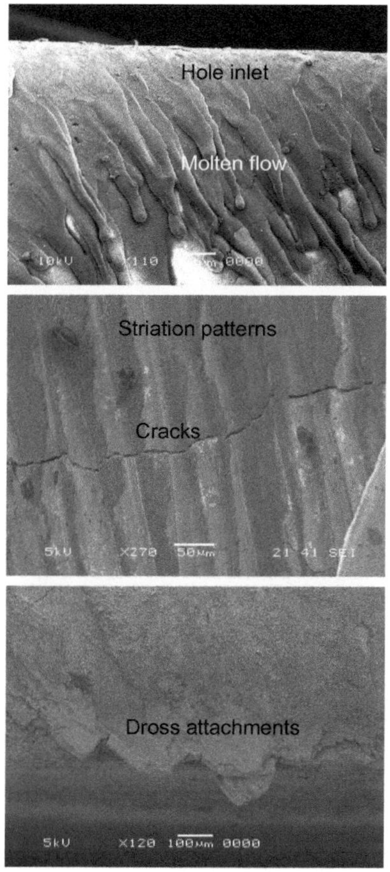

Fig. 5.37 SEM micrographs of laser cut-hole surfaces [4].

the high-cooling rates. It should be noted that high-cooling rates result in the formation of brittle and dense structures at the surface in the form of a recast layer. The fracture toughness of this layer becomes significantly lower than the as-received material fracture toughness. Consequently, a high residual stress formation in the recast layer causes the crack formation in this layer. However, the close examination of the SEM micrographs indicates that the cracks formed at the kerf surface are shallow and do not extend below the recast layer thickness. This can be attributed to the stress relaxation in this region. It should be noted that energy relaxation takes place through cracking in this region.

5.5.3 Rectangular laser cutting of alumina tiles

Laser cutting of rectangular geometry into alumina tiles introduces complicated stress fields in the corner region of the cut geometry. This is because of the excessive heating in this region, which, in turn, modifies the stress levels. In this section, the findings of

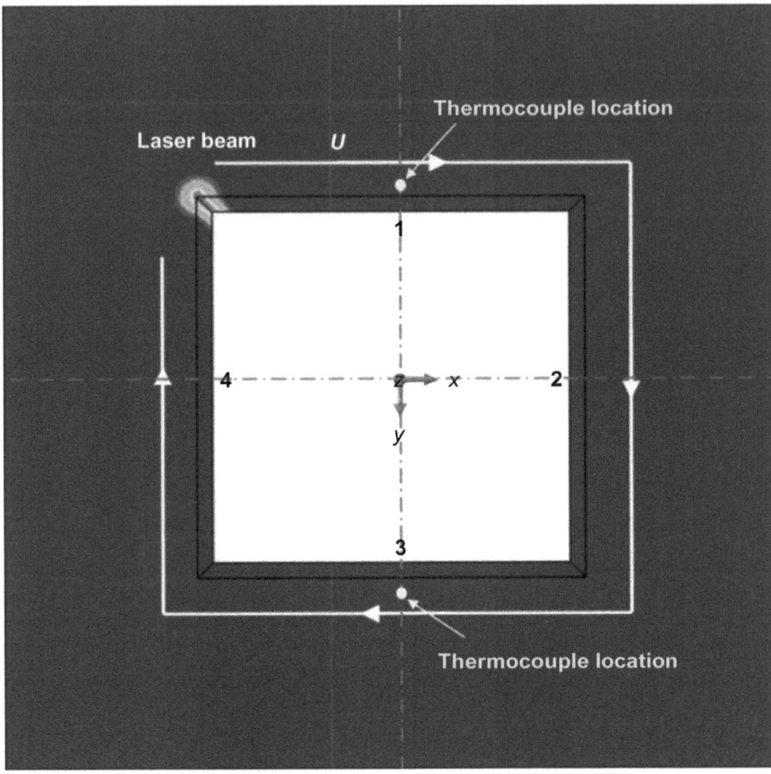

Fig. 5.38 A schematic view of the laser cutting of rectangular geometry, coordinate system, and the thermocouple locations [5].

the simulation and experimental data are presented in line with the previous work [5]. Fig. 5.38 shows the schematic view of the laser-cutting path in the tile.

Fig. 5.39 shows the temperature distribution predicted along the top and bottom circumferences of the rectangular cut geometry for different cooling periods while Fig. 5.40 shows temperature contours in the cutting section. The cooling period starts at 0.4 s in Fig. 5.39. The temperature remains high in the region where the laser beam is switched off at 0.4 s and the location of the laser beam is 0.04 m along the circumference of the rectangular geometry. However, the temperature decays gradually along the initially-cut sections, except in the corner region of the rectangular geometry. The attainment of low temperature in this region is associated with the heat conduction from this region to its neighbourhood. The area cut in the vicinity of the corner is larger than that of the straight cut. This, in turn, enhances the cooling rates through conduction heat loss in this region. Moreover, a temperature well in excess of the melting temperature occurs in the region of the last location of the laser spot at the circumference of the cut geometry. This indicates that the superheating of the liquid phase takes place in this region. In addition, the temperature remains in between the liquidus and solidus temperatures of the alumina in the neighborhood of the superheated liquid

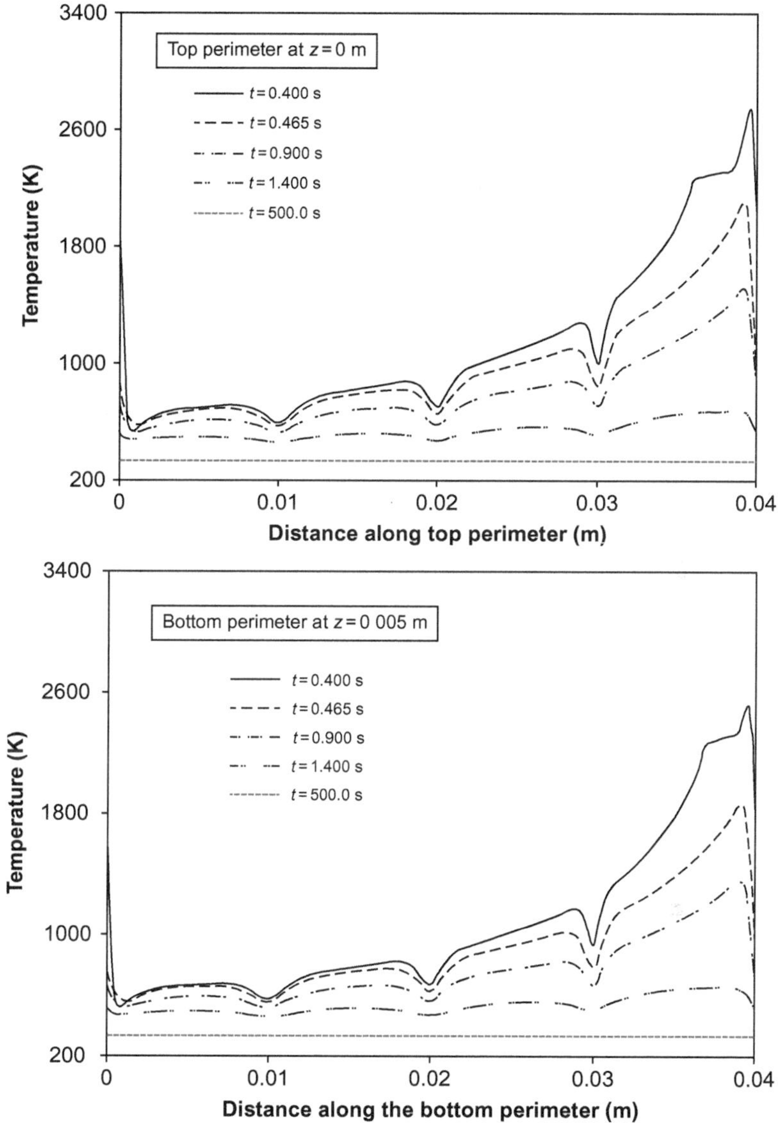

Fig. 5.39 The temperature variation along the top and bottom circumference of the rectangular cut for different cooling periods [5].

region. This shows the presence of the mushy zone in this region. When comparing temperature behaviors corresponding to the top and bottom circumference of the rectangular geometry, both temperatures behave similarly, provided that the maximum temperature at the bottom circumference is slightly lower than that of the top circumference. This is because of the attenuation of the penetrated laser beam during the

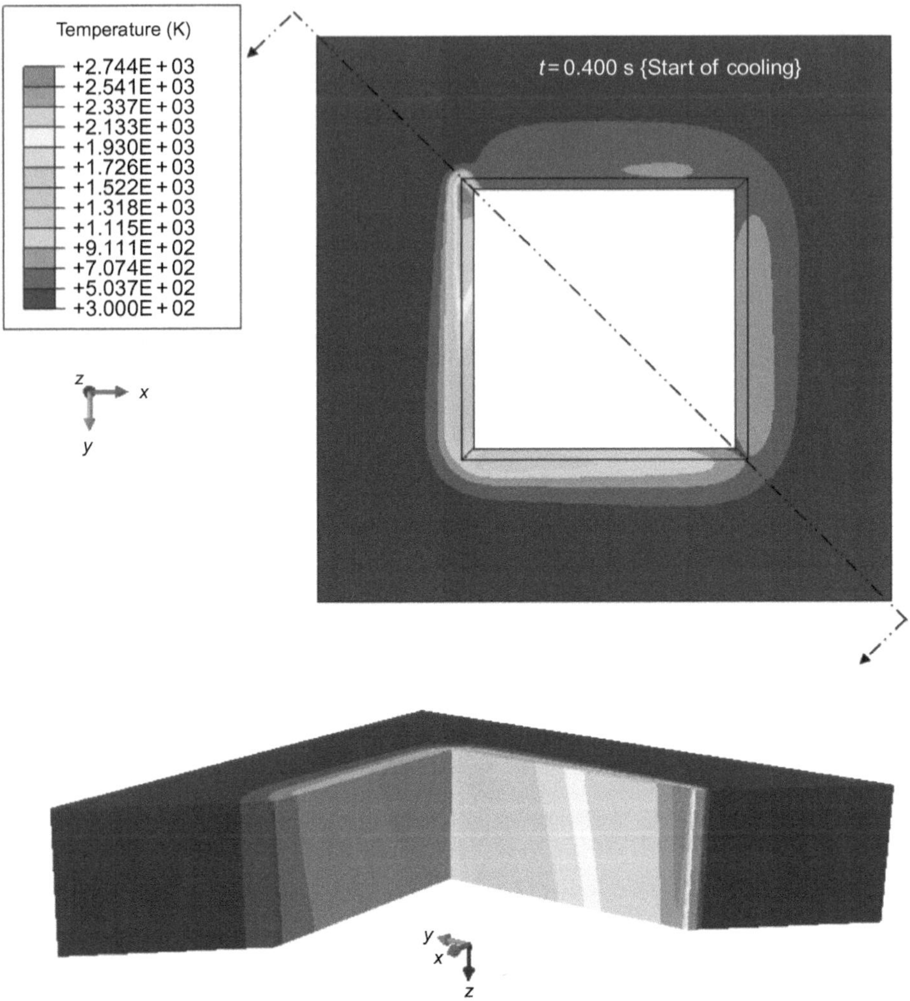

Fig. 5.40 The temperature contours in the cutting section onset of cooling period [5].

cutting process. In addition, the size of the mushy zone is smaller at the bottom circumference than at the top circumference. As the cooling period progresses, the temperature reduces and reaches the initial temperature at the end of the cooling period, which is at $t=500$ s.

Fig. 5.41 shows the von Mises stress distribution predicted along the top and bottom circumferences of the rectangular cut geometry for different cooling periods, as similar to those shown in Fig. 5.39, while Fig. 5.42 shows the von Mises stress contours in the cut section. The von Mises stress attains low values in the region where temperature is high. This is associated with temperature-dependent elastic modulus, which reduces with increasing temperature. However, at the initiation of the cooling

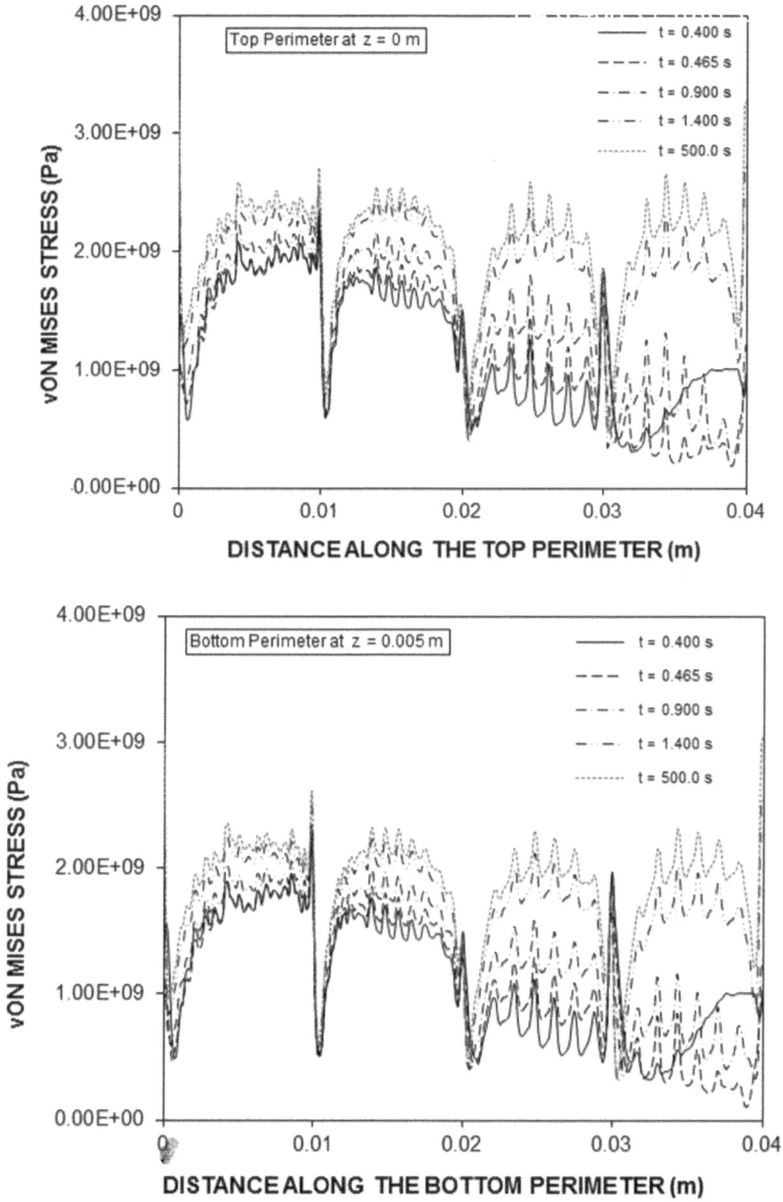

Fig. 5.41 The von Mises stress variation along the top and bottom circumference of rectangular cut for different cooling periods [5].

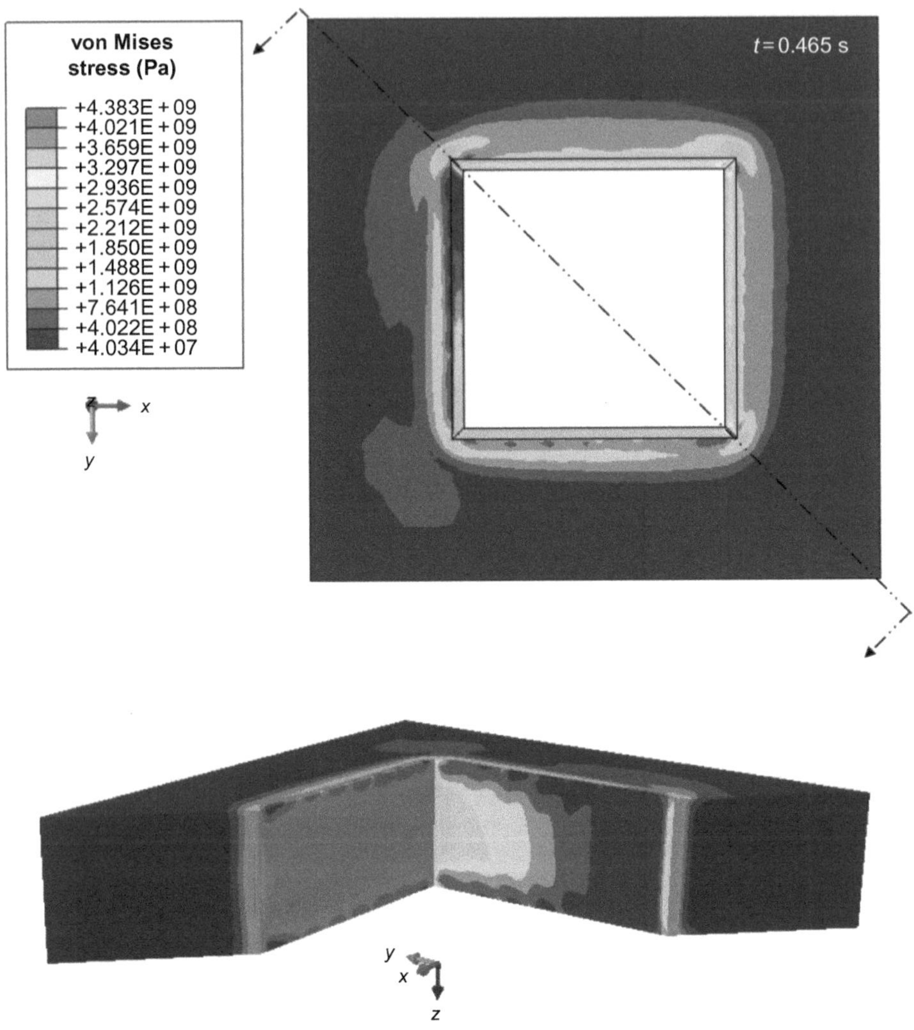

Fig. 5.42 The von Mises stress contours in the cutting section onset of cooling period [5].

period ($t = 0.4$ s), the von Mises stress attains high values in the close region of the corners of the rectangular geometry. This is associated with the high temperature gradients developed in this region during the early cooling period. However, the von Mises stress attains high values at the surfaces of the rectangular geometry toward the end of the cooling period ($t = 500$ s), and it remains low at the corners of the geometry for the all cooling periods. The same arguments are true for a von Mises stress distribution at the bottom circumference of the rectangular geometry. However, the magnitude of the von Mises stress is lower at the bottom surface as compared to its counterpart at the top surface. This behavior is related to temperature variation along the circumference, which attains relatively lower values at the bottom

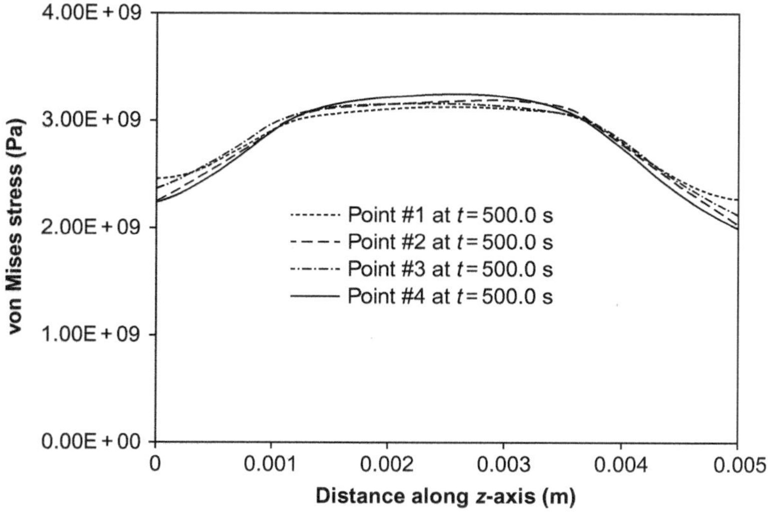

Fig. 5.43 von Mises distribution along the workpiece thickness: (A) onset of the cooling period and (B) at the end of the cooling period [5].

circumference. The maximum value of the von Mises stress is on the order of 2.6 GPa, which is considerably high.

Fig. 5.43 shows the von Mises stress along the z-axis at the onset of cooling period and at the end of the cooling period. It is evident that the von Mises stress attains low values at location #4 at the onset of cooling cycle, which is related to the reduced elastic modulus because of the attainment of high temperatures at this location. In general, the von Mises stress remains low at the top ($z=0$ m) and bottom ($z=0.005$ m) surfaces of the workpiece. This is associated with the free expansion of the surface during the cutting process. It should be noted, in line with the experimental conditions, the top and bottom surfaces of the workpiece are set as the stress-free boundary conditions. Because the surfaces expand freely during the heating cycle, they also contract freely in the surface region during the cooling cycle. Therefore, thermally-induced strains attain lower values in this region than those in the mid-thickness of the workpiece. The attainment of a considerably high strain causes the formation of high-stress levels in this region. Moreover, the stress levels remain high in this region at the end of the cooling period. The values of the von Mises stresses become almost the same at all locations around the circumference of the rectangular geometry, and the maximum magnitude of the von Mises stress is on the order of 3.1 GPa. Its location is almost at the mid-thickness of the workpiece.

Fig. 5.44 shows the optical photographs of the laser cut rectangular geometry while Fig. 5.45 shows the SEM micrograph of the laser cut surface. It should be noted that as the cutting speed reduces while keeping the laser output power for the optimal cutting conditions, thermal erosion partially deforms the sharp edges of the corners of the rectangular geometry. This can be observed from Fig. 5.44. However, for the optimum

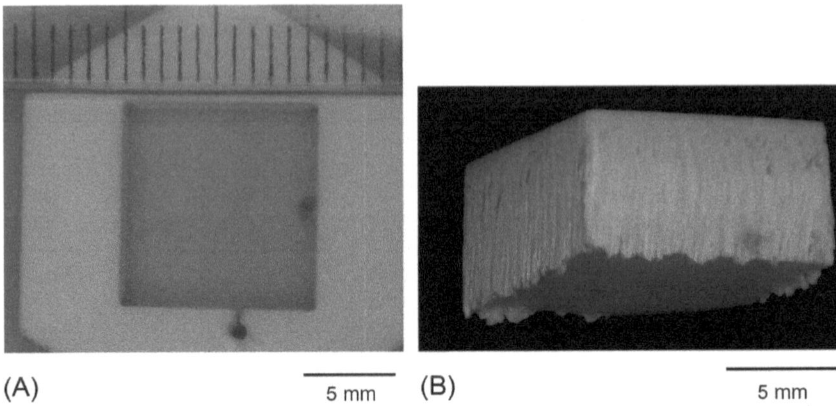

Fig. 5.44 Optical photographs of laser cutting sections: (A) top circumference and (B) bottom circumference [5].

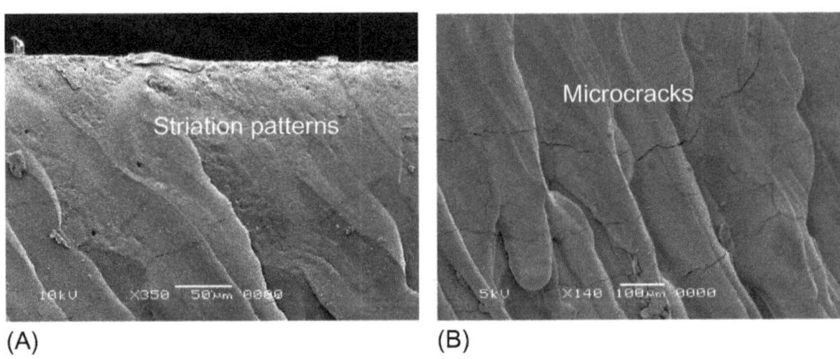

Fig. 5.45 SEM micrographs of the kerf surface: (A) the top surface and (B) the central region of the kerf surface [5].

cutting conditions, it is evident that the cut sections are free from large defect sides such as major cracks or large-scale cavities due to sideways burning. Moreover, close examination of the optical photograph and SEM micrograph reveals that striation patterns are formed at the edges of the cut section. This is attributed to the melt instability and assisting gas jet effect on the molten flow at the kerf surface. It should be noted that the assisting gas purges the molten material from the cut section. Therefore, it generates a drag force at the surface of the molten flow. Depending on the depth of the molten material, the rate of fluid strain changes, which in turn alters the amount of material purged from the kerf site. In addition, the melt instability causes the variation of the melt depth at the kerf surface and initiates the striation formation. The dross attachment is also evident at the surface. This is associated with the molten viscosity at the kerf edge, which increases due to the cooling effect of the assisting gas.

It should be noted that the viscosity of the molten material increases as the temperature reduces. The convection cooling of the molten surface by the assisting gas lowers the temperature of the molten flow. Consequently, high viscosity combined with surface tension force results in the attachment of dross at the kerf edges. The SEM micrograph shows the presence of microcracks at the surface. These cracks are shallow in depth and do not conform to the base material; in which case, their depth is limited with the recast layer thickness.

5.5.4 Laser cutting of aluminum foam

Aluminum foam is one of the metallic foams with superior mechanical properties. They provided significant advantages when the weight of the parts became important in the practical applications. The foam is a cellular structure consisting of metal containing large voids and pores filled with air. The high-volume ratio of voids and pores reduces the density of the foam, which enables it to be used in aerospace and other cooling applications. One of the problems encountered in manufacturing of parts from aluminum foam is net shaping for the desired geometries. Due to the cellular structures, conventional machining, in most cases, fails to produce net shapes for complicated geometries. Additionally, the process is also costly. Laser machining offers significant advantages over the conventional machining methods. Some of these advantages include fast processing, precision of operation, and low cost. However, the presence of air in the pores and voids triggers high-temperature exothermic oxidation reactions in the cut sections during the process while lowering the cutting quality and enhancing thermal stress developed in the cutting sections.

In the analysis, the solid body heat conduction with temperature-dependent conductivity, internal energy (including latent heat effects), and convection and radiation boundary conditions are considered. Although the metal foam is formed from porous cells, the governing equation of energy transfer for cutting needs to be modified to incorporate the presence of pores in the workpiece. Because the pores have irregular shapes and are randomly distributed in the aluminum solid, the modeling of the aluminum foam including air, which is trapped in pores and the solid part of the foam, is difficult; in which case, such predictions could result in large errors due to unknown porous cell geometry and their orientations. Therefore, the continuum conservation equations are used in the analysis with the assumption of effective properties for the aluminum foam. In the analysis, the dynamics effect of the assisting gas on the cutting process is omitted due to the complicated nature of the gas flow behavior in large cells in the foam; however, it is referred to in the previous studies [23–25]. Moreover, the convective cooling effect of the assisting gas in the cutting section is incorporated in the analysis through the convection boundary condition. In thermal analysis, temperature dependence of the material properties, namely, the density, specific heat, and thermal conductivity, are considered. The values of these equivalent properties for an aluminum porous metal are calculated from that of an aluminum metal without pores by the rule of mixture. Based on the assumption that the porosity of the lotus-type porous metal is constant with respect to the temperature change, and the equivalent density and equivalent specific heat of the lotus-type porous metals can be expressed using the same rule as that of conventional porous metals, their equivalent properties are expressed by the following equations [26]:

$$\rho_{eff}(T) = (1-\varepsilon)\rho_s(T) + \varepsilon\rho_f(T) \quad (5.18)$$

and

$$C_{peff}(T) = \frac{(1-\varepsilon)\rho_s(T)C_{ps}(T) + \varepsilon\rho_f(T)C_{pf}(T)}{(1-\varepsilon)\rho_s(T) + \varepsilon\rho_f(T)} \quad (5.19)$$

and

$$k_{eff}(T) = \frac{(1-\varepsilon)\rho_s(T)k_s(T) + \varepsilon\rho_f(T)k_f(T)}{(1-\varepsilon)\rho_s(T) + \varepsilon\rho_f(T)} \quad (5.20)$$

where $\rho_{sub}(T)$, $C_{psub}(T)$, and $k_{eff}(T)$ are the density, specific heat and the effective thermal conductivity of porous metal, respectively, and ε is the porosity expressed by the volume ratio of pores versus the total volume of the porous metal. The subscripts (sub) '*eff*', '*s*', and '*f*' indicate the effective property, property of the nonporous aluminum metal, and property of the gas in the pores, respectively.

To analyze the phase change problem, the enthalpy method is used [13]. The specific heat capacity is associated with the internal energy gain (U) of the substrate material; that is $Cp(T) = \frac{\partial U}{\partial T}$. Latent heat effects are significant in the laser melting of aluminum foam material due to the phase change and must be taken into account. The equivalent latent heat ($L_{eq} = \frac{L_s}{1 + \frac{\rho_f}{\rho_s} \in \left(1 + \frac{\in}{\in - 1}\right)}$, where L_{eq} is the equivalent latent heat of melting and L_s is the melting heat of aluminum) is used in terms of solidus and liquidus temperatures (the lower and upper temperature bounds of the phase change range), and the total internal energy associated with the phase change.

The metal foam behavior differs from that of solid metal, and classical plastic theory cannot be used to describe their behaviors. Deshpande and Fleck [27] developed a three-dimensional model primarily based on the experimental tests of aluminum foam, which has been built in the finite element package ABAQUS [13]. The model assumes similar behaviors in tension and compression, hence, isotropic hardening. This model has the simplest expression and has found wide-spreading applications. In this paper, three-dimensional finite element modeling analysis was carried out for the thermal and stress analysis using the dynamic explicit code ABAQUS/Explicit using the crushable foam model with isotropic hardening. This stress analysis is coupled with the previous thermal analysis to import the thermal history during the laser treatment process. The crushable foam plasticity model available in ABAQUS is based on the assumption that the resulting deformation is not recoverable instantaneously and can be idealized as plastic for short-duration events. In this model, the contribution of the mean stress on the yield function is realized through a material parameter known as a shape factor. It defines the aspect ratio of the elliptical stress.

The shape factor quantitatively distinguishes the plastic behavior of metal foams from solid metals. The yield function (F) is given by Deshpande and Fleck [27]:

$$F=\sqrt{q^2+\alpha^2p^2}-B=0 \tag{5.21}$$

where q is the von Mises equivalent stress and p is the mean stress. B defines the size of the yield ellipse. Here, α represents the shape of the yield ellipse in the p–q stress plane, B defines the size of the yield ellipse. The yield surface is the von Mises circle in the deviatoric stress plane, and an ellipse in the meridional plane. The shape factor, α, remains as a constant during the plastic deformation process. The evolution of the yield ellipse is controlled by an equivalent plastic, $\overline{\epsilon}^{pl}$. In the present work, compression yield stress ratio k is taken as unity.

The flow potential for the isotropic hardening model is chosen as Deshpande and Fleck [27]:

$$G=\sqrt{q^2+\beta^2p^2} \tag{5.22}$$

where β represents the shape of the flow potential in the p–q stress plane and is related to the plastic Poisson's ratio, v_p, by:

$$\beta=\frac{3}{\sqrt{2}}\sqrt{\frac{1-2\nu_p}{1+\upsilon_p}} \tag{5.23}$$

The plastic Poisson's ratio, v_p, which is the ratio of the transverse to the longitudinal plastic strain under uniaxial compression, should be defined by the user, and it must be in the range of -1 and 0.5. The upper limit $v_p=0.5$ corresponds to an incompressible plastic flow. In the current simulation, the value of the plastic Poisson's ratio of 0.34 is assumed.

The plastic strains are defined to be normal to a family of self-similar flow potentials parameterized by the value of the potential G [27]:

$$\dot{\varepsilon}^{pl}=\dot{\lambda}\frac{\partial G}{\partial \sigma} \tag{5.24}$$

where $\dot{\lambda}$ is the nonnegative plastic flow multiplier. The hardening of the foam is described through:

$$\sigma_c=\sigma_c\left(\overline{\varepsilon}^{pl}\right) \tag{5.25}$$

where $\overline{\varepsilon}^{pl}$ is the equivalent plastic strain. The evolution of $\overline{\varepsilon}^{pl}$ is assumed to be governed by the equivalent plastic work expression; that is:

$$\sigma_c\dot{\overline{\varepsilon}}^{pl}=\sigma:\dot{\varepsilon}^{pl} \tag{5.26}$$

The equivalent plastic strain is equal to the absolute value of the axial plastic strain in uniaxial tension or compression. The plastic flow is associative when the value of β is the same as that of α. In general, the plastic flow is nonassociated to allow for the independent calibrations of the shape of the yield surface and the plastic Poisson's ratio.

5.5.4.1 *Laser straight cutting of aluminum foam*

The findings of simulations and experimental data are presented below in line with the previous study [6].

Fig. 5.46A–B shows the temperature variation along the x-axis at the top and bottom surfaces of the cut section. Temperatures reach above the liquidus temperature of the substrate material at the point where the laser beam axis is located ($x=0.03$ m) at

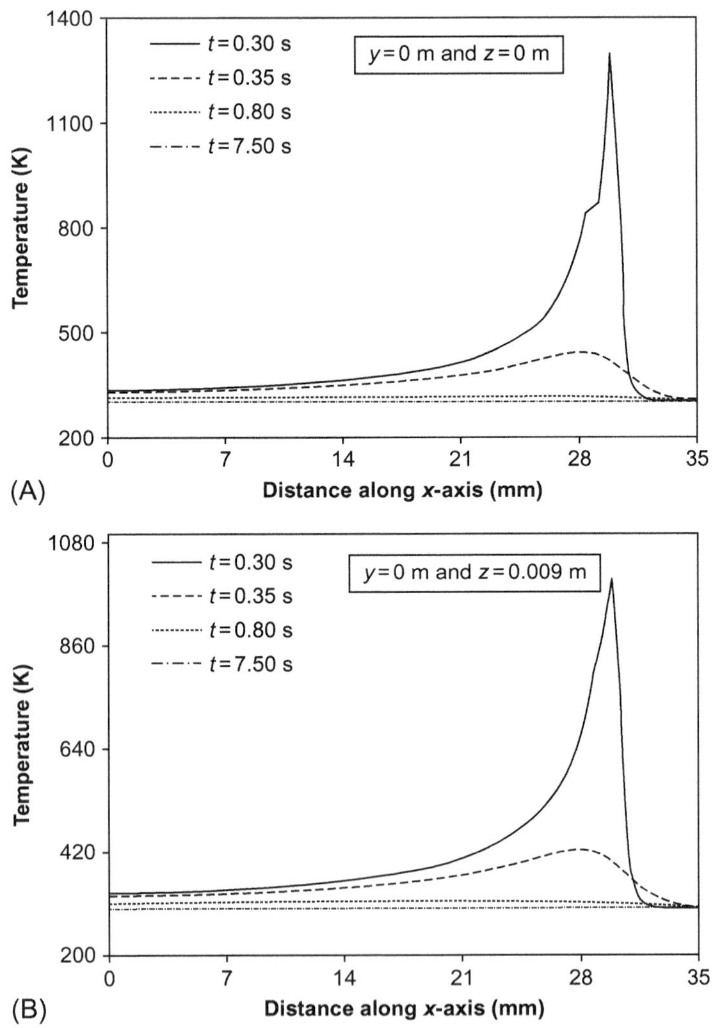

Fig. 5.46 (A) Temperature distribution along the x-axis for different cooling periods [6]. (B) Temperature distribution along the x-axis for different cooling periods at the bottom surface [6].

the onset of the cooling cycle. It should be noted that the laser beam cutting speed is 0.1 m/s and time taken to reach at $x = 0.03$ m is 0.3 s. Attainment of high temperatures at location $x = 0.03$ m results in superheating of the liquid phase at the irradiated spot surface. The temperature decays sharply in the close region of front and backsides of the irradiated spot; in which case, the temperature gradient attains high values in these regions. Moreover, temperature decay is rather gradual in the region $0 \leq x \leq 0.025$ m as compared to that in the irradiated spot. This region corresponds to the edge of the initially-cut section and convective as well as conduction cooling reduces the temperature in this region. However, the temperature decay next to the irradiated spot is high, which results in a self-annealing effect in the neighborhood region of the laser spot. As the cooling period progresses, the temperature decays sharply due to convection and conduction heat transfer from the top surface. When comparing temperature behavior between top and bottom surfaces of the cutting section, it is evident that temperature profiles along the x-axis are almost the same for both surfaces, but not identical, because the value of the temperature at the bottom surface is lower than that at the surface. This is attributed to the absorption of laser-irradiated power in the cutting section of the workpiece material, which decreases along the thickness according to the Beer-Lambert law.

Fig. 5.47A–B shows the von Mises stress distribution along the x-axis at the top and bottom surfaces of the cut section. The von Mises stress attains low values in the regions where temperature is high (Fig. 5.46A–B). This is attributed to the elastic modulus of the workpiece material, which reduces with increasing temperature. However, the von Mises stress attains high values in the region where the temperature gradient is high while the temperature is low. This corresponds to the front region of the laser spot. In addition, the von Mises stress also attains high values in the region $0.007 \leq x \leq 0.025$ m. This is because of the high elastic modulus at low temperatures. The von Mises stress attains low values in the neighborhood of the backside of the laser irradiated spot. This is associated with the self-annealing effect of the laser spot. As the cooling cycle ends, the von Mises stress attains the highest value in the front region of the irradiated spot. The maximum magnitude of the maximum von Mises stress is on the order of 1.7 MPa, which is less than the yielding limit of the substrate material; that is, the yielding limit is on the order of 3.5 MPa. When comparing the von Mises stress distribution along the top and bottom surfaces of the cutting section, the behavior of each is similar, but not the same, because the von Mises stress attains higher values at the top surface as compared to the bottom surface. This is attributed to temperature gradients; which are higher at the top surface. In this case, high-temperature gradients result in high thermal strain and thermal stress at the top surface of the cutting section.

Fig. 5.48 shows optical photographs of the laser cut section. Uniform and parallel-sided cuts are evident. The presence of large voids in the workpiece did not notably affect the cut section geometry. However, close examination of the surface reveals that some sideways burning takes place along the cut edges at the top plate of the workpiece. This results in kerf-width variation along the cutting section. The kerf width size along the cutting section is measured using a moving optical microscope. The kerf-width variation $\Delta w / w_{average}$ (where $\Delta w = w_{\max} - w_{\min}$ and $w_{average} = \frac{\sum_1^n w}{n}$)

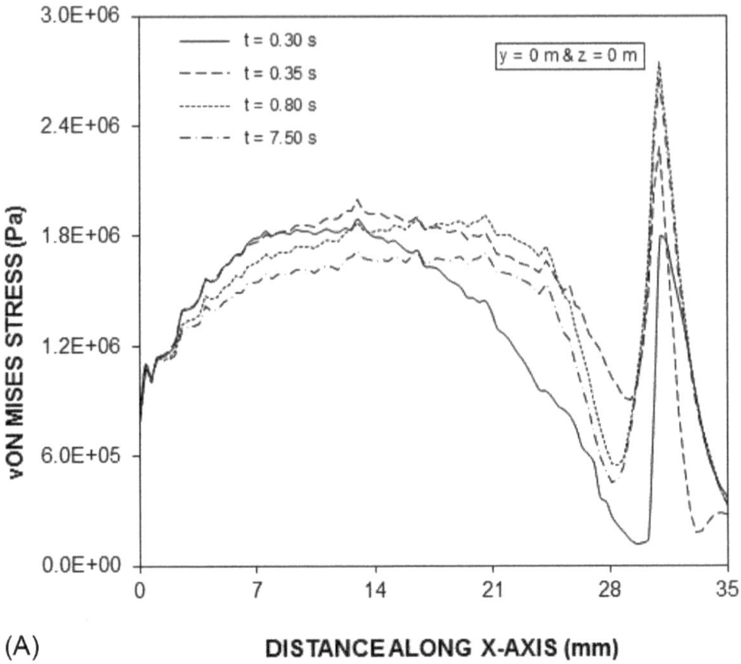

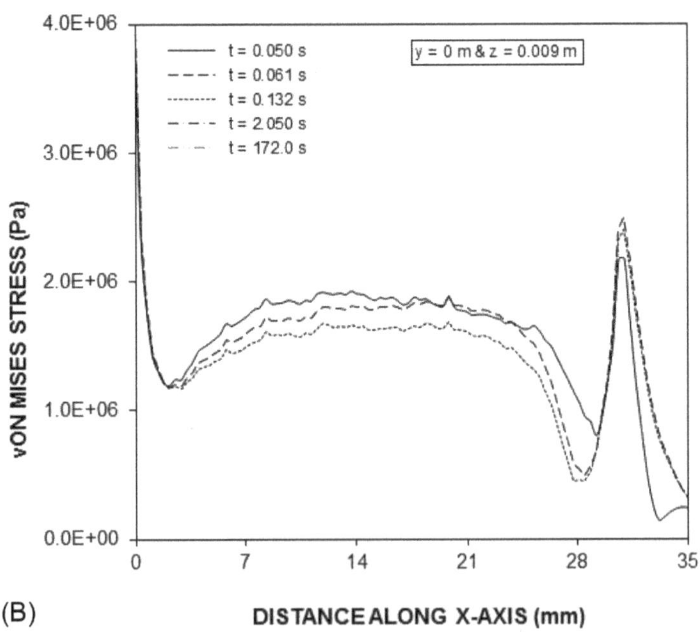

Fig. 5.47 (A) von Mises stress distribution along the x-axis for different cooling periods [6]. (B) von Mises stress distribution along the x-axis at the bottom surface for different cooling periods [6].

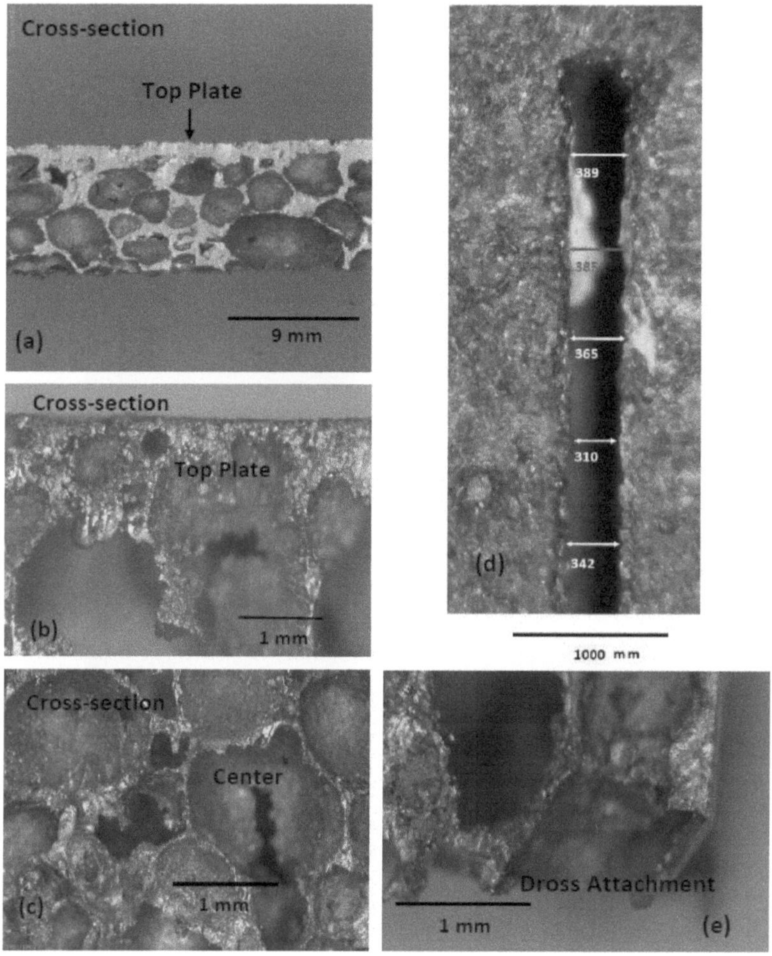

Fig. 5.48 Optical photograph of the laser cut sections: (A) a cross-section of the cut section in small magnification, (B) a cross-section close to the top surface, (C) a cross-section in the center of workpiece, (D) a top view of the cut, (E) a cross-section close to the bottom surface [6].

is found to be on the order of 7%, which is similar to those corresponding to solid metal cutting. The occurrence of the sideways burning is associated with the high temperature oxidation reactions taking place in the cutting section despite the fact that nitrogen at high pressure is used as the assisting gas. In this case, air trapped in the voids is responsible for the high temperature oxidation reactions taking place in the cutting section.

Fig. 5.49 shows SEM micrographs of the top and kerf surfaces of the cutting section. The close examination of the top surface reveals the presence of thermal erosion due to sideways burning taking place during the cutting process. However, the

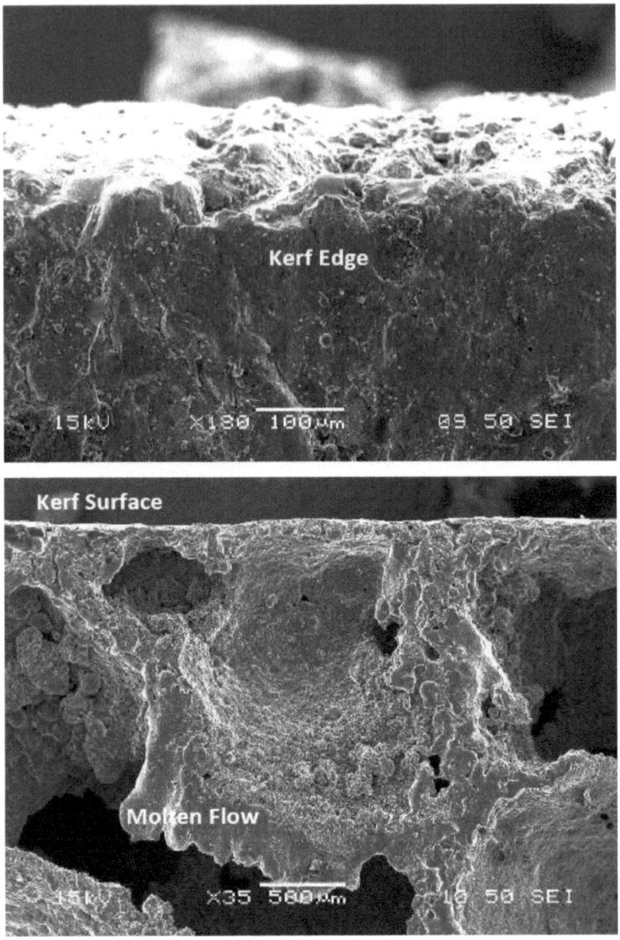

Fig. 5.49 SEM micrographs of the top and kerf surfaces of the cutting sections [6].

thermally-eroded region is limited to the cut edge, and the extension of the thermally eroded region toward the solid bulk is not observed at the surface. This is attributed to the use of nitrogen assisting gas at high pressure, which suppresses the high temperature oxidation reactions taking place in the region away from the cut edge. The close examination of the SEM micrograph corresponding to the kerf surface reveals that excessive molten flow occurs around the voids. This is because of one or both of the following reasons: (i) the impinging gas in the kerf width is redirected toward the large pores, while reducing its axial moment during suppression of the drag force on the molten material in the kerf. This reduces the molten ejection from the kerf, and (ii) oxidation reactions taking place at the molten surface, due to the presence of air in the pores. This results in the formation of aluminum oxides altering the viscosity of the molten layer while suppressing the molten removal rate from the kerf.

5.5.4.2 Laser circular cutting of aluminum foam

The findings of the simulation results and experimental data for laser circular cutting of aluminum foam are given in line with the previous studies [7].

Fig. 5.50 shows optical photographs of the front and bottom views of the laser cut hole. It can be observed that the laser cut hole does not have large size burr at the bottom side of the cut edge. It should be noted that air was trapped in the pores of the foam prior to the laser beam cutting process. Hence, during the cutting operation, oxygen in the air underwent exothermic reactions with aluminum molten metal in the cut section while modifying the viscosity of the molten flow. Because nitrogen at a high pressure (~550 kPa) is used as an assisting gas, the drag force generated in the kerf width due to the assisting gas purging the molten flow exiting the kerf. However, increased

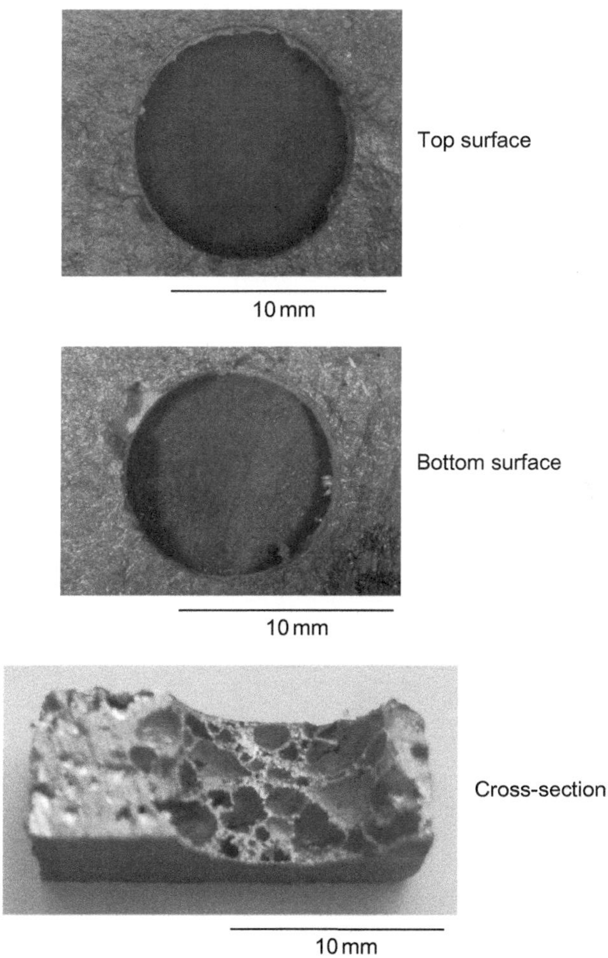

Fig. 5.50 Optical photographs of top and bottom surfaces of laser cut hole and its cross-section [7].

viscosity due to oxidation of the molten material via an exothermic reaction increases the resistance of the molten flow while causing some asperities at the kerf exit, which is mainly observed for metal cutting by a laser beam. It should be noted that viscosity of the molten metal increases after oxidation, which gives rise to high-shear stress at the interface between the molten flow and the solid surface. Therefore, the pressure force required to overcome the frictional drag force increases significantly. This causes dross attachments or the formation of some asperities such as small scale burrs at the kerf exit. Because the amount of molten metal is less because of the high porosity of the foam, the flow resistance does not result in excessive burr formation or a large-scale dross attachment at the hole exit. SEM micrographs of the laser cut section reveals that some small molten flow occurs at the surface of the cut section (Fig. 5.51). However, the molten flow does not substantiate to cover pores or all cut surfaces around the pores as observed from Fig. 5.51. In addition, no striation pattern is observed unlike that observed in metallic sheet laser cutting. This is attributed to the presence of large pores and a comparatively smaller area of solid structures

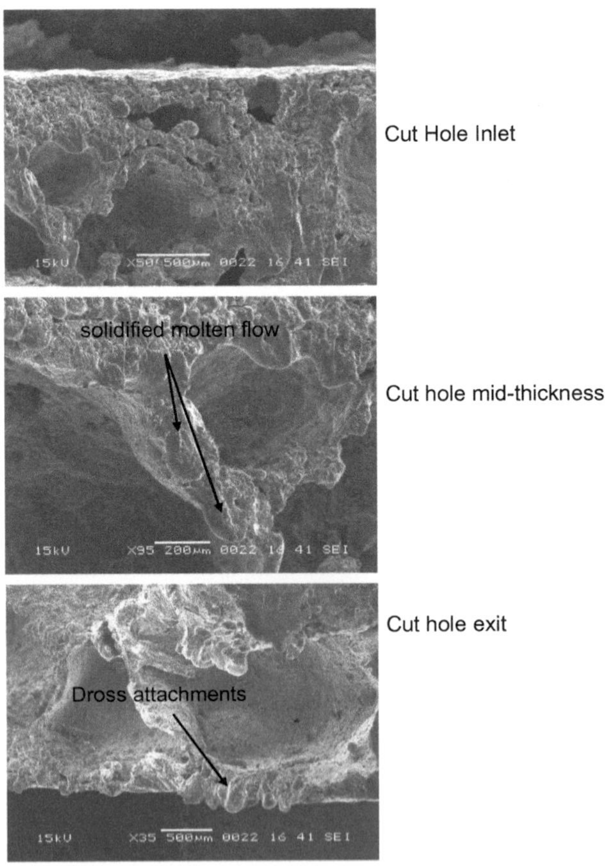

Fig. 5.51 SEM micrographs of laser cut surface at different locations along the hole thickness [7].

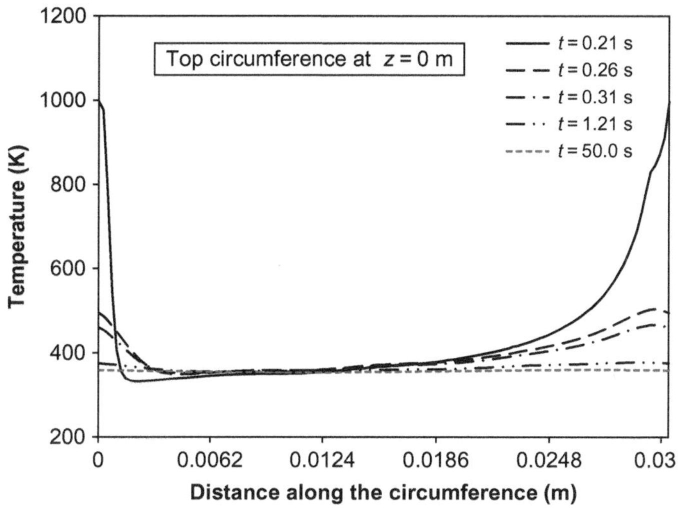

Fig. 5.52 Temperature variation along the circumference of the hole for different cooling periods [7].

around the pores. In the case of the hole exit, a small and locally-distributed dross attachment is evident (Fig. 5.51). This is attributed to resolidification of local molten flow. However, the size of dross attachment is small unlike that which occurs in laser metal sheet cutting.

Fig. 5.52 shows the temperature distribution along the hole circumference for different cooling periods. The cooling cycle starts at $t=0.21$ s, which corresponds to complete cutting of the hole; in which case; the laser power is ceased off at $t=0.21$ s. It should be noted that the radiation losses during laser cutting is found to be less than 10%. Temperature remains high at the location when the laser power is switched off. It is evident that the temperature exceeds the liquidus temperature of aluminum foam ($T_l=865$°K) at this location. This indicates the superheating of the liquid phase occurs in this region. However, the presence of a mushy zone is also visible from the figure where the temperature is between solidus ($T_s=835$ K) and liquidus temperature ($T_l=865$ K). However, the size of the mushy zone is very small. Temperature decays sharply in the vicinity of the heated region at the onset of cooling ($t=0.21$ s). This results in the formation of a high-temperature gradient in this region. As the cooling period progresses, the temperature along the circumference reduces sharply which is attributed to high cooling rates due to a convection cooling effect of the assisting gas and conduction heat transfer in the solid structure of the aluminum foam. As the cooling period progresses ($t=50$ s) the temperature reduces to initial temperature (300°K) around the circumference indicating the end of the cooling period.

Fig. 5.53 shows the von Mises stress along the hole circumference for different cooling periods. The von Mises stress attains low values in the region where temperature is high, and it becomes high where the temperature gradient is high in the initial

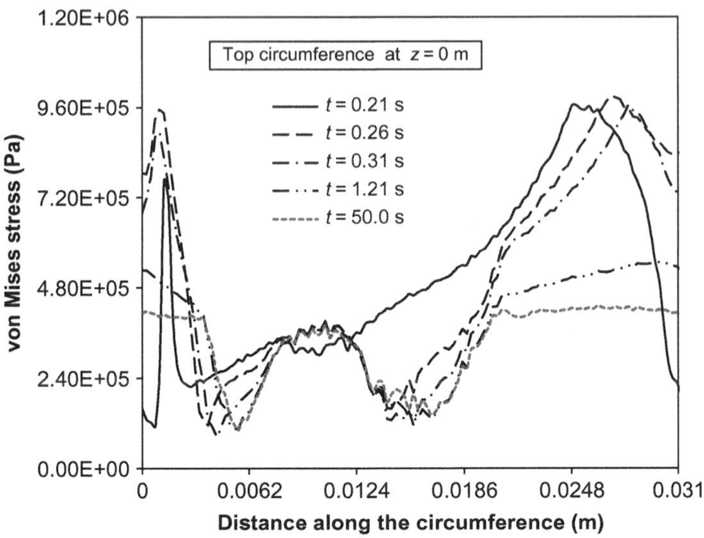

Fig. 5.53 The von Mises stress variation along the circumference of the hole for different cooling periods [7].

cooling period. This is attributed to the temperature dependent elastic modulus, which decreases with increasing temperature. It should be noted that the high-temperature gradient enhances heat conduction from the heated region to the solid bulk; which, in turn, results in high-cooling rates in this region. However, the von Mises stress first increases with a progressing cooling period and, later, it reduces with a progressive time toward the end of the cooling period. This is associated with the self-annealing effect of the initially cut hole section during the cutting process. In this case, initially, the cooling effect of the assisting gas results in a high temperature gradient in the cut surface around the hole circumference. However, the heat transfer from the neighborhood of the cut section from the aluminum foam bulk toward the cut surface reduces the temperature gradient while lowering the thermal strain in the vicinity of the cut section. This, in turn, lowers the thermal stress at the cut section. The maximum von Mises stress is on the order of 0.25 MPa at the end of cooling cycle, which is less than the yielding limit of aluminum. Consequently, no cracks are expected to form at the cut section, which is also evident from Fig. 5.50.

Fig. 5.54 shows the von Mises stress along the z-axis at different locations around the hole circumference for two cooling periods. The von Mises stress attains low values in the region of the front and back surfaces of the hole section and becomes high in the central region of the hole ($0.001\ \text{m} \leq z \leq 0.0075\ \text{m}$). The attainment of low stress at the front and back surfaces of the cut section is attributed to free expansion of both surfaces during the cooling process. Consequently, thermal strain developed in this region is low due to unconstrained surface effects. In the mid-section of the hole, thermal strain is high during the expansion and contraction of the hole wall, which does not freely expand unlike the front and back surfaces. The von Mises stress

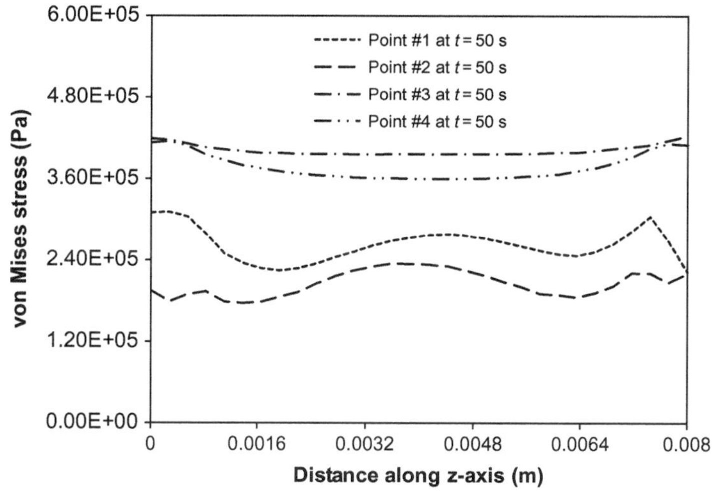

Fig. 5.54 von Mises stress distribution along the *z*-axis at different locations along the hole circumference onset of cooling initiation and at the end of the cooling period [7].

is low in the region of the high temperature during the early cooling period ($t = 0.21$ s) and attains high values when the temperature becomes low. This is because of the elastic modulus of the workpiece, which reduces with the increasing temperature. However, the von Mises stress remains high along the *z*-axis for all locations around the hole circumference at the end of the cooling period ($t = 50$ s) when the temperature reduces to the initial temperature in the cutting section. The von Mises stress becomes lower at locations #1 and #2. This is associated with the self-annealing effect of the laser beam at these locations on the cut section around the hole circumference. Consequently, slow cooling results in stress relaxation in the region where the laser beam is switched off around the hole circumference.

5.5.4.3 Laser triangular cutting of aluminum foam

The simulations results and experimental findings are presented in line with the previous study [8]. Fig. 5.55 shows the laser cutting direction of triangular geometry.

Fig. 5.56 shows the temperature distribution along the top perimeter of the triangle cut of two different sizes (namely A and B) for three different time periods of the cooling cycle. Triangle A represents a side length of a 4 mm equilateral triangle, and Triangle B corresponds to a side length of an 11 mm equilateral triangle. It should be noted that the distance along the perimeter of the triangular cut is normalized with its total length. This provides the comparison of two different sized triangle cuts in Fig. 5.56.

The cooling cycle starts immediately after the cutting process ends and it corresponds to $t = 0$ s. The temperature attains high values in the region where the laser power is switched off (laser power is set to zero), which is true for early cooling periods. However, the temperature exceeds the melting temperature of the substrate material at the center of the irradiated spot at the cooling cycle initiation; that is, time

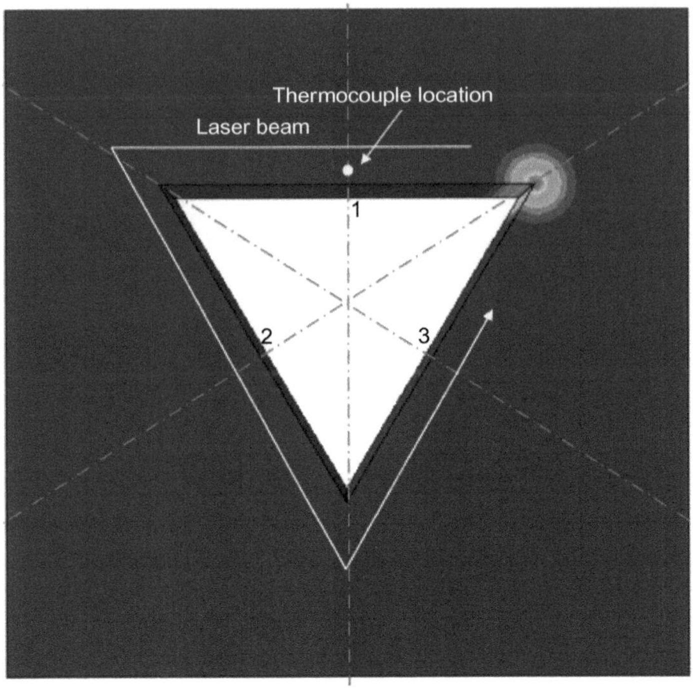

Fig. 5.55 A schematic view of laser cutting of equilateral triangular geometry and the location of thermocouple [8].

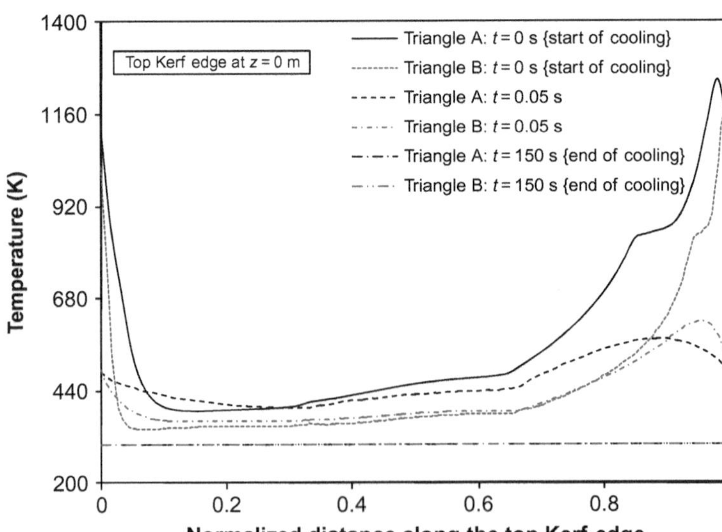

Fig. 5.56 Temperature distribution along the top edge circumference of the triangle cut for two sizes (A represents a side length of a 4 mm equilateral triangle, and B corresponds to a side length of an 11 mm equilateral triangle) [8].

corresponding to $t=0$ s. In this case, superheating at the liquid phase takes place in this region. The temperature remains almost in the range of solidus and liquidus temperatures in the region close to the irradiated spot center. This shows the phase change taking place in this region. It should be noted that the phase change takes place between the solidus and liquidus temperature of the substrate material. Temperature decay is gradual behind the laser spot while it is sharp in front of the spot. This is attributed to the initial cut sections, which is situated behind the irradiated laser spot. Consequently, heat transfer from the cut edges to the surroundings through conduction, convection, and radiation lowers the temperature gradually along the cut edges. The maximum temperature is high for a 4 mm triangle as compared to that obtained for an 11 mm triangle at the cooling initiation. This is associated with the rate of heat dissipation from the cut section to its surrounding; in which case, heat dissipation due to conduction in the direction of the cut along the cut edges is less for the 4 mm triangle than that of an 11 mm triangle. This is due to the fact that the temperature along the cutting direction is lower for a 4 mm triangle. It should be noted that the conduction heat transfer increases as the temperature gradient increases. As the time progress in the cooling cycle, the temperature around the perimeter reduces and it reduces to almost initial temperature after 150 s from the onset of the cooling cycle. Once the temperature reduces to the initial temperature, the cooling cycle ends.

Fig. 5.57 shows the von Mises stress distribution along the top perimeter of the cut section for two triangles and three cooling periods, as similar to those shown in

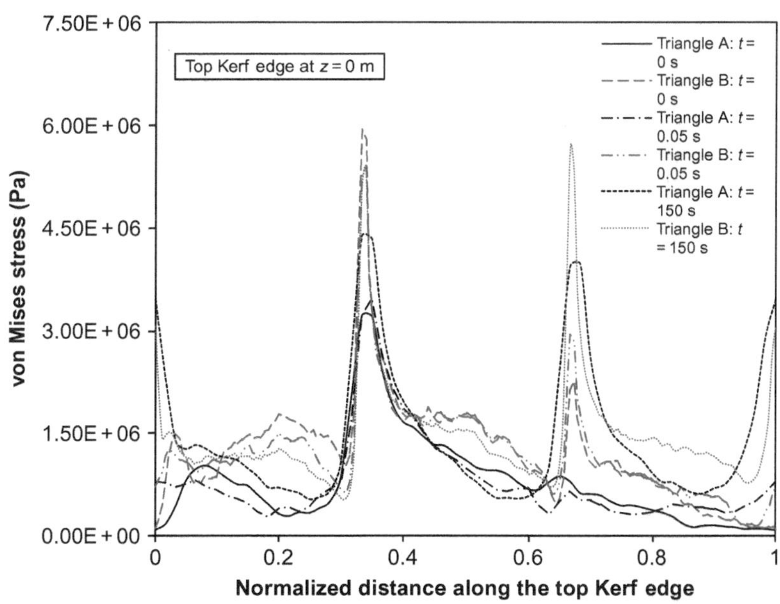

Fig. 5.57 von Mises stress distribution along the top edge circumference of the triangle cut for two sizes (A represents a side length of a 4 mm equilateral triangle and B corresponds to a side length of an 11 mm equilateral triangle) [8].

Fig. 5.56. The von Mises stress attains high values in the regions close to the corners of the triangles. This is attributed to the thermal strain formed at both edges of the triangle that meet at the corner and the strains are superimposed at the corners. Because the corners are not free to expand, high thermal strain results in the formation of high thermal stress in these regions. The thermal expansion coefficient is temperature-dependent and reduces with increasing temperature; the von Mises stress attains low values where temperature is high. Consequently, at the initiation of cooling cycle, the von Mises stress attains low values along he triangle edges. However, as the cooling period progresses, the von Mises stress attains high values particularly in the regions close to the corners. The maximum von Mises stress becomes maximum for 11 mm size triangle despite the fact that the maximum temperature is less for an 11 mm size triangle. This is attributed to the cooling rates; in which case, a large-sized triangle cools at a faster rate than that of a small-sized triangle. This, in turn, results higher stress levels for a large sized triangle. The maximum values of the von Mises stress is on the order of 1 MPa, which is less than the yielding limit of the aluminum foam.

Fig. 5.58 shows the von Mises stress distributions along the z-axis at different locations along the edges of the triangle for two sizes of triangle at $t=150$ s of cooling cycle. In general, the von Mises stress attains low values at $z=0$ (top edge) and $z=0.005$ m (bottom edge). This is because of the surfaces at which the workpiece is free to expand. This lowers the stress levels in this region. The stress distribution is not uniform along the z-axis at different locations around the cut edges. This is attributed to the thermal expansion and contraction of the workpiece along the z-axis. The free expansion of the surfaces and displacement constraints at mid-thickness of the workpiece modifies the thermal stress levels along the z-axis. In addition, a

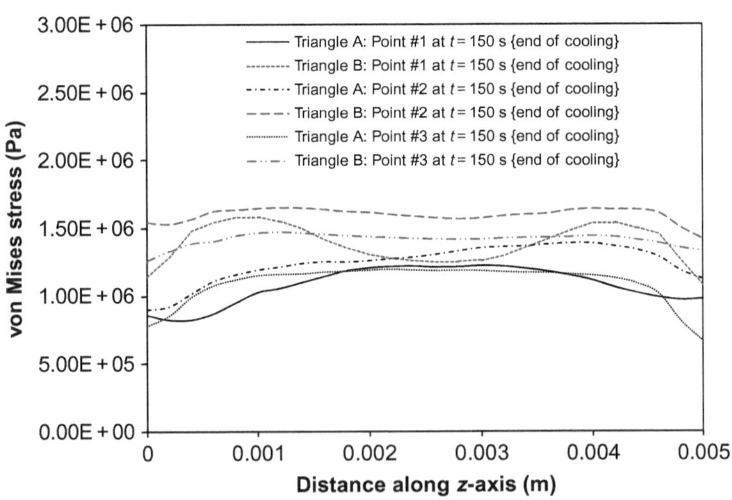

Fig. 5.58 von Mises stress distribution along the z-axis at different locations around the cut edges for two sizes (A represents a side length of a 4 mm equilateral triangle and B corresponds to a side length of an 11 mm equilateral triangle). The cooling period is 150 s [8].

different cooling rate along the cut edges of the triangle contributes to the formation of a nonuniform stress distributions along the z-axis.

Fig. 5.59 shows an optical photograph of the laser cut triangles. It can be observed that the cut section is free from large-size asperities such as surface debris, which results from the irregular melting around the cut edges. However, close examination of the photographs reveal that some small sideways burning occurs around the cut edges, particularly in the corner region. This is attributed to the presence of air in the closed pockets of the foam prior to cutting. During the cutting process, air is

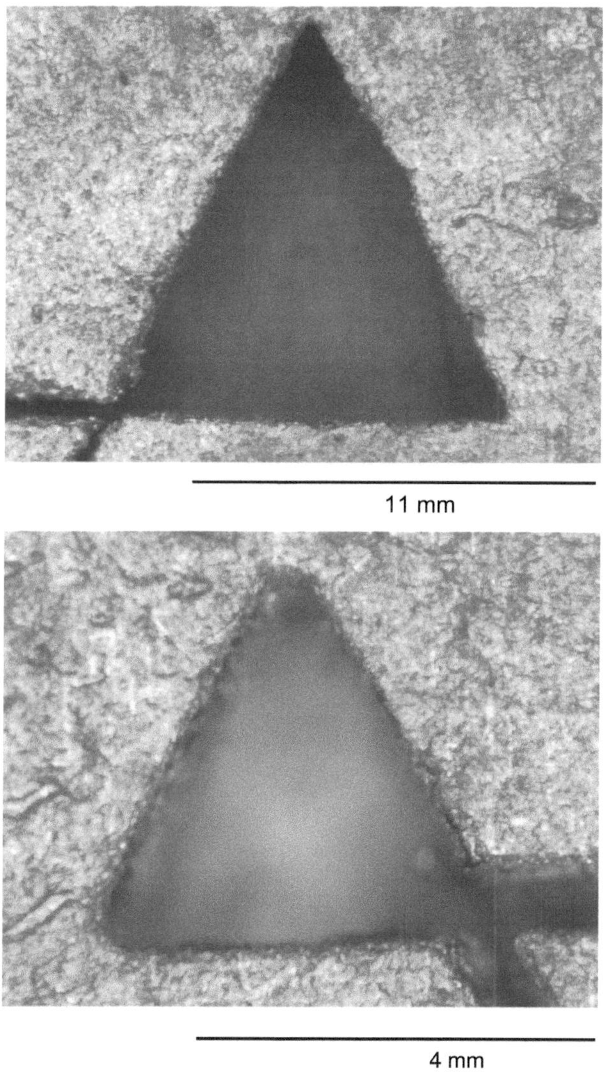

Fig. 5.59 Optical photograph of the laser cut triangles [8].

released and oxygen in the air undergoes the exothermic reactions with the molten material. This gives rise to excessive heat generation in the cutting section while causing the sideways burning at the cut edges. However, the size of the sideways burning is considerably small. This is attributed to the use of nitrogen assisting gas, which suppresses the rate of high temperature oxidation reactions in the cut section.

Fig. 5.60 shows SEM micrographs of laser cut sections. It is evident that the top surface of the cut is considerably clean and free from the striation patterns, which are usually observed for solid metal cutting. The formation of a striation pattern is associated with the transient nature of melt flow at the cut edges. Because the effective density of the aluminum foam is on the order of 250 kg/m^3, the amount of melt flow along the cut surface is negligibly small as compared to solid metals. However, some small flow of molten metal is observed at the cut surface. This is attributed to the presence of high-pressure assisting gas, which purges the molten metal in the kerf toward the cut exit. The pressure of the assisting gas reduces toward the exit of the cut section due to the frictional drag in the kerf. In addition, the presence of large voids in the cut section aids to develop turbulence flow in the kerf. This further influences the pressure drop at the kerf exit. Consequently, some fraction of the molten flow at the exit of the cut section could not be swept away by the assisting gas and it adheres to the kerf exit. Moreover, the viscosity of the molten flow reduces toward the kerf exit because of the temperature drop due to cooling in this region. This enhances the dross attachment at the kerf exit. Close examination of the micrographs reveal that few dross attachments

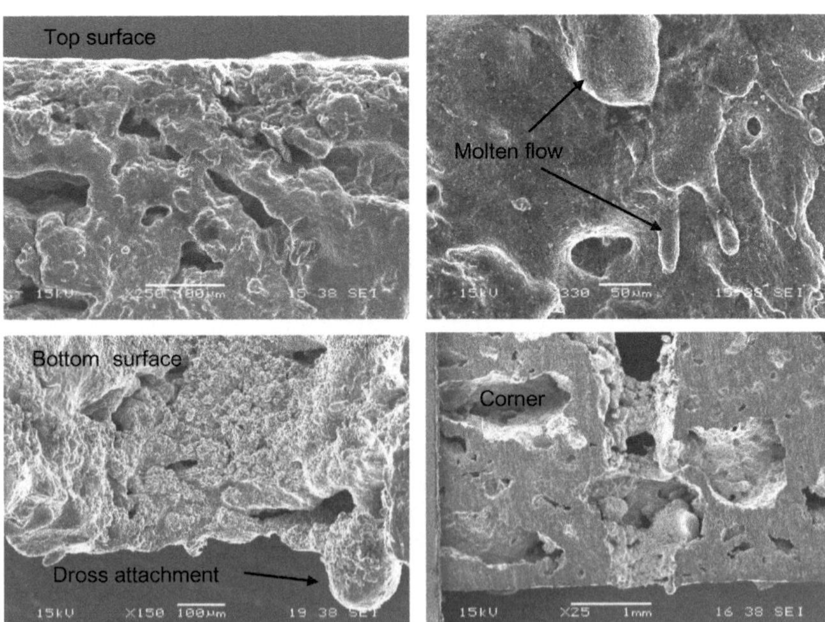

Fig. 5.60 SEM micrographs of laser cut sections [8].

are present at the kerf exit. The dross attachment does not extend over the edge of the cut section at the exit. This is because of the low density of foam; in which case, the molten flow rate at the kerf exit is low. SEM micrographs in the region of the corner of the triangle cut shows that cut section is free from geometric defects and a straight cut is observed in this region.

5.5.5 *Laser cutting of Kevlar*

Kevlar has unique properties, such as high tensile strength, high toughness, and chemical stability at high temperatures in aromatic polyamides. Kevlar is widely-used as a friction material in the automotive industry and a combustion protection material in the aerospace industry. The net shaping of Kevlar laminates is difficult using the conventional machine tools due to whiskers forming and loose fibers around the machine edges. In addition, the specific energy required for machining is higher than most of the engineering materials. Moreover, laser machining offers considerable advantages over the conventional techniques. Some of these advantages include a high-quality end product, precision of operation, and low cost. Laser cutting eliminates the whiskers emanating from the machined edges due to thermal effect. Although unwoven Kevlar laminates are easy to bend, the development of resolidified regions at the kerf edges and the formation of a heat-affected zone in the cut section increases brittleness of laser machined Kevlar laminates in this region. This, in turn, limits the practical application of the machined laminates.

The results obtained from laser cutting experiments associated with Kevlar cutting are presented below in light of the previous study [9].

Fig. 5.61 shows the variation of kerf width size with laser power intensity for different assisting gas pressures. It should be noted that the kerf width size is predicted from Eq. (3.231) in Chapter 3 and data used in the simulations are given in Table 5.5. Nitrogen was used as the assisting gas. The kerf width size increased with increasing laser power intensity, which appears to be more pronounced at high power intensities. In this case, increasing the laser power intensity enhances the rate of material removal from the kerf and causes small oxidation reactions to take place in the kerf, due to the air trapped in the voids. This enhances the sideways burning. The laminate consists of layers, as shown in Fig. 5.62A, in which air is trapped during the manufacturing of the laminate. During the cutting process, air is freed, which results in oxidation reactions in this region. Because the amount of oxidation reactions is small, enlargement of the kerf width size is not substantial as observed for metals. The influence of the assisting gas pressure on the kerf size is evident. The assisting gas enhances the mass removal rate from the kerf sites as well as preventing oxidation reactions. The cooling effect of assisting gas is expected to reduce the kerf width size slightly. However, enhancing a mass removal rate at high assisting gas pressure diminishes the cooling effect of the assisting gas, which in turn, increases the kerf width size with increasing assisting gas pressure. Some small discrepancies between the experimental findings and the theoretical predictions are attributed to the thermal properties of the laminate; that is, in the analysis an averaged thermal conductivity is considered while in actual experimental situations, thermal conductivity depends on the fiber orientation and spacing.

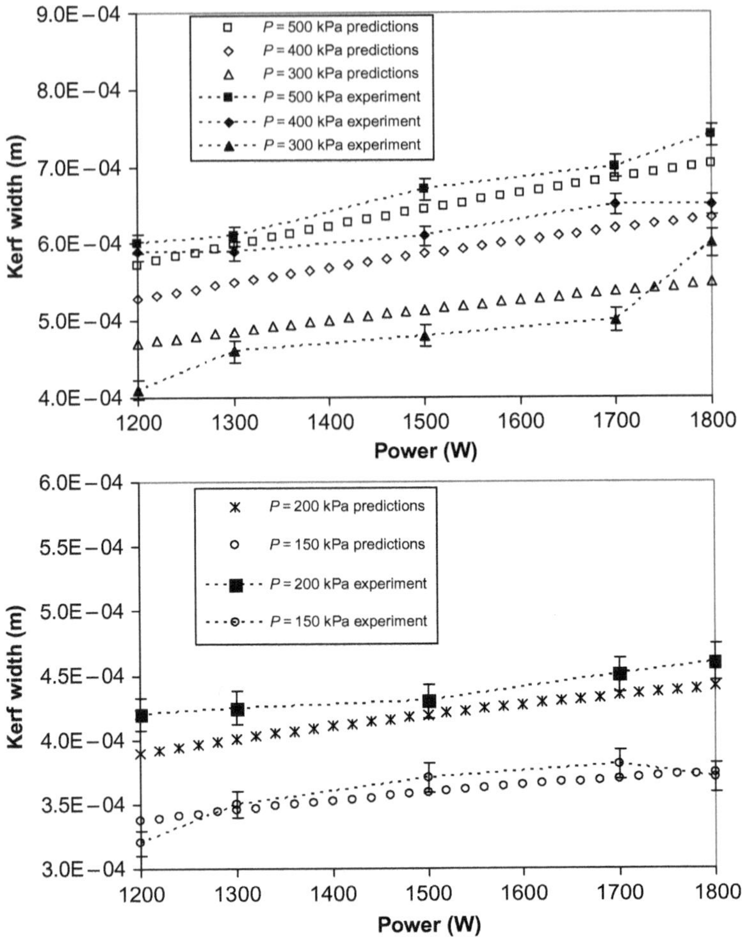

Fig. 5.61 Kerf width predicted and measured from the experiment for different power intensities as assisting gas pressure being variable [9].

Table 5.5 Kevlar and assisting gas properties used in the simulations

Source of variation	Value	Units
Density of assisting gas	2.786	kg/m^3
Density of workpiece	1340	kg/m^3
Fraction of evaporation contribution (β)	0.1	–
Fraction of pressure drop (f)	0.5	–
Energy coupling factor (A)	0.6	–
Super heating factor in the melt front (η_u)	0.01	–
Heat of reaction	26.92×10^3	J/kg
Molecular diameter of nitrogen	0.3176	Å
Assisting gas pressure (P)	150–500	kPa
Specific heat capacity of workpiece	1400	J/kg K
Superheating factor	0.02	–
Thermal conductivity of workpiece	0.14	W/m K
Combustion temperature	2800	K

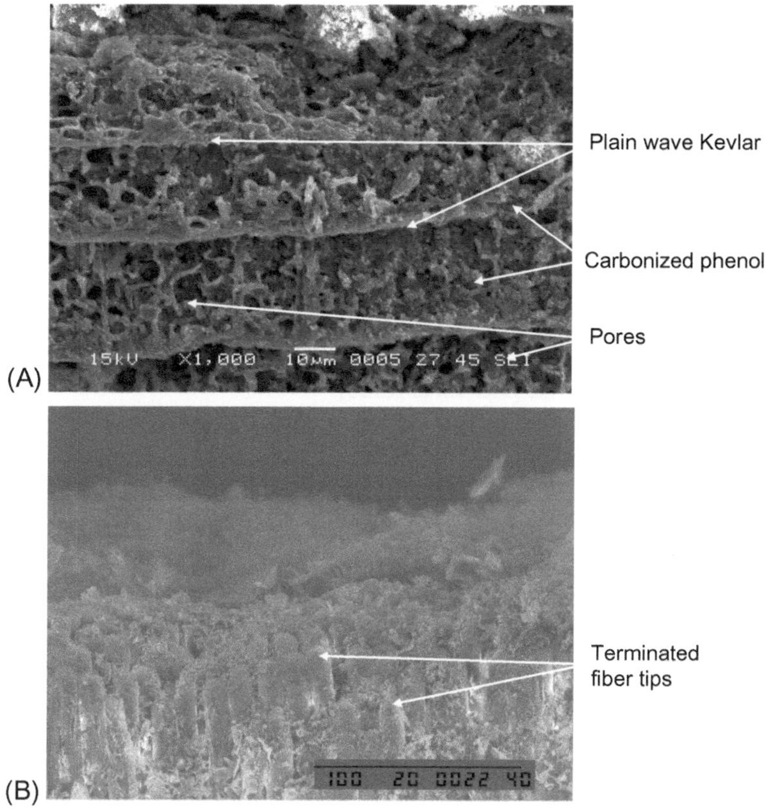

Fig. 5.62 (A) A cross-section of laminate used in the cutting experiments [9]. (B) Laser cut edges of laminate and thermally terminated fiber tips at the cutting edge (laser power = 1800 W and $P = 500$ kPa) [9].

Consequently, nonuniform thermal conductivity results in some small variation in the kerf width size. Moreover, heat-affected zones around the cutting edges are not observed clearly. In this case, plain wave fibers are terminated thermally at the cutting edges by a laser beam (Fig. 5.62B). In addition, due to the low thermal conductivity of Kevlar, the heat diffusion into the fibers is small at the cutting edges. This, in turn, lowers the heat-affected zone size unlike in the case of fiber reenforced plastics.

Fig. 5.63 shows the surface roughness profile for two different assisting gas pressures. The average surface roughness is on the order of 25 μm for assisting gas pressure of 200 kPa, while it is on the order of 17 μm for a 500 kPa assisting gas pressure. It can be observed that surface roughness reduces for high assisting gas pressure; in which case, the sideways burning due to oxidation reactions becomes minimum due to the shielding effect of the assisting gas at high pressure. Fig. 5.64 shows the ratio of the kerf width obtained from $\Delta\text{Kerf width ratio} = \frac{(\text{Maximum Kerf width} - \text{Minimum Kerf width})}{\text{Averaged Kerf widh}}$ using the experimental data. The kerf width ratio is determined from the ratio of the

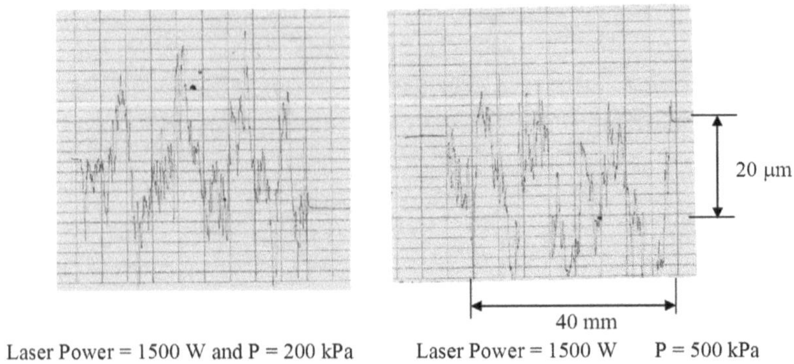

Fig. 5.63 Surface profiles of the laser cut edges [9].

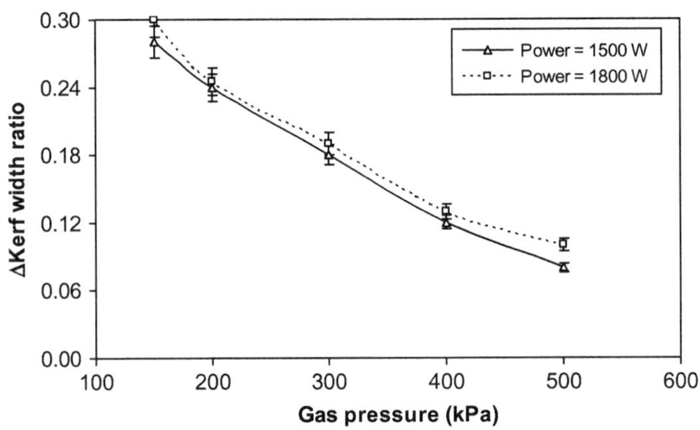

Fig. 5.64 Kerf width ratio with different gas pressure for laser output power of 1500 and 1800 W [9].

difference in the kerf width size along the cut edges to the average kerf width size along the same cut edges. The influence of assisting gas pressure is significant on the kerf width ratio while the laser output power influences the kerf width ratio slightly. In this case, increasing the assisting gas pressure lowers the kerf width ratio while alternatively increasing the laser output power enhances the kerf width ratio slightly. This suggests that the sideways burning due to air trapped in the layers is a significant contributor to the kerf width size. High assisting gas pressure enhances the rate of material removal from the kerf and suppresses the oxidation reaction taking place in the cutting section via reducing the oxygen concentration in air. In this case, nitrogen acts as a shielding gas in the cutting section, suppressing the oxidation reactions in this region. This, in turn, results in improved cut quality. This improvement is significant at high assisting gas pressures (500 kPa).

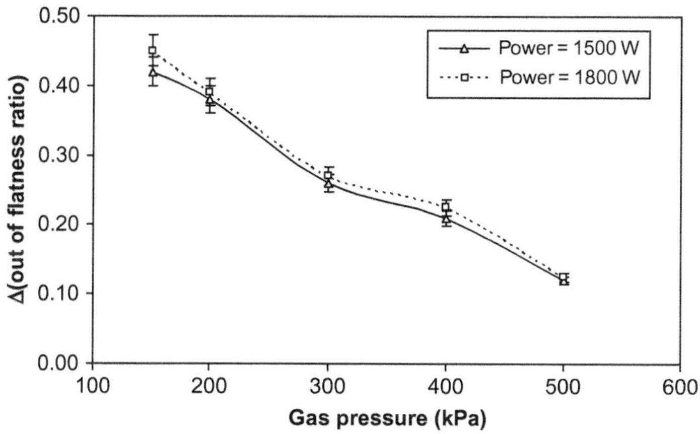

Fig. 5.65 Out-of-flatness ratio with gas pressure for power intensities of 1500 and 1800 W [9].

Fig. 5.65 shows the ratio of out-of-flatness, obtained from $\Delta(\text{Out of flatness}) = \frac{(\text{Maximum out of flatness} - \text{Minimum out of flatness})}{\text{Averaged out of flatness}}$ using the experimental data, with assisting gas pressure for laser output power intensities of 1500 and 1800 W. It should be noted that the data presented in Fig. 5.65 is a nondimensional form (it is the ratio of out-of-flatness differences over the average value of the out of flatness). The out-of-flatness ratio decreases significantly with increasing assisting gas pressure for all power intensities. This is because nitrogen prevents high-temperature oxidation reactions in the cut region. Although nitrogen prevents oxidation reactions, which take place at high temperatures in the cutting section, air, which is trapped in the laminate, is released during the cutting process and causes local sideways burning. Despite the fact that the sideways burning is local, it reduces the cut quality via enhancing the out-of-flatness ratio. Consequently, nitrogen at high pressures prevents the sideways burning partially in the kerf. This, in turn, improves the end product quality. The out-of-flatness ratio increases with increasing laser output power, provided that the change in the out-of-flatness is smaller as compared to the assisting gas pressure variation. This may be attributed to the shielding effect of the assisting gas, which influences the out-of-flatness ratio while suppressing the oxidation reactions during the cutting process.

Fig. 5.66 shows SEM micrographs of laser cut sections for three assisting gas pressures. Parallel side-cut edges are evident from the micrographs. In the case of high assisting gas pressure, the out-of-flatness ratio is considerably low at the cut edges. This can be explained in terms of air being trapped in the laminate structure. Air is released during the cutting and it mixes with the assisting gas, nitrogen. However, this mixing does not result in a significantly low concentration of oxygen in the air at low assisting gas pressures. Consequently, the releasing of air at high temperatures causes exothermic reactions within the substrate material leading to the sideways burning while enhancing the out-of-flatness ratio in the cut section. The thermal damage on the workpiece surface is more pronounced for the workpiece having fiber axis in

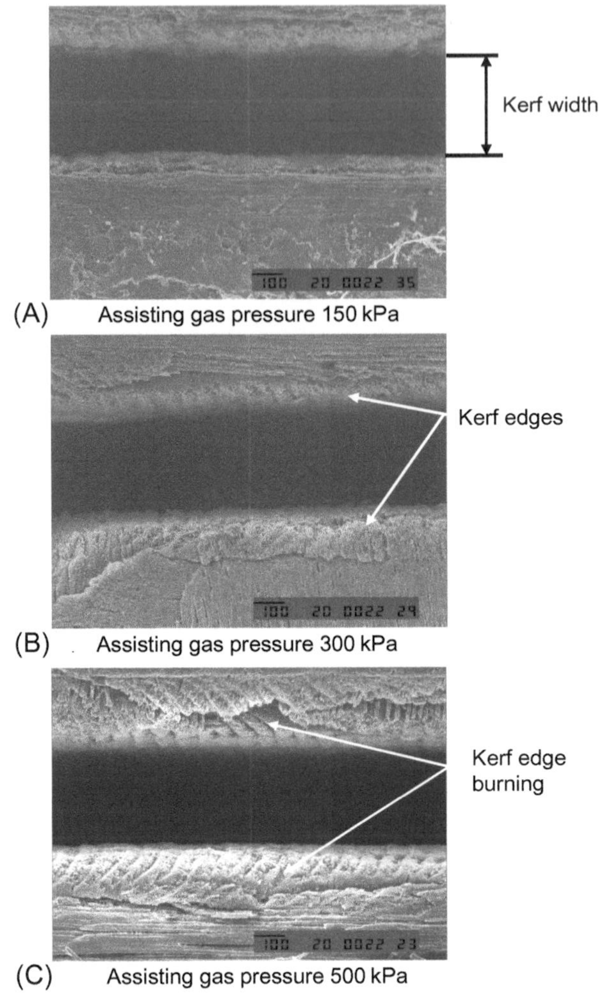

Fig. 5.66 SEM micrographs of the top view of cut sections for three different assisting gas pressure. The laser output power is 1500 W. The scale bar in micrographs is in micrometers [9].

the cutting direction. However, the workpiece with a fiber axis in the plane normal to the workpiece motion is less damaged due to low-bulk thermal conductivity.

5.5.6 Laser cutting of a carbon/carbon multilamelled plain-weave structure

Laser cutting of composites find wide application in industry due to improved end-product quality, low cost, and short processing time. In a laser cutting operation, an assisting gas is used to protect the cutting section from the exothermic reactions. In this case, an inert gas emerging from the nozzle coaxially with a laser beam and impinging onto the workpiece surface is employed. Impinging gas has a shielding effect, protecting the cutting section from the oxidation reactions. The end product quality is influenced significantly by the high-temperature combustion reactions taking place in the cutting section. Moreover, when cutting thick workpieces, the assisting gas may not have enough pressure to shield the cutting edges from the combustion reactions when it reaches the bottom of the workpiece. This, in turn, results in sideways burning at the bottom edge of the workpiece. Laser cutting of a carbon/carbon multilamella structure introduces the challenges of high-temperature combustion and sideways burning, which causes the low quality of cut sections. The present section provides useful information on the laser cutting of carbonic materials. The experimental data findings are given in line with the previous study [10].

Fig. 5.67 shows the front view of the cut cross section of the workpiece. The kerf width size remains almost the same in the laser entry region, and as the depth increases the kerf width size reduces. This may occur because of the absorption of the laser beam energy from the cut surfaces. Unlike metals, carbon acts as an absorbent to the internally-reflected laser beam from the kerf sites. Consequently, laser beam energy reaching the kerf bottom becomes less, due to no contribution of reflected laser beam energy from the kerf sides, reducing the kerf width. Although the workpiece is in a lamella-type structure, the orientation of the carbon fibers remains the same in each layer, and an uneven cut surface is not observed, provided that at a high-power intensity, some undulations at the cut surface are evident (Fig. 5.68). In this case, the size of the kerf width becomes slightly larger for the layer composing of carbon fibers with their axis being normal to the axis of the workpiece motion than that which is parallel to the axis of workpiece motion. This could be explained in terms of thermal conductivity, which reduces for fibers having their axis normal to the axis of workpiece motion. Consequently, less radial heat conduction along the fiber axis, due to low thermal conductivity in this direction, enhances the energy available in the cutting section; that is, the size of kerf width increases in this region.

Fig. 5.69 shows top views of cutting sections for three laser power intensity settings. Almost parallel-sided cut edges are observed, provided that some small variation in kerf width occurs along the cutting direction. This is because of the absorbed laser energy at the surface, which varies due to some small surface plasma generated at the surface. In this case, the surface plasma generated acts as a heat source, due to its high temperature, enhances the kerf width, and partially blocks the incident reaching the workpiece surface through absorption. In this case, plasma absorption has an adverse effect on the kerf width size. Consequently, the surface plasma has two coupling effects on the kerf formation. Once the coupling of heating and absorption effects of the surface improves, the kerf width size increases. It should be noted that the surface plasma is transiently hot, and its temperature increases via absorbing the incident laser energy. As the surface plasma absorption progresses, plasma temperature and pressure increase. This, in turn, results in plasma expansion; that is, the density of surface plasma reduces. The less dense surface plasma becomes less opaque to the incident beam, and the amount of laser beam power reaching the workpiece surface increases. This improves the mass removal rate and kerf size in the cutting section. Moreover, the transient behavior of the surface plasma introduces instability in a mass removal rate and kerf width size, because its effect varies with time. This appears as striation formation at kerf sides.

Fig. 5.67 Micrographs of laser cut cross-sections for two laser power settings [10].

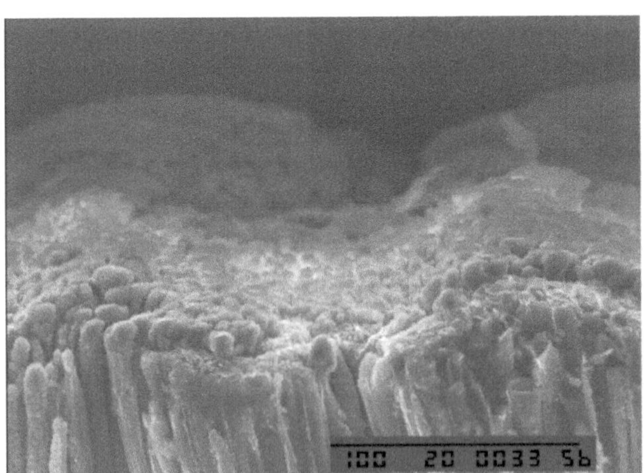

Fig. 5.68 Micrograph of laser cut surface. Power = 1200 W [10].

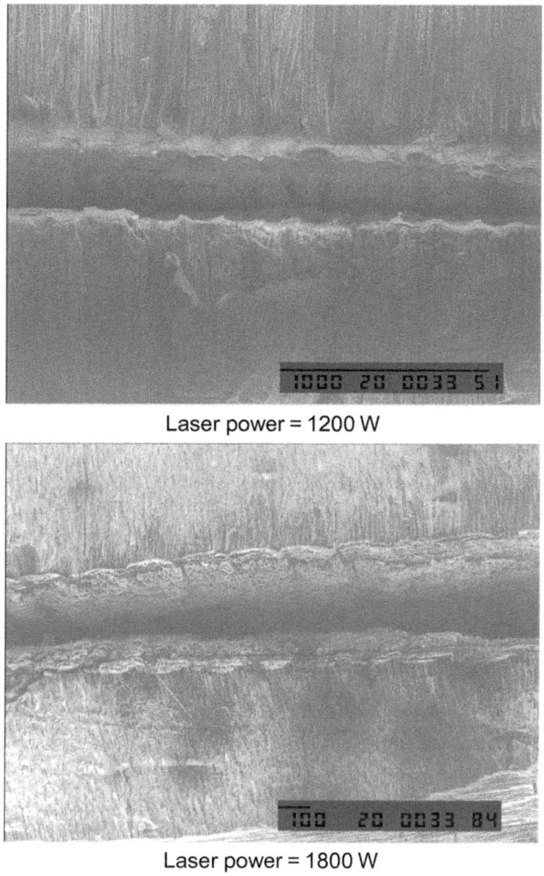

Fig. 5.69 Micrographs showing the top view of laser cut surfaces [10].

Fig. 5.70A–B shows a SEM micrograph of two striation patterns formed for two different orientations of carbon fiber axis referenced to the axis of the workpiece motion. Because nitrogen at high pressure (250 kPa) is used as an assisting gas, oxidation reactions resulting in sideways burning are not achieved, that is, no excessive deformation is observed. The cyclic behavior of the surface plasma results in slightly more deep and elongated striation for carbon fibers with their axis normal to the direction of workpiece motion than those lying in the same direction of workpiece motion. This is because of the change in thermal conductivity with reference to the axis of the carbon fiber and the workpiece motion during the cutting. The depth and length of the striation enhance with increasing power intensity.

Fig. 5.71 shows SEM micrographs of carbon fiber ends after the cutting process. The globule-like formations at the fiber ends is evident, particularly fibers with their axis normal to the direction of workpiece motion. This suggests that due to low thermal conductivity, excessive heat transfer into the neighboring cut surfaces of the fiber is suppressed and this elevates the fiber temperature at the cut surface. This, in turn,

(A)

(B)

Fig. 5.70 Micrographs of striation patterns obtained when (A) workpiece motion normal to fiber principle axis during cutting [10] and (B) workpiece motion parallel to fiber principle axis during cutting [10].

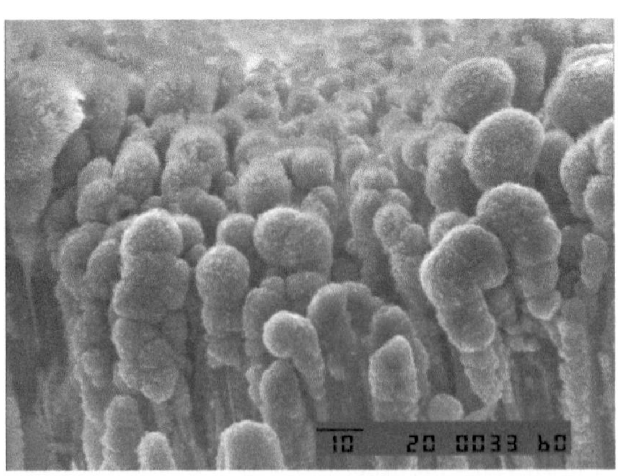

Fig. 5.71 Micrograph of fiber tips after the laser cutting process [10].

results in softening and partial melting of the fiber in the region of the fiber tip. The surface tension and gravitational forces are responsible for the formation of the globule-like structures in this region. Moreover, it is also possible that although nitrogen is used as the assisting gas to prevent the oxidation reactions in the cutting section, the water content in the workpiece may cause limited oxidation reactions to take place in the cutting section. This results in scattered, irregular formations of structures around the fiber tips at the cut edges.

5.5.7 Wedge cutting of mild steel

Laser cutting of inclined metallic panels is interesting because of the flexibility of the laser cutting in various applications. Although the kerf size and cut section are influenced by the tilting of the panel, the cut section can be free from large-size asperities including sideways burning and dross attachments. Therefore, in this section, laser cutting of stainless steel sheets at an inclination angle of 18°, as well as normal to the laser beam axis, is presented. The cut quality is assessed by the international standards for thermal cutting. The findings of the experimental data are presented in line with the previous study [11].

Fig. 5.72 shows photographs of the top view of the laser cut surfaces due to wedge and normal cutting situations. In the case of a wedge-cutting situation, the flow of

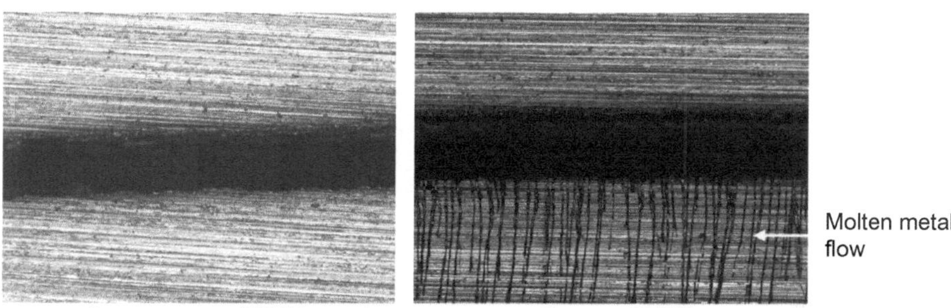

Laser power = 229 W Power = 133 W

Photographs of laser cut sections at the top of the workpiece surface for two different power settings of wedge cutting situation. Inclination is in downward direction (↓).

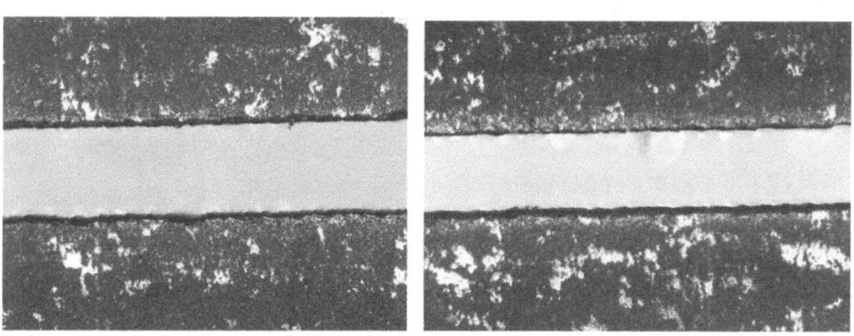

Laser power = 229 W Power = 133 W

Photographs of laser cut sections at the top of the workpiece surface for two different power settings of normal cutting situation.

Fig. 5.72 Photographs of laser cut surfaces due to wedge and normal cutting situations [11].

molten metal occurs on the top surface of the workpiece. This forms the surface debris and lowers the cutting quality. The flow of molten metal is more pronounced at low laser power intensities. In this case, the time required for penetration through the workpiece is larger than that of the high-power intensities. Consequently, the liquid metal generated in the kerf during laser melting of the workpiece drag away from the kerf sides by the assisting gas as well as the recoil pressure developed in the kerf vicinity. Although recoil pressure diminishes once the through penetration of the workpiece is achieved, the assisting gas pressure in the kerf pushes the molten metal flowing to the top surface of the workpiece. It should be noted that the flow of the molten metal is toward the down side of the inclination; that is, it flows only on the inclined side of the cut surface. The possible explanation for the flow situation is because of the gravity effect; in which case, gravity-driven flow along the top surface of the workpiece occurs. Moreover, the flow of molten metal along the cut edge is not continuous, but contains some distance interruptions. This indicates that the molten metal flow from the kerf is pulsative in nature along the cutting site. This can be associated with the striation formation along the cut surface during the cutting process. As the cutting progresses, thermal erosion of the cut sides pulsates due to molten instability during the cutting process. This gives rise to the striation formation at the cutting surface (Fig. 5.73). In the case of high-laser power intensities, the absorption of the

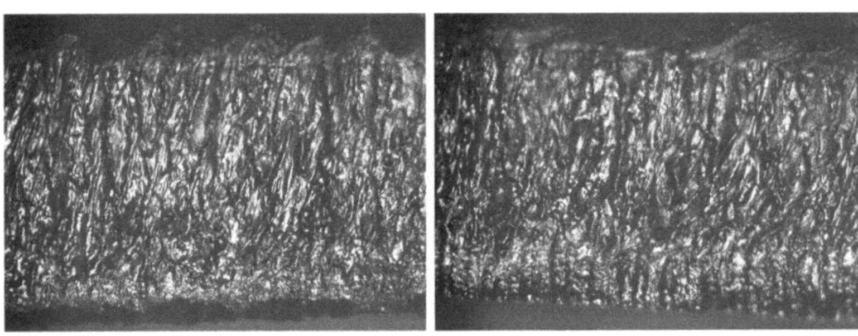

Laser power = 133 W Power = 229 W

Photographs of striations at the side of cut surfaces for two different power settings of wedge cutting situation.

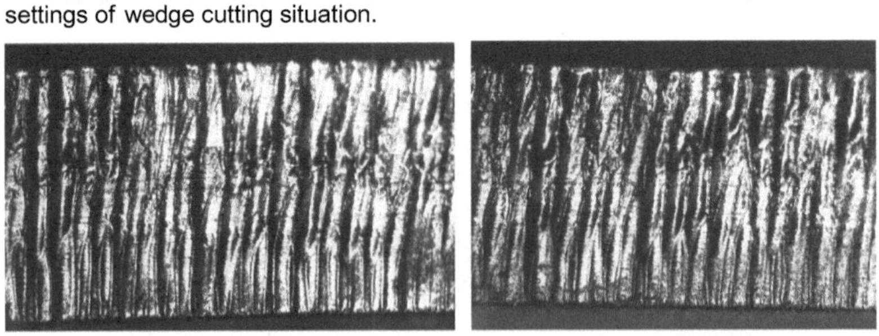

Laser power = 133 W Power = 229 W

Photographs of striations at the side of cut surfaces for two different power settings of normal cutting situation.

Fig. 5.73 Photographs of side of cut surfaces for wedge and normal cutting situations [11].

laser beam at the molten surface improves significantly; that is, exothermic reactions at the molten surface enhances the energy available in the cut section. This increases the penetration speed. In this case, molten metal is pushed away from the bottom of the workpiece during the cutting process, that is, the molten metal is ejected from the cutting section as dross leaving the bottom surface of the workpiece. Moreover, thermal erosion at the bottom side of the workpiece lowers the cutting quality due to considerably high-temperature molten ejection from this region. However, in the case of a normal cutting situation, molten flow at the workpiece surface is not observed; in which case, parallel-sided cut sections result. In addition, regular striation patterns occur at the cut edges.

Fig. 5.74 shows the variation of normalized dross height corresponding to the top and bottom surface of the cut edges. The errors are based on the five times repeat of the measurements, which are indicated as bars at each data point. The dross height is the result of the striation formed at the cut surface due to molten flow instability occurring during the cutting process. In the case of a wedge cutting situation, the dross height attains high values for low power intensities at the front surface of the workpiece. However, in the case of high power intensities, the dross height is low at the front surface and it becomes high at the back surface of the workpiece. This is because of the dross ejection from the bottom surface of the workpieec during the cutting process. Consequently, high-temperature molten material causes thermal erosion from the kerf sides in the region of the bottom surface of the workpiece. This is more pronounced for high-power intensities. Consequently, low-power intensity cutting of the wedge (inclined) workpiece gives rise to poor cutting quality at the top surface of the workpiece, while increasing the power intensity causes deformation of kerf due to thermal erosion, lowering the cut quality at the bottom surface of the workpiece.

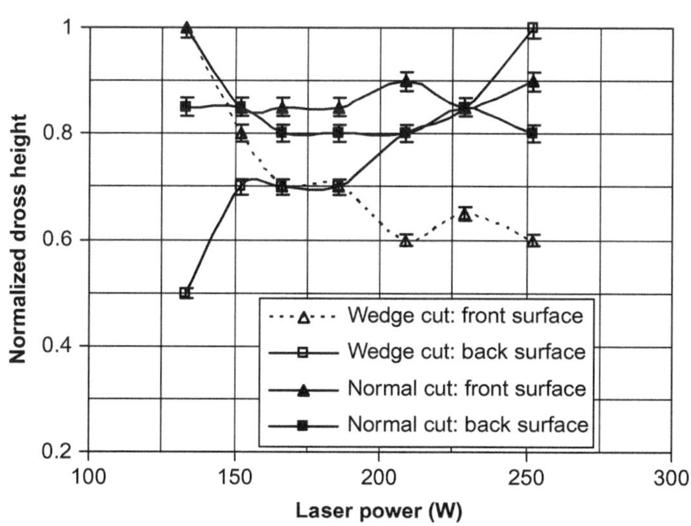

Fig. 5.74 Normalized dross height corresponding to the front and back side of the workpiece surface due to wedge and normal cutting situations [11].

Moreover, at a power intensity of 160 Watts, both front and bottom surfaces of the workpiece have small dross heights. This suggests that the cutting quality improves at certain power intensities. In the case of a normal cutting situation, dross height attains low values, and the dross height ratio becomes more than 0.8 for front and back surfaces of the workpiece. This indicates that the good quality of cut edges is resulted for normal cutting situation.

Fig. 5.75 shows normalized out-of-flatness of the cut edges with laser power intensity for wedge and normal cutting situations. The errors are based on the five times repeat of the measurements, which are indicated as bars at each data point. In the case of a wedge cutting situation, the out-of-flatness is high along the edges at the bottom surface of the workpiece than at the top surface. In addition, the out-of-flatness is poor for the top surface of the workpiece at low-power intensities while it is the opposite for high-power intensities, as similar to the observation made for the dross height. Moreover, the out-of-flatness varies along the cut edges due to flow-induced instabilities during the cutting process. The out-of-flatness along the upper edge of the inclined surface of the workpiece is higher than that at the lower edge of the inclined surface. This occurs because of the laser power intensity distribution at the irradiated workpiece surface; that is, the irradiated power absorbed in the region close to the upper edge is higher than the lower edge and the molten metal in the kerf does not contribute to the enhancement of the size of the melt in this region. Consequently, out-of-flatness increases in this region. In the case of a normal cutting situation, out-of-flatness becomes low and the normalized out-of-flatness shows a steady trend within the range of 0.9–0.95. This indicates that the quality of the cut edge is higher than that of the wedge cutting situation.

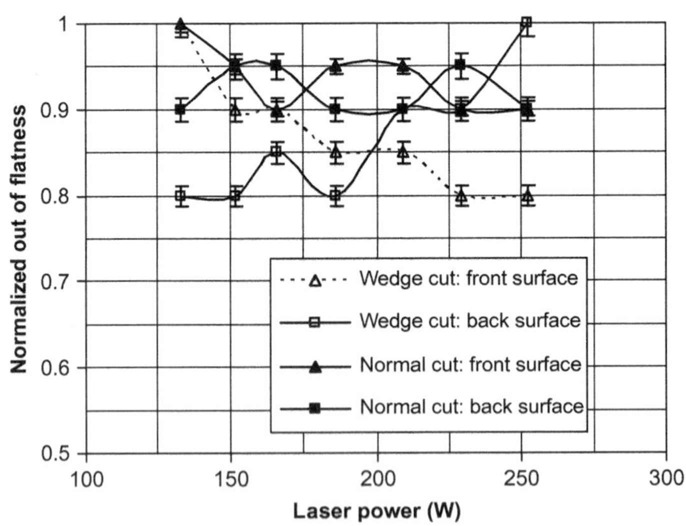

Fig. 5.75 Normalized out-of-flatness corresponding to the front and back side of the workpiece surface due to wedge and normal cutting situations [11].

5.5.8 Laser cutting of titanium alloy

The oxidation reactions in the cutting section results in excessive heat release during the high-temperature exothermic reactions. This modifies the kerf width because of the sideways burning and thermal erosion while lowering the cut quality. This situation is particularly true for metals having high oxygen affinity such as titanium and its alloys. Therefore, cutting of such sheet metals requires special attention to secure the end-product quality. On the other hand, titanium and its alloys have low thermal diffusivity (3.64×10^{-6} m^2/s), as compared to steel (1.17×10^{-5} m^2/s), which causes less energy to be transferred from the cutting edge to the solid bulk through the thermal conduction. This causes a temperature increase at the kerf surface and results in the high-temperature gradient formation in the vicinity of the kerf surface. The high-temperature gradient, in turn, causes the development of high-stress fields around the cutting section, which is particularly true for the hole cutting into a titanium alloy sheet. In this section, laser straight cutting of titanium alloy (Ti6Al4V) is presented in light of the previous studies [28,29]. The parameters used in the simulations are given in Tables 5.6 and 5.7.

5.5.8.1 Laser straight cutting of titanium alloy

Fig. 5.76 shows temperature distribution along the x-axis for different cooling periods. The x-axis represents the distance along the laser cut edge. It should be noted that the cooling period starts after the cutting ends, which corresponds to $t = 0.08$ s in the

Table 5.6 Mechanical properties of Ti-6Al-4V alloy and cutting parameters used in the simulations

(A) Mechanical properties

Temperature (K)	Elastic modulus (GPa)	Poisson's ratio	Expansion coefficient $\times 10^6$ (1/K)	Yield stress (MPa)
300	113	0.34	8	14.37
373	108	0.34	8.6	12.81
473	103	0.34	9	11.2
589	95	0.34	9.2	10.5
698	85	0.34	9.4	7
803	70	0.34	9.7	5
973	50	0.34	1.05	2.25

(B) Laser cutting parameters

Gaussian parameter (m)	Hole radius (m)	Angular velocity (rad/s)	Tangential velocity (m/s)	Power intensity (W/m^2)
0.0003	0.006	25	0.15	2.8×10^{12}

Table 5.7 **Thermophyscial properties of Ti-6Al-4V alloy used in the simulations**

(A) Density and melting properties of Ti-6Al-4V alloy			
Density (kg/m^3)	**Latent heat of melting (J/kg)**	**Solidus temperature (K)**	**Liquidus temperature (K)**
4430	4.2×10^5	1878	1933

(B) Thermal properties of Ti-6Al-4V alloy

Temperature (K)	***Cp* (J/kg K)**	***k* (W/m K)**
300	403	6.5
573	505	7
673	520	9
773	530	11
873	545	12
1073	620	17

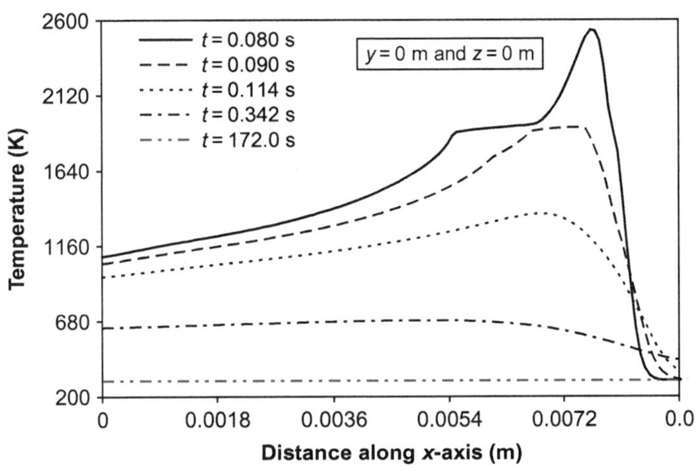

Fig. 5.76 Temperature distribution along the x-axis for different cooling periods [28].

figure. Because the laser cutting speed is 0.1 m/s, the laser beam is at $x=0.008$ m where temperature is well above the melting temperature of the substrate material and the laser beam is switched-off at this location. It is evident that behind the laser spot ($x\leq0.007$ m), temperature remains the same at the melting temperature of the workpiece. This indicates that the cut edge is at a melting temperature in this region. However, temperature drops below the melting temperature along the cut edge $0\leq x\leq0.005$ m, which is associated with the conduction, convection, and radiation cooling of the initially cut edges. Because the thermal diffusivity of the alloy is low, temperature decay is gradual. Moreover, attainment of a considerably-high temperature at the irradiated spot indicates the presence of superheating in the liquid phase. As the cooling period progresses ($t > 0.08$ s), the temperature reduces along the cut edges; however, temperature decay in the irradiated spot is gradual due to superheating of the liquid phase. The temperature gradient is high in the front region of the laser irradiated spot ($x > 0.0082$ m). This is attributed to a large temperature difference between the irradiated spot and the base material, which is at a low temperature. Consequently, the difference between a superheated liquid temperature and a low base material temperature results in a high temperature gradient in this region.

Fig. 5.77 shows the von Mises stress along the x-axis for different cooling periods, which are similar to those shown in Fig. 5.76. The von Mises stress attains low values in the regions where temperature is high. This is associated with temperature-dependent elastic modulus, which reduces with increasing temperatures (Table 5.6). Moreover, the von Mises stress increases as the temperature along the cut edge reduces. In addition, the von Mises stress attains high values in the front region of the laser spot where temperature gradient is high. This is more pronounced in the early cooling period ($t = 0.08$ s). The von Mises stress reduces in the region where superheating of liquid takes place at the onset of the cooling cycle. This is attributed to the self-annealing effect of the laser-irradiated region due to high temperature. In this case, slow cooling results in the stress relaxation in this region. As the cooling period ends ($t = 172$ s) and the temperature reduces to the initial

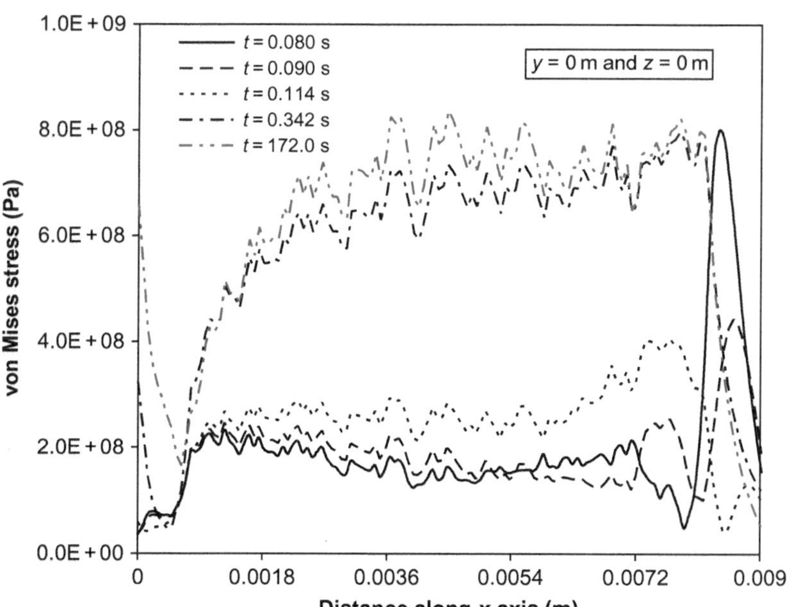

Fig. 5.77 von Mises stress distribution along the x-axis for different cooling periods [28].

temperature, the von Mises stress remains high. The maximum value of the von Mises stress along the cut edge is on the order of 800 GPa.

Fig. 5.78 shows temperature distribution along the *y*-axis for different cooling periods. It should be noted that the *y*-axis is normal to the laser cutting direction at the surface. The temperature remains high in the vicinity of the cut edge and decays sharply as the distance from the cut edge vicinity increases along the *y*-axis. Moreover, temperature decay in the vicinity of the cut edge is gradual, which can be associated with the superheating of the liquid phase. The gradual decay of temperature results in the low temperature gradient, while a sharp decay generates the high-temperature gradient along the *y*-axis. The temperature decay rate is lower in the vicinity of the cut edge. This is attributed to the occurrence of superheating in the liquid phase in this region. Fig. 5.79 shows the von Mises stress distribution along the *y*-axis for different cooling periods similar to those shown in Fig. 5.78. The von Mises stress attains low vales in the region where temperature is high, particularly in the early period while it increases in the region of the high-temperature gradient. This is associated with the elastic modulus of the workpiece (Table 5.6), which reduces with increasing temperature. As the cooling period progresses, the maximum von Mises stress moves toward the edge of the cut section. The maximum von Mises stress is on the order of 750 MPa and the high-stress region extends 80 μm along the *y*-axis, whereas the magnitude of the von Mises stress is on the order of 180 MPa at 80 μm and beyond. However, as the distance along the *y*-axis increases further, the von Mises stress remains the same low value.

Fig. 5.80 shows the temperature distribution along the *z*-axis for different cooling periods. It should be noted that the *z*-axis extends along the thickness of the workpiece. Consequently, $z = 0$ m represents the free surface where the laser beam interacts with the substrate material during cutting while $z = 3$ mm corresponds to the back surface of the workpiece. The temperature decays gradually along the thickness of the cut section. The gradual decay of temperature is associated with the cutting surface where the

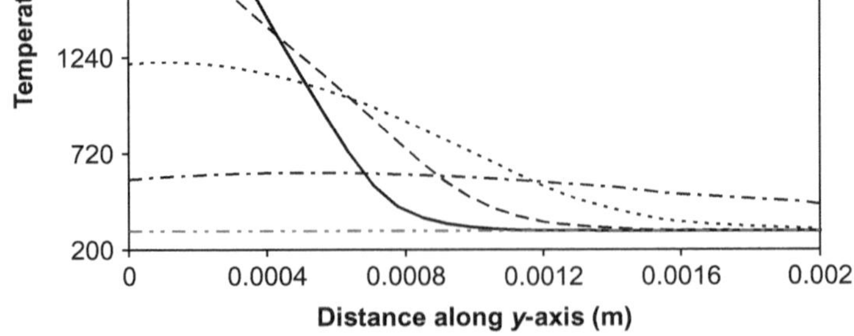

Fig. 5.78 Temperature distribution along the *y*-axis for different cooling periods [28].

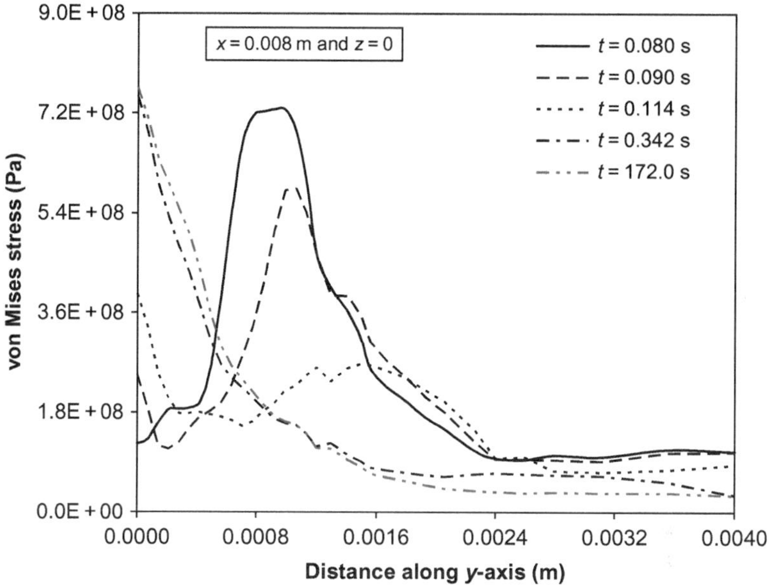

Fig. 5.79 von Mises stress distribution along the *y*-axis for different cooling periods [28].

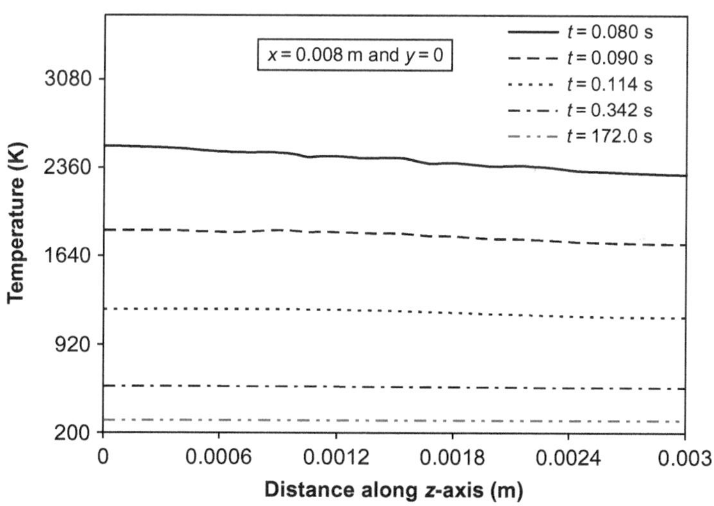

Fig. 5.80 Temperature distribution along the *z*-axis for different cooling periods [28].

temperature remains above the melting temperature of the substrate material. It should be noted that $x=0$ and $y=0$ corresponds to the kerf (cut edge) surface. As time progresses, temperature decays almost uniformly along the thickness at the kerf surface. This is attributed to the convection, conduction, and radiation heat losses from the kerf surface.

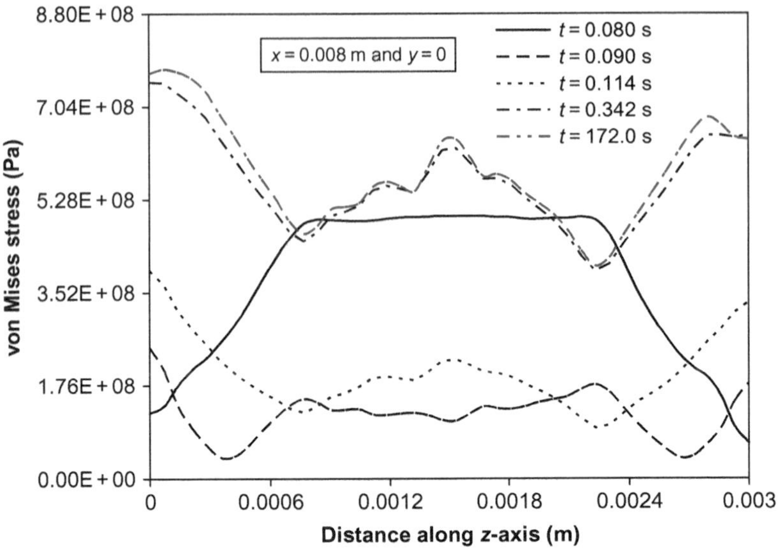

Fig. 5.81 von Mises stress distribution along the z-axis for different cooling periods [28].

Fig. 5.81 shows the von Mises stress distribution along the z-axis for different cooling periods similar to those shown in Fig. 5.80. The von Mises stress attains low values at high temperatures during the initial cooling period and it increases considerably as the cooling period progresses. The von Mises stress reaches its maximum in the vicinity of the top and bottom surfaces of the workpiece despite the fact that the top and bottom surfaces of the workpiece are free to expand. The attainment of high von Mises stress in these regions is attributed to thermal strain formation. Moreover, the von Mises stress increases in the mid-thickness of the workpiece; however its value is less than the maximum stress. This behavior is associated with the compressive stress, which is formed in this region. Because the top and bottom surfaces expand freely, the mid-thickness region undergoes compression due to free expansion in the workpiece surface region.

Fig. 5.82 shows optical photographs of the top and bottom surfaces of the laser cut sections. It is evident that the laser cut sections are parallel and free from excessive defects such as large area sideways burnings. It can be observed that some small sideways burning occur around the kerf edges. This can be attributed to the presence of oxygen resulting in oxidation of the workpiece surface prior to laser cutting. In this case, the presence of oxygen in the cutting section initiates high-temperature oxidation reactions during the cutting process while generating some small amount of sideways burning around the cut sections despite the fact that nitrogen, at high pressure, is used as the assisting gas. At the bottom surface of the laser cut section, some small dross attachments are observed. This occurs because of the molten flow which cools down with the presence of the assisting gas upon ejection from the kerf. Although the temperature remains high at the bottom surface of the workpiece at a location where the

Top surface of laser cut section

Top surface of laser cut section

Fig. 5.82 Close view of top and bottom surfaces of cut sections [28].

laser beam passes, in the region behind the laser spot, the temperature reduces at the cut section. This increases the viscosity of the molten flow, and the drag force generated in the kerf, acting onto the molten flow due to assisting gas, is not sufficiently high, resulting in total ejection of the molten material from the kerf. Nevertheless, the dross size is small and it is randomly distributed along the cut edges at the bottom side of the workpiece.

5.5.8.2 Laser circular cutting of titanium alloy

Fig. 5.83 shows temperature distribution along the top circumference of the laser cut edges for different cutting periods. The temperature increases gradually with progressing time in the region before the melting temperature of the substrate material, and the increase becomes sharp around melting temperature. However, the temperature remains almost the same across the melting region, because the liquidus and soliudus temperatures are very close to each other (Table 5.7). The temperature rises sharply in the liquid phase, resulting in compressed liquid heating to reach its maximum value. In the neighborhood of its peak value, the temperature decays sharply with progressing time. The sharp decay occurs at the frontal region of the laser beam in the direction of the cutting. This situation is true where the laser beam is located around the circumference for all the cutting periods. The sharp decay of temperature results in the development of high-temperature gradients, and the gradual decay in temperature gives rise to the gradual decay of the temperature gradient around the circumferences.

Fig. 5.84 shows the von Mises stress along the top circumference of the hole cut for different cutting periods similar to those shown in Fig. 5.83. The von Mises stress attains relatively low value when the temperature variation is gradual, and it becomes significantly high when the temperature gradient is high. Consequently, the von Mises stress is less in the early cut region and it becomes high in the vicinity of the frontal region of the laser beam during the cutting process. As cutting continues along the edges, the von Mises stress reduces in the region previously cut. This situation applies to all locations around the circumference during the cutting process. This is associated with the thermal stress relaxation via the annealing effect of the laser heated region during the cooling cycle; that is, annealing of a particular location once the laser beam passes over this location. A similar situation is observed for the bottom circumference

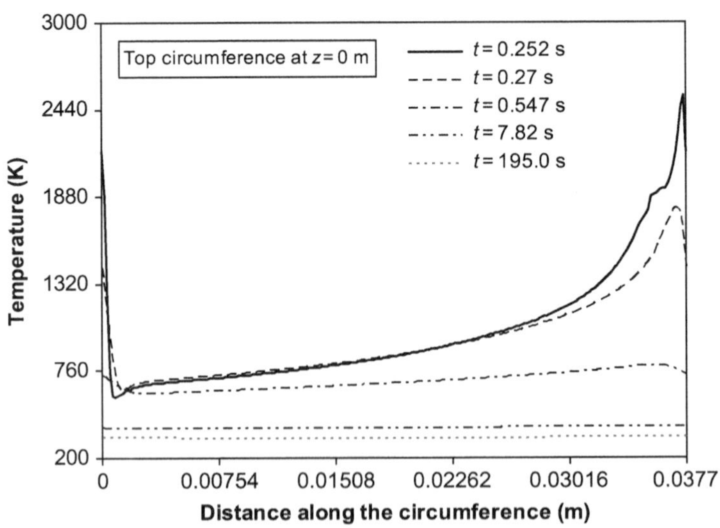

Fig. 5.83 Temperature distribution along top circumference at different cooling periods [29].

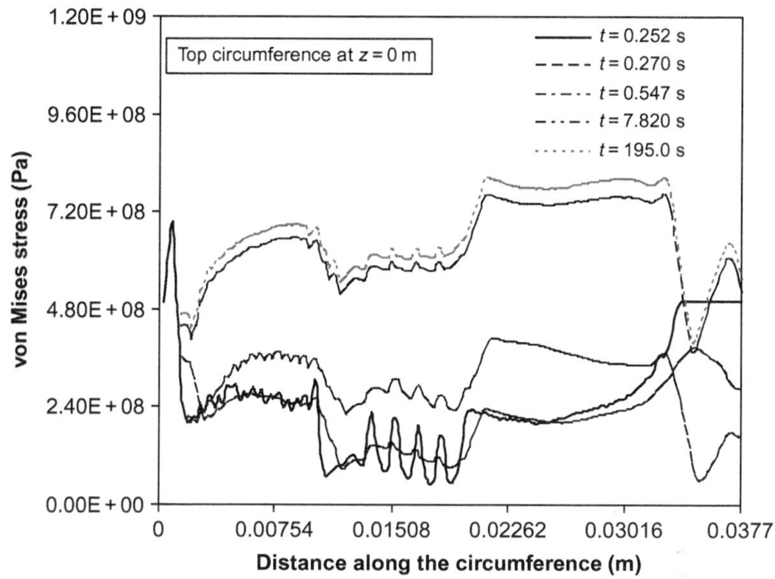

Fig. 5.84 von Mises stress distribution along top circumference at different cooling periods [29].

of the hole cut. Moreover, the small differences in temperatures between the top and bottom circumferences result in small differences in von Mises stress values between the circumferences.

Fig. 5.85 shows the temperature distribution along the z-axis after the laser beam passes the location at point #4 (Fig. 5.86) and the time corresponding to the completion of the circular cutting. It should be noted that $z = 0$ m represents the top surface and $z = 0.003$ m corresponds to the bottom surface of the workpiece. The temperature is slightly lower at the top surface as compared to the distance below the top surface ($z = 0.0002$ m). This is because of the convective and radiative cooling of the top surface during the cutting process, despite the fact that absorbed intensity is highest at the surface. As the distance increases beyond the point of the maximum temperature toward the workpiece bottom surface, temperature decays monotonically toward the bottom surface. The gradual decay in temperature is associated with the absorption of the intensity, which reduces with increasing depth. However, the temperature remains low at other locations because these locations are cut by the laser beam prior to reaching the #4 location. In this case, convection, conduction, and radiation cooling of the cut surface is responsible for the attainment of low temperatures at these locations. The temperature gradient becomes high in the vicinity of the top and bottom surfaces of the workpiece due to convection and radiation cooling.

Fig. 5.87 shows the von Mises stress distribution along the z-axis after the laser beam passes the location #4 for top and bottom circumferences of the hole cut. The von Mises stress attains low values in the vicinity of top and bottom surfaces. This occurs because of the expansion of the free surface, despite the fact that the

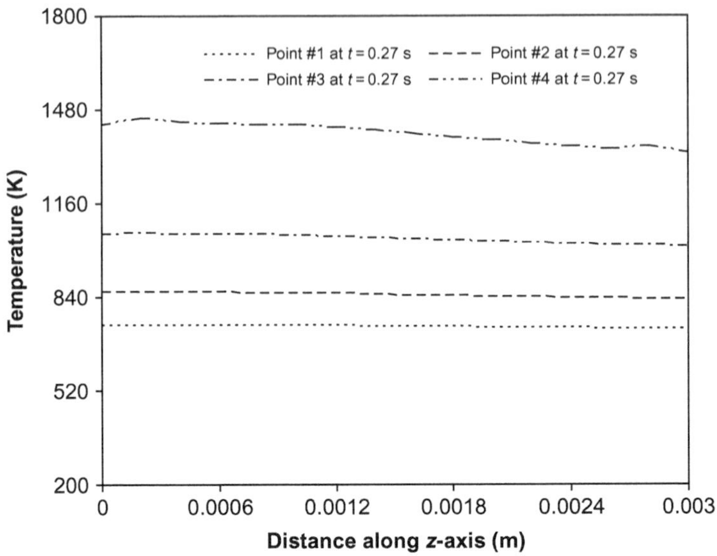

Fig. 5.85 Temperature distribution along the z-axis for different locations around the hole circumference for cooling period $t=0.27$ s [29].

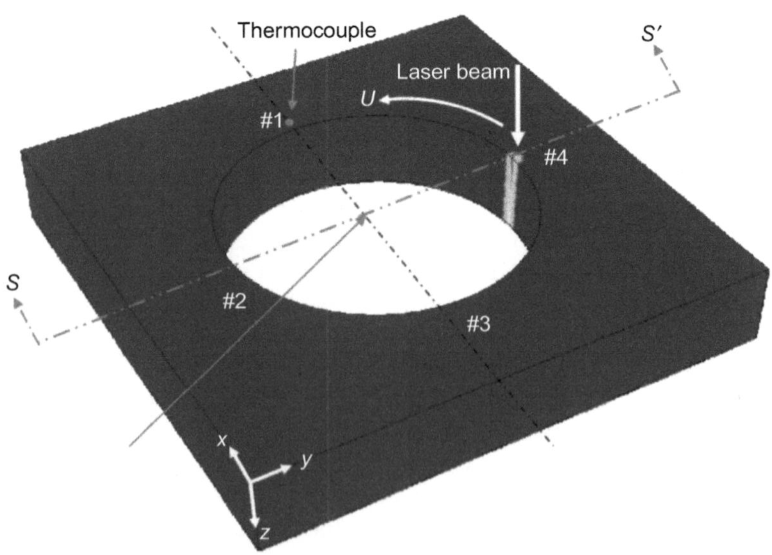

Fig. 5.86 A schematic view of laser cut hole and the coordinate system [29].

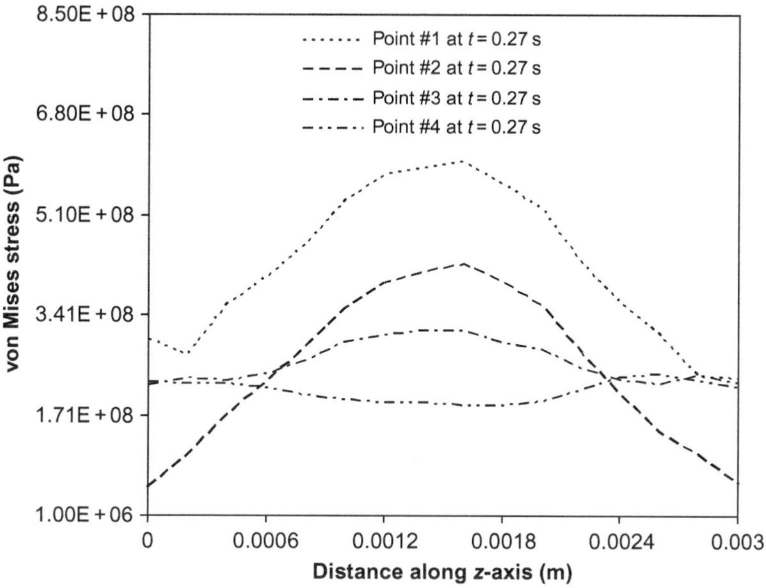

Fig. 5.87 von Mises stress distribution along the z-axis for different locations around the hole circumference for cooling period $t=0.27$ s [29].

temperature gradient is high in this region. Moreover, the von Mises stress attains high values in the region at some distance away from the top surface. The gradual decay of temperature along the z-axis results in an almost uniform von Mises stress distribution along the z-axis at location #4. However, the von Mises stress reaches its peak value at the mid-height of the workpiece thickness at other locations, provided that the magnitude of stress is the highest at location #1 among the locations #1, #2 and #3. This is attributed to the temperature at location #1, which is lower than at locations #2 and #3 (Fig. 5.86). It should be noted that the elastic modules of the workpiece reduces with increasing temperature (Table 5.6).

Fig. 5.88 shows SEM micrographs of the kerf surface. It is evident from the optical photographs that the top surface is clear from resolidified material attachment; however, some small dross attachments are observed at the bottom surface cut edges. The presence of dross is associated with: (i) the drag force generated in the kerf by the assisting gas, which may not be large enough to force out the molten metal from the kerf exit, and (ii) the presence of some oxygen in the region of the bottom surface, which initiates the exothermic oxidation reactions taking place in this region. This, in turn, results in thermal erosion and doss attachment in this region. Nevertheless, the dross attachment is not substantially large. The SEM micrograph shows that irregular striation patterns are formed at the kerf surface. The irregularity is associated with the melting and subsequent solidification of the alloy under the high cooling rates at the kerf surface. Nonetheless, no microcracks are observed at the surface. The influence of assisting gas pressure on the striation pattern is evident in the region of the top surface of the kerf; that is, the striations are slightly bended in this region.

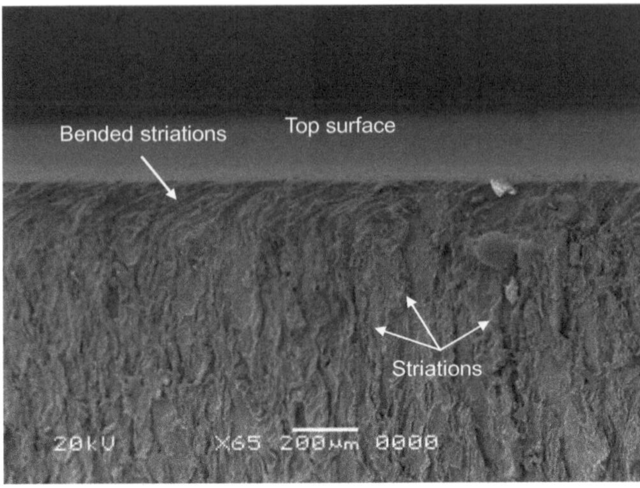

Fig. 5.88 SEM micrograph of the kerf surface [29].

References

[1] B.S. Yilbas, S. Akhtar, C. Karatas, Laser straight cutting bronze sheets: thermal stress analysis and experiment, Mach. Sci. Technol. 16 (1) (2012) 20–39.

[2] B.S. Yilbas, S.S. Akhtar, C. Chatwin, Laser hole cutting into bronze: thermal stress analysis, Opt. Laser Technol. 43 (7) (2011) 1119–1127.

[3] B.S. Yilbas, S.S. Akhtar, C. Karatas, Laser straight cutting of alumina tiles: thermal stress analysis, Int. J. Adv. Manuf. Technol. 58 (2012) 1019–1030.

[4] B.S. Yilbas, S.S. Akhtar, O. Keles, Laser cutting of different diameter holes in alumina: thermal stress analysis, Int. J. Adv. Manuf. Technol. 133 (2) (2011) 024503.

[5] B.S. Yilbas, S.S. Akhtar, C. Karatas, Laser cutting of rectangular geometry into alumina tiles, Opt. Lasers Eng. 55 (2014) 35–43.

[6] B.S. Yilbas, S.S. Akhtar, O. Keles, Laser cutting of aluminum foam: experimental and model studies, ASME J. Manuf. Sci. Eng. 135 (5) (2013), 051018. https://doi.org/10.1115/1.4025009.

[7] B.S. Yilbas, S.S. Akhtar, O. Keles, Laser cutting of large diameter holes into aluminum foam, Mach. Sci. Technol. 17 (4) (2013) 524–544.

[8] B.S. Yilbas, S.S. Akhtar, O. Keles, Laser cutting of triangular geometries in aluminum foam: effect of cut size on thermal stress levels, Opt. Laser Technol. 48 (2013) 523–529.

[9] F. Al-Sulaiman, B.S. Yilbas, M. Ahsan, C. Karatas, CO_2 laser cutting of Kevlar laminate: influence of assisting gas pressure, Int. J. Adv. Manuf. Technol. 45 (1–2) (2009) 62–70.

[10] F.A. Al-Sulaiman, B.S. Yilbas, M. Ahsan, CO_2 laser cutting of a carbon/carbon multi-lamelled plain-weave structure, J. Mater. Process. Technol. 173 (2006) 345–351.

[11] B.S. Yilbas, C. Karatas, I. Uslan, O. Keles, Y. Usta, Z. Yilbas, M. Ahsan, Wedge cutting of mild steel by CO_2 laser and cut quality assessment in relation to normal cutting, Opt. Lasers Eng. 46 (10) (2008) 777–784.

[12] S.Z. Shuja, B.S. Yilbas, The influence of gas jet velocity in laser heating-a moving workpiece case, Proc. Inst. Mech. Eng. Part C: J. Mech. Eng. Sci. 214 (2000) 1059–1078.

[13] ABAQUS Theory Manual, Version 6.2, ABAQUS Inc., Pawtucket, USA, 2015.
[14] J.O. Kristiansson, Thermo-mechanical behavior of the solidifying shell within continuous casting billet molds—a numerical approach, J. Therm. Stresses 7 (1984) 209–226.
[15] J.M. Risso, A.E. Huespe, A. Cardona, Thermal stress evaluation in the steel continuous casting process, Int. J. Numer. Methods Eng. 65 (9) (2006) 1355–1377.
[16] G.E. Mase, G.T. Mase, Continuum mechanics for engineers, second ed., CRC Press, Boca Raton, FL, 1999.
[17] C. Li, B.G. Thomas, Thermo-mechanical finite-element model of shell be-havior in continuous casting of steel, Metal Mater. Trans. B 35B (6) (2004) 1151–1172.
[18] L. Anand, Constitutive equations for the rate dependant deformation of metals at elevated temperatures, ASME J. Eng. Mater. Technol. 104 (1982) 12–17.
[19] A.M. Lush, G. Weber, L. Anand, An implicit time-integration procedure for a set of integral variable constitutive equations for isotropic elasto viscoplasticity, Int. J. Plast. 5 (1989) 521–549.
[20] A. Mendelson, Plasticity: Theory and Applications, Krieger Pub. Co., New York, 1983.
[21] Young Modulus of Elasticity for Metals and Alloys and Coefficients of Linear Expansion, 2015, http://www.engineeringtoolbox.com/young-modulus-d_773.html&www.engineering toolbox.com/linear-expansion-coefficients-d_95.html.
[22] ETEC Gesellschaft für Technische Keramik GmbH, Germany, 2015, *www.etec-ceramics.de*.
[23] L.D. Scintilla, L. Tricarico, Estimating cutting front temperature difference in disk and CO_2 laser beam fusion cutting, Optics Laser Technol. 44 (2012) 1468–1479.
[24] S. Guo, H. Jun, L. Lei, Z. Yao, Numerical analysis of supersonic gas-dynamic characteristic in laser cutting, Opt. Lasers Eng. 47 (2009) 103–110.
[25] A. Ottoa, H. Koch, K. Leitz, M. Schmidt, Numerical Simulations—a versatile approach for better understanding dynamics in laser material processing, Phys. Procedia 12 (2011) 11–20.
[26] T. Mukarami, T. Tsumura, T. Ikeda, H. Nakajima, K. Nakata, Ansitropic fusion profile and joint strength of lotus-type porous magnesium by laser welding, Mater. Sci. Eng. A 456 (2007) 278–285.
[27] V.S. Deshpande, N.A. Fleck, High strain rate compressive behavior of aluminum alloy foams, Int. J. Impact Eng. 24 (2000) 277–298.
[28] B.S. Yilbas, S.S. Akhtar, O. Keles, Laser Straight Cutting of Ti-6Al-V: Temperature and Stress Fields, in: J. Paulo Davim (Ed.), Materials and Surface Engineering: Research and Development, Mechanical Engineering Series No. 2, Woodhead Publishing/ WOODHEAD/CHANDOS, Cambridge/Oxford, ISBN 0-85709-151-4, 2012.
[29] B.S. Yilbas, S.S. Akhtar, C. Karatas, Laser hole cutting into Ti-6Al-4V alloy and thermal stress analysis, Int. J. Adv. Manuf. Technol. 59 (9–12) (2012) 997–1008.

Concluding remarks

6

Laser cutting is one of the important laser machining applications in practice. Although the studies associated with the laser machining of materials have been active for over 30 years, the research on the improvement of the process is still current. This is because of the complicated physical phenomena influencing the end-product quality. In addition, improvements in laser technology and new materials boost the research in the laser machining area. There are many challenges associated with the laser-cutting process, some of which include high-speed processing, machining to difficult machine parts, end-product quality improvement, low cost, etc. In laser cutting applications, cutting thickness, laser output power and the quality, repetition rate of pulses, workpiece thickness, and assisting gas type and pressure are the main factors affecting the end-product quality. However, the interactions of these parameters are not linear and the combined effects become important to achieve good quality cutting. In addition, thermal stress fields developed around the cut section also play an important role in the practical applications of the laser cut sections. High-stress fields cause failure to laser machined parts in some applications such as under high-cyclic fatigue or high-thermal load. Nevertheless, laser cutting offers significant advantages over conventional cutting methods. Some of these include high-speed processing, precision of operation, cost-effective processes, and local treatment. The findings of the analytical and numerical studies related to laser cutting, as well as practical applications of the laser cutting process for difficult-to-cut materials, are given below under the appropriate subheadings.

6.1 Analytical treatment of the laser cutting process

Although laser cutting is involved with complicated physical processes, the useful assumptions enable the formulation of the cutting process in terms of temperature and stress fields as well as kerf width and cutting efficiency. These assumptions enable the correlation of the laser parameters with temperature rise and stress development in terms of analytical functions. This makes it easy to determine the maximum temperature and stress in the heated region of the irradiated workpiece during laser processing. Moreover, some of these assumptions can be justified because of the localized effects generated during the processing. For example, the assumption of a one-dimensional body can be justified through knowing that the axial heat conduction overtakes the radial conduction due to the one-directional absorption of the incident beam (Lambert's Bear law) and considerably smaller absorption depth of the irradiated material as compared to its thickness.

In general, the exponential laser pulse, the laser pulse parameter (β), together with the heat transfer coefficient at the surface, affect the temperature rise and its distribution inside the substrate material. However, the effect of heat transfer becomes critical for large Biot numbers; in which case, the heat transfer coefficient increases to

The Laser Cutting Process. https://doi.org/10.1016/B978-0-12-812982-1.00006-0

10^7 W/m^2 K. This situation is mostly seen when a phase change takes place, such as melting or evaporation. In the case of nonconduction limited heating situations, the recession velocity obtained from constant temperature evaporation at the surface results in a low magnitude of recession velocities at low power intensities, particularly in the early period of evaporation; however, it predicts substantially higher recession velocities for high laser power intensities. The equilibrium-based recession velocity (steady heating situation) gives rise to high recession velocity at low laser power intensities. This suppresses the rise of surface temperature after the evaporation process is initiated. Therefore, the surface temperature drops well below the evaporation temperature of the substrate material. Although evaporation initiation cools down the substrate surface, the temperature at the surface remains at around the evaporation temperature of the substrate material. Consequently, recession velocity determined from the steady heating situation wrongly predicts the surface temperature rise immediately after the evaporation process. Therefore, constant temperature evaporation at the surface estimates recession velocity correctly in the early evaporation period, and recession velocity based on steady evaporation occurs as the evaporation period progresses further. In addition, there exists an evaporation period when recession velocities obtained from the constant temperature evaporation at the surface and the steady heating situation become almost identical. In laser solid phase heating prior to key-hole formation, the temperature gradient approaches zero, indicating that energy balance takes place among the internal energy gain, energy conducted, and energy absorbed by the substrate material from the irradiated field. However, this situation changes with time because of the transient nature of the heating. The thermal stress field generated in the substrate material is mostly compressive at some depth below the surface. This is mainly associated with the thermal expansion of the substrate, which is suppressed by the expanded material in the surface region. Thermal stresses demonstrate the wave behavior in the substrate material under the thermal load generated during the laser pulse heating.

The analytical formulation of kerf width provides useful information on the kerf size determination in terms of surface absorptivity, laser pulse power, assisting gas pressure, and type of the assisting gas. In addition, it gives insight into the kerf width variation with the various levels of these parameters. For example, the influence of a small variation in laser output power, which resembles the laser output power oscillations during cutting, kerf width variations can be observed prior to the experiments. This argument is also true for the laser cutting speed as well as assisting gas type and its pressure. In the practical application of laser cutting, the regulation of the assisting gas pressure is important to achieve successful cutting with small kerf size and free from asperities such as waviness, dross attachment, and out-of-flatness. Moreover, assisting gas in the kerf creates multiple effects including dragging of the molten material from the kerf surface, cooling of the molten surface through convection heat transfer, and purging the dross formed at the kerf exit toward the kerf exit. Therefore, it has significant effect on the quality of the resulting cut geometry.

The neural network introduced to assess the striation patterns is successful in identifying the patterns according to the control chart. The findings revealed that no standard pattern is identified for the striations; however, the normal pattern happens to be the winner for all the striation patterns. This reveals that the striation formation, in general, is random, and depends on the transient nature of the physical processes generated in the kerf such as surface plasma, oxidation reactions, molten flow, assisting gas, and thermal properties of the material. Nevertheless, neural network is found to be useful tool to assess the cut section properties.

6.2 Numerical analysis associated with laser cutting

Because the assisting gas has a significant effect on the cutting process, its effects in the kerf, in terms of skin friction (momentum loss) and heat transfer (thermal energy loss), are important. In addition, increasing the frictional losses enhances the molten material attachment to the kerf surface while lowering the end product quality. On the other hand, increasing heat transfer losses gives to low thermal efficiency of the cutting process. Therefore, examination of the assisting gas effect on the skin friction and heat transfer becomes essential for improved laser cutting process. Because the process is involved with three-dimensional heating in the kerf and multiphysics of the process, the analytical methods become almost impossible to provide close form solution to the problem. Therefore, the numerical method incorporating three-dimensional effects and the multiphysics of the problem becomes necessary to tackle the problem. In the numerical analysis, the jet temperature at the nozzle exit is selected as 300 K and the kerf wall temperature is kept at 1500 K to resemble the laser-cutting situation. The kerf width is fixed while the kerf thickness is varied in the simulations. Different gasses are used as the working fluid resembling the influence of different assisting gas types on the heat transfer and skin friction at the kerf surface. The recent findings of the numerical predictions related to the assisting-gas effect are given below.

The radial expansion of the assisting gas jet occurs in the region close to the kerf inlet. This is more pronounced for high kerf wall wedge angle (6 degree), which is associated with the increasing blockage effect of the kerf with increasing kerf wall wedge angle. The shear layer developed at the jet outer boundary, in between the nozzle exit and the kerf surface, results in the circulation cell. The orientation and size of the circulation cell changes with increasing the kerf wall wedge angle. Static pressure increases in the kerf; however, as the depth below the kerf surface increases toward the kerf exit, static pressure reduces due to flow acceleration. The Nusselt number increases in the region of the kerf inlet because of enhanced cooling of the kerf wall surface in this region. The radial acceleration of the jet is responsible for the Nusselt number enhancement. The second peak of the Nusselt number occurs in the kerf, which is associated with the flow acceleration in the kerf toward the kerf exit. The influence of the kerf wall wedge angle on the Nusselt number is notable; in which case, increasing the kerf wall wedge angle increases the second peak of the Nusselt number. The skin friction decreases with increasing depth toward the kerf exit. However, increasing the rate of fluid strain due to velocity enhancement results in the formation of the second peak of the skin friction in the kerf. The skin friction reduces toward the kerf exit, which is more pronounced for the high kerf wall wedge angle.

It is also found that the jet expands upon leaving the nozzle and the shear layer developed at the outer boundary of the jet results in a circulation cell in this region. The axial momentum of the jet reduces in the region of the stagnation zone, and the presence of the kerf results in the radial expansion of the jet in the entry kerf region. This, in turn, increases the velocity magnitude after the entry region of the kerf. However, the pressure buildup, due to a kerf blockage effect and the thermodynamic pressure in the kerf wall vicinity, suppresses the flow velocity toward the kerf exit. However, the pressure differential across the kerf inner region and the kerf exit accelerates the flow at the kerf exit. The Nusselt number increases significantly in the region of the kerf inlet where the radial acceleration takes place. Moreover, flow acceleration at the kerf exit results in a second peak in the Nusselt number in the kerf toward the kerf exit. This is influenced by the kerf thickness; in which case, increasing kerf thickness lowers the value of the Nusselt number peak. The skin friction at the kerf inlet is high due to high rate of fluid strain in this region. Increasing kerf thickness has an adverse effect on the skin friction because of the reduced rate of fluid strain in the kerf due to thickening of the thermal boundary layer and the flow deceleration in the kerf.

The shear flow at the expanding jet boundary results in the formation of the circulation cells in the neighborhood of the jet emerging from the nozzle. The flow velocity reduces at the kerf inlet due to the blockage effect of the kerf. Because of the pressure drop toward the kerf exit, flow acceleration is observed toward the kerf exit. The boundary layer thickens toward the kerf exit because the high-wall temperature of the kerf suppresses the velocity rise at the kerf exit region. Because the density of helium is low, its velocity is very high in the jet. This alters the behavior of pressure in the kerf; in which case, pressure increases significantly toward the kerf exit for helium. The Nusselt number increases sharply at the kerf inlet while it decreases gradually with increasing distance along the kerf wall. This behavior is attributed to the balance between the amount of convection heat transfer due to flow velocity and temperature increase in the flow. The Nusselt number is high due to low fluid temperature at the kerf inlet despite the low flow velocity. Fluid temperature and velocity are high in the kerf at some distance away from the kerf inlet; in which case, the Nusselt number reduces. The influence of assisting gas type on the Nusselt number is notable. The Nusselt number is the highest for nitrogen, followed by air and helium. The skin friction increases sharply at the kerf inlet due to high rate of fluid strain developed in this region. The skin friction is the highest for helium, followed by air and nitrogen. The attainment of high skin friction for helium is due to its high velocity in kerf, which results in a high rate of fluid strain and shear stress along the kerf wall. Therefore, for the practical application of the laser cutting process, the use of helium as an assisting gas is preferable. This is because helium gives rise to less heat transfer from the kerf surface, due to a low Nusselt number, which delays the resolidification at the surface, and a high skin friction coefficient results in a large drag force enhancing the liquid metal flow from the kerf surface while reducing the amount of dross attachment at the kerf exit.

The circulation cells are formed above the kerf surface due to the frictional flow in the vicinity of the jet's outer boundary and the radial acceleration of the jet impinging upon the kerf surface. The flow first accelerates in the kerf inlet and then decelerates toward the kerf exit. This is because of the pressure buildup in the kerf. Moreover, the blockage effect, due to the kerf wedge angle, contributes to the flow behavior in the kerf. The Nusselt number increases sharply at the kerf inlet and decreases gradually for a kerf wall wedge angle of 0 degree. However, the Nusselt number increases slightly toward the kerf exit due to pressure buildup and the density increase of the fluid in this region. Increasing the mean jet velocity at the nozzle exit enhances the values of the Nusselt number, which is more pronounced for a kerf wall wedge angle of 4 degree. The skin friction increases sharply at the kerf inlet and decreases sharply toward the kerf exit for a kerf wall wedge angle of 0 degree, which is more pronounced for the low-averaged jet velocity at the nozzle exit. However, the skin friction decreases in the region of the kerf inlet first, and then, increases toward the kerf exit. This is because of the rate of fluid strain, which is modified by the pressure buildup in the kerf toward the kerf exit region.

6.3 Laser cutting of difficult-to-cut materials

Although the practical applications of laser cutting spans a wide area of engineering, laser cutting of difficult-to-cut materials limits its applications. Therefore, the research to develop an efficient and end-product quality improvement is still in progress. However, the findings of numerical and experimental results of laser cutting of such materials are given below, under the appropriate subheadings.

6.3.1 Laser cutting of bronze

Laser cutting of brass is challenging because of its high thermal conductivity, which diffuses the localized energy gain from the irradiated field toward the bulk of the material during processing. The findings of laser straight-cutting of bronze show that temperature increases beyond the melting temperature of the substrate material at the cut edges, and the heating of the liquid phase, enhances the temperature increase above the melting temperature in the vicinity of the cut edges. Temperature decay is sharp in the frontal region of the laser beam once the cutting process is terminated. This, in turn, results in development of the high-temperature gradients in this region, which give rise to the formation of two stress peaks along the cutting direction: one in the recast region, and the other in the frontal region of the laser beam. Temperature decay along the y-axis is gradual in the cut edge surface vicinity and it becomes high at some distance away from the surface. This causes the formation of high von Mises stress in this region. The maximum magnitude of a von Mises stress at the end of the cooling cycle is higher than the yielding limit of the workpiece material, which results in plastic deformation in this region. The striations with small depth are formed at the kerf surface. The striation patterns changes toward the kerf exit due to the change in the viscosity of the molten metal and the momentum loss of the assisting gas in the kerf. The top surface of the laser cut is free from the sideways burning and resolidification. However, dross attachment at the kerf exit takes place. This is associated with the low viscosity and less drag force acting on the molten metal surface in the kerf because of the momentum loss of the assisting gas toward the kerf exit. The cast layer formed at the surface of the kerf is on the order of 2.5 μm and the size of the heat-affected zone below the cast layer is on the order of 10 μm. The elemental analysis at the kerf surface reveals that phosphorous depletion occurs at the surface. In addition, Cu_3N and Cu_2O compounds are formed at the kerf surface due to the presence of nitrogen in the cutting region and the exothermic reactions taking place in the cutting section during the cutting process.

On the other hand, laser cutting of circular geometry in bronze demonstrates that the maximum temperature along the hole circumference exceeds the melting temperature of bronze; hence superheating in the liquid phase takes place in the laser-irradiated region. The maximum temperature remains higher along the top circumference of the hole than that of the bottom circumference. The conduction and convection cooling of the kerf surface lowers the temperature along the hole circumference once the cutting process is completed. However, the temperature remains high and the cooling rate is slow at the location where the cutting terminates. The maximum residual stress is on the order of 200 MPa, which is almost uniformly distributed along the hole edges. The high residual stress region extends almost 0.0025 m along the hole circumference in the substrate material. The optical photographs reveal that the cutting is free from irregularities. The close view of the kerf surface shows that striation patterns are formed at the hole inlet. However, the striation patterns change toward the hole exit. This is attributed to the frictional drag force in the molten flow. In this case, attainment of the high viscosity in the melt flow due to temperature drop is responsible for the bending of the striation patterns. Dross attachment is observed at the hole exit, which indicates the solidification of molten flow upon exiting the hole. The high thermal conductivity of bronze, and the low assisting gas pressure in the kerf at the hole exit, cause the dross attachment. The cross-sectional examination of the hole section indicates the cast layer is formed at the kerf surface, which becomes thicker at the hole exit. The heat-affected zone in the substrate material close to the kerf surface is large. The EDS results reveal that oxidation reactions are taking place during the cutting process, despite the use of nitrogen as an assisting gas in the cutting section.

6.3.2 Laser cutting of alumina tiles

The findings of laser straight cutting of alumina tiles are presented. Temperature and stress fields developed along the cut edges are simulated in line with the experimental conditions and parameters. The finite element method is incorporated to model the thermal stress field, thus the ABAQUS code is used in this regard. The residual stress at the cut section is obtained from the XRD technique and compared with the predictions. The geometric features of the cut section are examined using the optical and scanning electron microscopes. It is found that temperature at the cutting edge exceeds the melting temperature of the substrate material during the cutting process. This, in turn, results in annealing effect on the cutting edge lowering the cooling rate in the cooling cycle. Consequently, the von Mises stress attains relatively lower value in this region as compared to other regions around the cutting edge. Sharp decay of temperature in the region next to the surface vicinity results in high thermal strain and stress formation in this region, particularly after the end of cooling cycle. In the early cooling period, the von Mises stress attains low values in the region where temperature is high, which is associated with the elastic modulus of the workpiece, that is, it reduces with increasing temperature. The von Mises stress attains its maximum at mid-height of the thickness on the kerf surface due to the thermal contraction in the cooling cycle. On the other hand, the von Mises stress reduces at the top and bottom surfaces of the cutting edge because of free expansion and contraction of the surface. The von Mises stress at the surface vicinity predicted from the simulations agree with that obtained from the XRD technique. Striations are found at the kerf surface and the depth of striations is shallow because of the high melting temperature and low thermal conductivity of the workpiece. Dross attachment is observed at the kerf exit, which is associated with low viscosity of the molten flow and weak drag force generated by the assisting gas due to blockage effect in the kerf. Microcracks form in the recast layer at the kerf surface because of high cooling rates at the kerf surface.

The findings of laser circular cutting of aluminum tiles are presented. The findings revealed that temperature decay is sharp around the hole circumference for a 15 mm diameter hole as compared to that of a 10 mm diameter hole. This behavior is attributed to the internal energy gain of the substrate material in the region of the hole circumference and heat transfer losses from this region to the solid bulk. The von Mises stress shows similar behavior around the hole circumference for both hole diameters. Heat conduction from the region where the laser spot is located to the location where cutting started causes a self-annealing effect in the initial region where the cutting started. This, in turn, results in low stress levels in this region at the end of the cooling period. The residual stress predicted is compressive and it is in the order of -2.9 GPa, and the residual stress measured using the XRD technique is in the order of -2.65 ± 0.08 GPa. Consequently, both results appear to be in very good agreement. In addition, surface temperature predicted agrees well with the thermocouple data. The morphological examination of the cut sections reveal that striation patterns are formed at the kerf surface, provided that the stria depth is shallow. Temperature drop toward the kerf exit causes the increase in viscosity of the molten flow in this region. The pressure drop, due to the friction effect in the narrow kerf depth, lowers the pressure drag of the assisting gas acting on the molten flow. These, in turn, reduce the molten ejection from the kerf and formation of dross attachments at the kerf exit. High cooling rates at the kerf surface causes the formation of a brittle recast layer, which endures multimicro cracking, provided that the cracks are shallow and do not extend into the substrate material.

Temperature and stress fields are predicted for the laser cutting of a rectangular geometry in thick alumina tiles. The findings demonstrate that the predictions of surface temperature and residual stress are validated through the thermocouple and XRD data. It is found that predictions of surface temperature and the residual stress agree well with their experimental counterparts in the cut section. Temperature well exceeds the liquidus temperature of the substrate material in the irradiated region, which, in turn, results in superheating of the liquid phase. The presence of the mush zone in the neighborhood of the superheated liquid zone is evident from the temperature profiles. Temperature attains low values in the corner regions of the rectangular geometry due to the conduction cooling effect. The von Mises stress attains low values in the region where temperature is high due to the consideration of temperature-dependent elastic modulus. The von Mises stress attains high values in the region of the mid-thickness of the workpiece. The maximum value of the von Mises stress is in the order of 3.1 GPa. The striation patterns are formed at the cut surface, which are associated with the melt instabilities and the assisting gas pressure in the kerf. The viscosity of the molten flow increases toward the kerf exit due to the convection cooling of the assisting gas. The combination of high viscous and surface tension forces results in the formation of the dross attachment at the kerf edge. The use of nitrogen at high pressure as an assisting gas causes the formation of AlON at the kerf surface.

6.3.3 Laser cutting of aluminum foam

Net shaping of metallic foams, particularly aluminum foams, by conventional techniques is difficult. In addition, thermal processing introduces complicated physical processes during the cutting. However, localized laser intensity enables the ability to cut aluminum foams with minimum asperities. The findings of the experimental and model study for laser straight cutting of aluminum foam reveal that the von Mises stress attains high values in the region where the temperature gradient is high and temperature is low. This is related to temperature-dependent elastic modulus, which reduces with increasing temperature. The maximum value of the von Mises stress is close to the yielding limit of the substrate material. The laser cutting resulted in parallel-sided edges of the kerf. The kerf width variation is on the order of 7% indicating narrow and quality cuts. Some small thermal erosion occurs along the kerf edges due to high-temperature oxidation reactions taking place during the cutting process. Although nitrogen at high pressure is used as the assist gas in the cutting section, the presence of air, which is trapped in large pores in the workpiece, is responsible for the high-temperature oxidation reactions. The depth of the local thermally eroded region along the kerf edge is narrow and does not extend into the solid bulk of the workpiece. Molten flow occurs on the pore sides, which is associated with the viscosity changes during oxidation reactions and low drag force due to axial momentum loss of the assisting gas in the pore sites. The dross attachment at the kerf exit is small and locally scattered. The small size of the dross attachment at the kerf is associated with less molten material flow at the kerf because of the presence of pores in the aluminum foam. It is evident that laser cutting resulted in burr-free and parallel-sided cuts into 9-mm-thick aluminum foam without damaging the foam's cell pockets; which may not be the case with the mechanical cutting process.

In the case of circular cutting of aluminum foam, the findings show that cut sections are free from large burrs and dross attachments. In addition, no defects, such as sideways burning, are observed at the hole circumference. Thermal stress analysis reveals that the maximum thermal stress formed around the hole circumference is less than the yielding limit of the substrate material. Consequently, no cracks are observed at the cut section. The presence of air, which is trapped in pores prior to the laser cutting process, undergoes exothermic oxidation reactions in the cut, section. This results in oxide compound formation at the kerf surface. A small, locally-scattered dross attachment is observed at the hole exit, which is associated with the local melt flow in this region. In this case, formation of oxide compounds alters the viscosity of the molten flow, which contributes to the dross attachment at the hole exit. The von Mises stress attains high values in the region where the temperature gradient is high; it becomes low when temperature is high. This is due to temperature-dependent elastic modulus incorporated in the simulations, which reduces with increasing temperature. The self-annealing effect of the laser heated region causes stress relaxation in the region where the laser beam is switched off. The von Mises stress attains high values in the region next to the free surfaces and in the mid-thickness of the workpiece, and remains low at the front and back surfaces of the cut section. This is attributed to the free expansion of both surfaces while lowering the stress levels in the free surface region.

Laser cutting of triangular shapes into aluminum foam is found to be interesting, because laser cutting of sharp corners, particularly into foam material, brings challenges to reduce the excessive heating in the corners of the triangular geometry. The findings reveal that the temperature predicted attains high values in the close region of the laser spot and it exceeds well above the melting temperature of the substrate material. Because the cooling rates around the cut edges are different for the small triangle cut as compared to the large triangle cut, the maximum temperature becomes higher for the small triangle than that of the larger one. The von Mises stress attains low values in the region of high temperature because of temperature-dependent elastic modulus. The von Mises stress is slightly lower for small triangle cuts as compared to large triangle cuts. This is attributed to the cooling rates, which is higher for the large triangle cut. The von Mises stress along the z-axis and around the cut edges is not uniform, which may be attributed to contraction and expansion of the cut surface. The von Mises stress attains low values in the region of the top and bottom edges of the cut surface. Optical photographs and SEM micrographs show that a laser cut surface is free from large scale defects; however, some small, local sideways burning is observed at the cut edges. This is attributed to air trapped in the large voids in the foam, which is released during the cutting, and it undergoes oxidation reactions locally with the molten metal. The small-scale dross attachment is observed at the kerf exit. This is associated with: (i) reduced viscosity of the molten flow due to cooling and (ii) reduced pressure of the assisting gas toward the kerf exit due to the frictional loss. The amount of dross attachment is small because of the low density of the foam, which results in a low rate of molten flow at the kerf surface.

6.4 Laser cutting of Kevlar

Conventional machining methods are, however, known to offer low-end product quality due to the abrasive reinforcement and inhomogeneous structure of Kevlar. Consequently, laser machining of Kevlar offers considerable advantages over the conventional machining methods. In laser machining of Kevlar, a shielding gas is used to eliminate the high-temperature oxidation reactions, which occur through combustion of carbon compounds in Kevlar at high temperature. This, in turn, results in sideways burning while reducing the machining quality through irregular cut surfaces. Consequently, conducting an investigation into laser cutting of Kevlar and the influence of assisting gas pressure and laser output power levels on the cut quality will constitute an important contribution to laser machining of Kevlar. Inert gas laser cutting is applicable for polymer and wood processing, because sideways burning is avoided by minimizing the exothermic reactions. Moreover, in some situations, high-temperature oxidation reactions cannot be avoided because of existing oxygen in the chemical composition. Kevlar is a material that burns at high temperatures. Because laser cutting is involved with high temperature processing, care must be taken during laser cutting of Kevlar. The findings of the experimental study pertinent to laser cutting of Kevlar is given herein. Increasing cutting speed improves the end-product quality, provided that the end-product quality becomes low when the cutting speed exceeds 0.4 m/s. In this case, a high cutting speed results in small kerf width and the amount of assisting gas passing through the kerf lessens which, in turn, prevents the suppression of the high-temperature oxidation reaction in the region close to the kerf exit. Consequently, some side burning and formation of char residue are observed in this region. On the other hand, at a low cutting speed (0.05 m/s), loose fibers appear in the region of the kerf entrance and exit sections, which lower the cut quality. The influence of thermal conductivity on the kerf size is significant. In this case, at high cutting speeds, a discrepancy between predictions and experimental results is observed. Therefore, nonuniform structure of the fibers in the substrate material (un-woven fabric) results in nonuniform thermal conductivity in the substrate material at high cutting speeds. Because the rate of energy available for cutting and radial heat conduction is highly dependent upon the cutting speeds, nonuniform thermal conductivity appears to enhance the kerf width size at high cutting speeds; that is, measured kerf width size becomes larger than those corresponding predictions at high-cutting speeds. It should be noted that in the mathematical model, a uniform thermal conductivity is assumed in the substrate material.

6.5 Laser cutting of composites

Laser cutting of composites finds wide application in industry due to improved end-product quality, low cost, and short processing time. The end-product quality is influenced significantly by the high-temperature combustion reactions taking place in the cutting section. Moreover, when cutting thick workpieces, the assisting gas may not have enough pressure to shield the cutting edges from the combustion reactions when it reaches the bottom of the workpiece. This, in turn, results in sideways burning at the bottom edge of the workpiece. Therefore, investigation into the end-product quality of laser cutting of composites is necessary. The findings of laser cutting of a workpiece composed of 64 layers of plain-weave carbon/carbon fibers with 0^0 orientation carbon/carbon fibers structured are presented herein. The fiber orientation in the workpiece with reference to the workpiece motion during the cutting process has significant effect on the kerf formation. In this case, fiber axis normal to the axis of the workpiece motion results in relatively large kerf width size due to the influence of thermal conductivity; that is, energy available for mass removal in the kerf is enhanced by less diffused energy in the radial direction to the solid bulk. In general, parallel-sided cuts are obtained, provided that as the penetration depth in the kerf increases, small tapering of the cut cross section is observed. This may be associated with less laser power available in this region, that is, kerf-side surfaces act as an absorbent to the laser irradiation, suppressing the side-wall reflections in the entry region of the kerf. The transient behavior of the surface plasma affects the striation formation on the kerf sides. Striation width and depth vary slightly with increasing laser power. This suggests that a coupling effect of the surface plasma is significant, particularly in the entry region of the kerf. The globule-like structure is observed at the tip of the fiber cut edges, which is particularly true at high laser powers. Moreover, the odd shapes of globule-like structures suggest that some local high-temperature oxidation reactions take place due to water content in the workpiece.

6.6 Laser cutting of wedges in steel

In the laser cutting process, laser parameters such as laser output power and pulsating frequency, cutting speed, assisting gas pressure, and focal distance determine the end-product quality. The sideways burning along the cutting paths results in poor cutting quality; particularly wedge cuts forming in sheet metal. In order to improve the cutting quality for such situations, the laser parameters should be readjusted to minimize the molten flow toward the wedge surface during the laser cutting process. The findings of the research study reveal that cutting quality is affected significantly by the laser output power intensity, particularly in a wedge cutting situation. In the case of a low power setting, molten metal flow on the top surface of the workpiece is observed. This reduces the cutting quality. On the other hand, for wedge cutting situations, the dross ejection from the bottom of the kerf causes thermal erosion of the cutting edges at the bottom surface of the workpiece. This also reduces the cutting quality. The dross height varies with low-power intensity while the variation of out of flatness is smaller than its counterpart corresponding to the dross height. It is observed from the neural network output that the normal pattern of striation is dominant over the other patterns. This suggests that molten flow induced instability occurs randomly along the cut edges. The cutting quality improves for a specific laser output power setting.

Index

Note: Page numbers followed by *f* indicate figures, and *t* indicate tables.

N

O

P

Q

R

S

T

FSC
www.fsc.org
MIX
Papier | Fördert
gute Waldnutzung
FSC® C083411

Kolibri 360 GmbH
Ferdinand-Jühlke-Straße 7
99095 Erfurt, Deutschland
produktsicherheit@kolibri360.de